Waves in Periodic and Random Media

CONTEMPORARY MATHEMATICS

339

Waves in Periodic and Random Media

Proceedings of an AMS-IMS-SIAM
Joint Summer Research Conference
on Waves in Periodic and Random Media
June 22–28, 2002
Mount Holyoke College, South Hadley, MA

Peter Kuchment
Editor

American Mathematical Society
Providence, Rhode Island

This volume contains proceedings of an AMS-IMS-SIAM Joint Summer Research Conference on Waves in Periodic and Random Media held at Mount Holyoke College, South Hadley, MA on June 22–28, 2002, with support from the National Science Foundation, grant DMS-9973450.

2000 *Mathematics Subject Classification.* Primary 35Pxx, 35Qxx, 78Axx, 81Qxx, 81Uxx, 81Vxx, 82Dxx.

Library of Congress Cataloging-in-Publication Data

AMS-IMS-SIAM Joint Summer Research Conference on Waves in Periodic and Random Media (2002 : Mount Holyoke College)

Waves in periodic and random media : proceedings of an AMS-IMS-SIAM Joint Summer Research Conference on Waves in Periodic and Random Media, June 22–28, 2002, Mount Holyoke College, South Hadley, MA / Peter Kuchment, editor.

p. cm. — (Contemporary mathematics, ISSN 0271-4132 ; v. 339)

Includes bibliographical references.

ISBN 0-8218-3286-7 (alk. paper)

1. Wave-motion, Theory of—Congresses. 2. Photons—Congresses. 3. Crystal optics—Congresses. 4. Order-disorder models—Congresses. I. Kuchment, Peter, 1949– II. Title. III. Contemporary mathematics (American Mathematical Society) ; v. 339.

QA927.A486 2002

2003063578

---

10 9 8 7 6 5 4 3 2 1 08 07 06 05 04 03

Contents

Preface

Sciences and engineering have been great sources of problems and inspiration for generations of mathematicians. This is probably clearer now than ever, when numerous challenges in sciences and technology are met by mathematicians. One of these challenges is studying propagation of acoustic, electromagnetic, and electron waves in systems of complex structure. Among the recently studied types of such media are photonic crystals (artificially created optical analogs of semi-conductors) and other composite materials and circuits of quantum wires (or more general mesoscopic systems, some dimensions of which are reduced to nanoscale), as well as older media like pure and doped crystals and waveguides. The Joint Summer Research Conference in the Mathematical Sciences "Waves in Periodic and Random Media" that convened June 22 through June 28, 2002 at Mount Holyoke College in South Hadley, MA gathered a strong group of mathematicians and physicists interested in wave propagation in complex media.

The Organizing Committee consisted of David Dobson (University of Utah), Alex Figotin (University of California, Irvine), Peter Kuchment (Texas A&M University), and Stephanos Venakides (Duke University). Specific topics of the conference included photonic crystals, Anderson Localization, wave propagation in linear and non-linear periodic media, waves in mesoscopic media (a crucial area for nanotechnology), and surface waves. These topics were addressed from analytic, numerical, and physics perspectives in the following invited lectures delivered by prominent experts:

- M. Aizenman (Princeton University) "Moment methods for continuum Schrödinger operators with random potentials"
- H. Ammari (École Polytechnique) "Guided waves in a photonic bandgap structure with a linear defect"
- A. Babin (UC Irvine) "Multilinear spectral decomposition for nonlinear Maxwell operators"
- E. Bonnetier (École Polytechnique) "Radiation properties of a planar antenna mounted on a photonic crystal substrate"
- J.-M. Combes (Univ. de Toulon, France) "Spectral properties of absorptive photonic crystals"
- D. Dobson (Univ. of Utah) "Optimal mode localization in heterogeneous media"
- P. Exner (Nuclear Institute, Czech Republic) "Magnetic transport in presence of perturbations"
- A. Figotin (UC, Irvine) "Basic properties of nonlinear Maxwell equations in inhomogeneous media"

- L. Friedlander (Univ. of Arizona, Tucson) "Application of Dirichlet-to-Neumann operators to spectral theory"
- R. Grigorchuk (Steklov Math. Institute, Moscow, Russia; starting fall 2002, Texas A&M University) "On spectral problems and two-dimensional dynamical systems associated with some self-similar structures"
- D. Hundertmark (Caltech) "A variational bound for eigenvalues of discrete Schrödinger operators with potentials of arbitrary sign"
- Yu. Karpeshina (Univ. Alabama, Birmingham) "Periodic magnetic Schrödinger operators. The isoenergy surface and density of states"
- A. Klein (Univ. of California, Irvine) "A characterization of the Anderson metal-insulator transport transition"
- F. Klopp (Univ. Paris-Nord, France) "Regularity properties for the density of states of random Schrödinger operators"
- T. Kriecherbauer (Universität München, Germany) "Multi-phase waves in discrete media"
- A. Krokhin (Instituto de Fisica, Universidad Autónoma de Puebla, Mexico) "Homogenization of photonic and phononic crystals"
- P. Kuchment (Texas A&M University, College Station) "Graph models of thin media"
- A. Maradudin (UC Irvine) "Designer surfaces"
- C. Nedelec (École Polytechnique) "Preconditioning for harmonic integral equations for Maxwell"
- B. Pavlov (Auckland University, New Zealand) "Dirichlet-to-Neumann map and mathematical modelling of nano-devices"
- J. Sipe (University of Toronto) "Gap and Bragg solitons in artificially structured materials"
- A. Sobolev (University of Sussex, UK) "Lattice points and periodic operators"
- T. Suslina (St. Petersburg State University, Russia) "Absolute continuity of the spectrum of periodic operators of mathematical physics"
- B. Vainberg (University of North Carolina, Charlotte) "Schrödinger operators with matrix potentials. Transition from absolute continuous to singular spectrum"
- S. Venakides (Duke University) "Resonances and surface waves in photonic crystal slabs"
- M. Weinstein (Bell Labs, Lucent Technologies) "Stopping light on a defect: Nonlinear waves in periodic structures with defects"

There were 44 participants of the conference, including one undergraduate and nine graduate students and five recent recipients of Ph.D. degrees. Two successful poster sessions were organized, where students and junior faculty could present their results. An important feature of the conference was a mix of mathematicians and physicists in all groups of participants (from students to invited speakers). This created an atmosphere of active and fruitful discussions during lectures, breaks, and in spare time. The organizers are grateful to the participants, who made the Conference such an exciting event. They feel that the conference facilitated further progress and brought the community of researchers in this area of research closer together.

This volume contains contributions from some of the conference participants. The articles included into this volume of proceedings reflect the topics of the conference. Among the issues discussed one finds photonic crystals (J.-M. Combes, B. Gralak, and A. Tip, P. Kuchment and B. Ong, S. E. Golowich and M. I. Weinstein), Anderson localization and spectral theory of random operators (F. Germinet and A. Klein, J.-M. Combes, P. Hislop and F. Klopp, V. Yurinsky), waveguide theory (L. Friedlander, P. Kuchment and B. Ong, S. E. Golowich and M. I. Weinstein), mesoscopic systems (P. Exner, B. Pavlov and K. Robert), spectral theory of periodic differential operators (L. Friedlander, Yu. Karpeshina, A. Sobolev, T. Suslina), and designer random surfaces (T. A. Leskova, A. A. Maradudin, E. R. Méndez, and J. Muñoz-Lopez). We hope that these articles will be useful to researchers interested in these active and promising areas of mathematical physics.

The conference was extremely well organized by the AMS, in particular thanks to a flawless job of the AMS Conference Coordinator Mr. Wayne Drady. Organizers and participants are grateful to the NSF and AMS for the financial and logistical support in organizing and running the conference.

The editor is indebted to Ms. Christine M. Thivierge and the technical staff of the AMS for their patience and efficient work in putting these proceedings together.

Peter Kuchment
College Station, TX
August 25th, 2003

Contemporary Mathematics
Volume **339**, 2003

Spectral properties of absorptive photonic crystals

J. M. Combes, B. Gralak, and A. Tip

ABSTRACT. We discuss some mathematical problems raised by the possibility to design photonic crystals using dispersive and absorptive materials. We review the results from A. Tip, A. Moroz and J. M. Combes, J. Phys. A: Math. Gen. **33**, 6223 (2000), showing in particular that, rigorously speaking, band gaps do not exist when absorption is present. Nevertheless we conclude that many physical features of absorptive systems can be obtained from perturbation theory applied to a suitable non-absorptive limit.

1. Introduction

Photonic crystals, i.e., dielectric devices with a dielectric function (dielectric constant, permittivity, electric permeability) $\varepsilon(\mathbf{x})$, possessing some form of spatial periodicity, have attracted considerable interest in the last 10 years due to their potential technological applications. With modern fabrication techniques devices can now be constructed with lattice parameters corresponding to optical wavelengths (i.e., in the region of visible light). Typical examples are 3D lattices of non-overlapping dielectric spheres (often called scatterers), centered on lattice sites in a background with different ε and 2D lattices of parallel cylinders or square rods. Producing frequency gaps in the optical range is one of the most important challenges in this field. It has turned out that a large contrast (ratio between the two ε's in a medium made up from two different dielectrics) is crucial. This is confirmed mathematically by A. Figotin and P. Kuchment [**1**], who showed that gaps open, e.g., for 2-dimensional square cylindrical structure in the limit of large dielectric contrast and filling ratio. However, it seems hard to reach such regimes with conventional dielectrics even with more sophisticated geometries such as, e.g., 3D wood pile structures [**2**]. Recently it was shown by Moroz [**3**], that these difficulties can be overcome using dispersive materials, having a dielectric function with Drude-like behavior. It turns out that for such systems the dielectric contrast can be arbitrarily large in some frequency windows, including the visible range, for

2000 *Mathematics Subject Classification.* Primary 78A25; Secondary 81Q10, 81V80.
Key words and phrases. Maxwell's equations, unitary time evolution, photonic crystals.
One of us (JMC) wants to thank the organizers of this conference for their invitation to present this talk. The work of BG and AT is part of the research program of FOM, which is financially supported by NWO. We also thank H. van der Lem and A. Moroz for many stimulating discussions.

some metals and also some new artificial structures. Although numerical results are promising [**4**], they certainly need to be confirmed by mathematical investigations about their interpretation. Here we present some steps in this direction. In this respect our first result is quite negative since it states that frequency gaps cannot exist for realistic models of dielectric functions with non-zero imaginary part, satisfying the Kramers-Kronig relations (which are basically a causality requirement). This clearly indicates that the limit of vanishing absorption requires a careful perturbative analysis of spectral quantities such as the density of states (which needs redefinition, the usual eigenvalue counting definition breaking down), whose smallness in a given frequency range has as much physical significance as the existence of a frequency gap. During this lecture we shall see that an efficient perturbative framework emerges quite miraculously from the introduction of auxiliary fields, initiated by one of us [**5**]. It can be considered mathematically as a constructive dilation method for certain time evolutions involving a time-convolution, allowing to recover a conservative system from an open one. This construction provides a natural non-absorptive limit as a starting point for an "almost regular" perturbation theory, allowing in particular, under realistic assumptions, the use of classical techniques such as complex scaling to study the fate of frequency gaps occurring in this limit. This talk summarizes the content of [**6**] and some preliminary results on these perturbative aspects.

2. Maxwell's equations for absorptive media

For conservative dielectrics, i.e., systems with a dielectric function $\varepsilon(\mathbf{x})$ only depending on the position variable $\mathbf{x}$ but not on frequency, band gaps Δ are just gaps in the spectrum of the Helmholtz operator, or, equivalently, intervals Δ, such that no stationary solutions of Maxwell's equations with frequency $\omega \in \Delta$ exist. This definition certainly needs to be revised if absorption is present, i.e., $\varepsilon = \varepsilon(\mathbf{x}, \omega)$ depends on the frequency ω and has a non-negative imaginary part. We shall only consider linear absorptive dielectrics for which Maxwell's equations read

$$\partial_t \boldsymbol{D}(\mathbf{x}, t) = \partial_{\mathbf{x}} \times \boldsymbol{B}(\mathbf{x}, t), \quad \partial_t \boldsymbol{B}(\mathbf{x}, t) = -\partial_{\mathbf{x}} \times \boldsymbol{E}(\mathbf{x}, t), \quad \partial_{\mathbf{x}} \cdot \boldsymbol{B}(\mathbf{x}, t_0) = 0, \tag{2.1}$$

where t_0 is some arbitrary initial time and with

$$\boldsymbol{D}(\mathbf{x}, t) = \boldsymbol{E}(\mathbf{x}, t) + \int_{t_0}^{t} ds\, \chi(\mathbf{x}, t-s) \boldsymbol{E}(\mathbf{x}, s). \tag{2.2}$$

Here $\chi(\mathbf{x}, t)$ is the electric susceptibility, for photonic crystals it has the spatial periodicity of a lattice Γ. With $\chi(\mathbf{x}, 0) = 0$ (no initial current surges) and in the absence of external currents, the electromagnetic energy,

$$\mathcal{E}_{em}(t) = \frac{1}{2} \int d\mathbf{x} \{ \boldsymbol{E}(\mathbf{x}, t)^2 + \boldsymbol{B}(\mathbf{x}, t)^2 \}, \tag{2.3}$$

satisfies

$$\partial_t \mathcal{E}_{em}(t) = -\int d\mathbf{x} \int_{t_0}^{t} ds\, \chi'(\mathbf{x}, t-s) \boldsymbol{E}(\mathbf{x}, s) \cdot \boldsymbol{E}(\mathbf{x}, t), \tag{2.4}$$

where $\chi'(\mathbf{x}, t) = \partial_t \chi(\mathbf{x}, t)$. Writing

$$\chi'(\mathbf{x}, t) = \int_{-\infty}^{\infty} d\lambda \exp[i\lambda t] \nu(\mathbf{x}, \lambda), \tag{2.5}$$

we then have

$$
\begin{aligned}
&\mathcal{E}_{em}(t) - \mathcal{E}_{em}(t_0) \\
&= -\int d\mathbf{x} \int_{t_0}^{t} ds \int_{t_0}^{t} du \int d\lambda \exp[i\lambda(s-u)]\nu(\mathbf{x},\lambda)\boldsymbol{E}(\mathbf{x},s)\cdot\boldsymbol{E}(\mathbf{x},u) \\
&= -\int d\mathbf{x} \int d\lambda\, \nu(\mathbf{x},\lambda)\Big|\int_{t_0}^{t} ds \exp[i\lambda s]\boldsymbol{E}(\mathbf{x},s)\Big|^2. \qquad (2.6)
\end{aligned}
$$

indicating a monotonous decay, provided ($\mathcal{M}$ is some set with positive Lebesgue measure)

$$
\nu(\mathbf{x},\lambda) > 0, \quad (\mathbf{x},\lambda) \in \mathcal{M}, \qquad (2.7)
$$

the condition for the system to be absorptive (dissipative). Note that the situation is quite different from that of C_0-semigroups of dissipative operators, where the dissipativity is generated by non-selfadjointness of the generator, rather than a time-convolution as is the case here. For $\chi(\mathbf{x},t)$ with fast decay in t, the former case can be obtained as an approximation [**5**]. Introducing Laplace transforms according to

$$
\widehat{f}(z) = \int_0^{\infty} dt \exp[izt]f(t), \quad \mathrm{Im} z \geqslant 0, \qquad (2.8)
$$

and setting $t_0 = 0$, it follows from Maxwell's equations that

$$
[z^2\varepsilon(\mathbf{x},z) - \mathsf{H}_0]\hat{\boldsymbol{E}}(\mathbf{x},z) = iz\boldsymbol{E}(\mathbf{x},0) - \partial_{\mathbf{x}} \times \boldsymbol{B}(\mathbf{x},0), \quad \mathrm{Im} z > 0, \qquad (2.9)
$$

where

$$
\varepsilon(\mathbf{x},z) = 1 + \hat{\chi}(\mathbf{x},z), \quad \mathrm{Im} z > 0, \qquad (2.10)
$$

and (here Δ is the Laplacian and not an interval)

$$
\mathsf{H}_0 = -\Delta \mathsf{P}^{\perp} = -\partial_{\mathbf{x}}^2 \mathsf{I} + \partial_{\mathbf{x}}\partial_{\mathbf{x}} = \mathbf{p}^2\mathsf{I} - \mathbf{pp}, \quad \mathbf{p} = -i\partial_{\mathbf{x}}, \qquad (2.11)
$$

where $\mathsf{P}^{\perp}$ is the projector upon transverse vector fields, i.e., solutions of $\partial_{\mathbf{x}} \cdot \boldsymbol{\psi} = 0$ and I is the unit 3×3-matrix. The object

$$
z^2\varepsilon(\mathbf{x},z) - \mathsf{H}_0 \qquad (2.12)
$$

is the Helmholtz operator. Note that the real and imaginary components of $\hat{\chi}(\mathbf{x},z)$, $\mathrm{Im} z \geqslant 0$, satisfy a Kramers-Kronig relation. Provided they exist, define

$$
\mathsf{R}_e(z) = [z^2\varepsilon(\mathbf{x},z) - \mathsf{H}_0]^{-1}, \quad \mathsf{R}_e(\omega) = \lim_{\delta\downarrow 0} \mathsf{R}_e(\omega + i\delta), \ \omega \in \mathbb{R}. \qquad (2.13)
$$

Then, if $\tilde{\boldsymbol{E}}(\mathbf{x},\omega)$ is the Fourier transform of $\boldsymbol{E}(\boldsymbol{x},t)$, we obtain formally, with $z = \omega + i\delta$,

$$
\begin{aligned}
\tilde{\boldsymbol{E}}(\omega) &= \lim_{\delta\downarrow 0}(2\pi)^{-1}[\{\mathsf{R}_e(z) - \mathsf{R}_e(z)^*\}\{i\omega\boldsymbol{E}(0) \\
&\quad -\partial_{\mathbf{x}} \times \ \boldsymbol{B}(0)\} - \delta\{\mathsf{R}_e(z) + \mathsf{R}_e(z)^*\}\boldsymbol{E}(0)] \\
&= \lim_{\delta\downarrow 0}(2\pi)^{-1}\{\mathsf{R}_e(z) - \mathsf{R}_e(z)^*\}\{i\omega\boldsymbol{E}(0) - \partial_{\mathbf{x}} \times \ \boldsymbol{B}(0)\}, \qquad (2.14)
\end{aligned}
$$

the last line holding if the limit in (2.13) makes sense. Here we note that, unless $t_0 = -\infty$, $\boldsymbol{E}(\boldsymbol{x},t)$ is so far not defined for $t < t_0$ but this can be remedied using the auxiliary field setup, discussed later on. In view of the above, let us take as a:

DEFINITION 2.1. $\Delta_+ = (\omega_1, \omega_2)$, $0 < \omega_1 < \omega_2$, is a complete band gap if $\mathsf{R}_e(\omega) = \mathsf{R}_e(\omega)^*$ for $\omega \in \Delta_+$.

According to (2.14) this means that for all allowed $\boldsymbol{E}(\boldsymbol{x},0)$, $\boldsymbol{B}(\boldsymbol{x},0)$ there are no propagating field modes with frequencies in $\Delta = \Delta_+ \cup (-\Delta_+)$.

REMARK 2.2. Here and in the following Fourier transforms are defined according to

$$\tilde{f}(\omega) = (2\pi)^{-1} \int_{-\infty}^{+\infty} dt \exp[i\omega t] f(t), \quad \omega \in \mathbb{R}. \tag{2.15}$$

REMARK 2.3. Existence of the boundary value operators in (2.13) in a suitable topology is one of the difficulties in the present situation. In the conservative case it is solved via a variant of the so-called "limiting absorption principle". However, for periodic absorptive dielectric functions, as considered below, existence can be established by means of a Floquet-Bloch decomposition (see Sections III and IV) At this point we note that, at least formally,

$$\mathsf{R}_e(\omega) - \mathsf{R}_e(\omega)^* = -2i\omega^2 \mathsf{R}_e(\omega)^* \mathrm{Im}\,\varepsilon(\mathbf{x},\omega) \mathsf{R}_e(\omega), \quad \varepsilon(\mathbf{x},\omega) = \lim_{\delta\downarrow 0} \varepsilon(\mathbf{x},\omega+i\delta), \tag{2.16}$$

so we do not expect band gaps to exist if $\mathrm{Im}\,\varepsilon(\mathbf{x},\omega) \neq 0$.

3. Absence of band gaps for absorptive systems

We will consider here only situations where $\chi(\mathbf{x},t)$ is constant as a function of $\mathbf{x}$ over the absorptive medium and zero in the non-absorptive surroundings. A typical situation is that of non-overlapping absorptive spheres centered on lattice sites. Denoting by Ψ_a and Ψ_b the characteristic functions of these components, we have

$$\varepsilon(\mathbf{x},\omega) = \Psi_a(\mathbf{x})\varepsilon_a(\omega) + \Psi_b(\mathbf{x}), \tag{3.1}$$

where $\Psi_{a,b}(\mathbf{x})$ have the periodicity of a lattice of dimension d, $1 \leqslant d \leqslant 3$. As noticed above, an important quantity is $\mathrm{Im}\,\varepsilon_a(\omega)$, or

$$n(\omega) = \frac{\omega}{\pi} \mathrm{Im}\,\varepsilon_a(\omega). \tag{3.2}$$

We make the following assumptions:

A_1 : $\chi(\mathbf{x},t) = \Psi_a(\mathbf{x})g(t)$ with $g(t) = 0$ if $t \leqslant 0$ and g possesses a Laplace transform $\hat{g}$ having boundary values $\hat{g}(\omega + i0)$, $\omega \in \mathbb{R}$. Then

$$\varepsilon_a(\omega) = 1 + \hat{g}(\omega + i0), \quad \omega \in \mathbb{R}. \tag{3.3}$$

A_2 : $g \in \mathsf{C}^1(\mathbb{R})$ and g' has a representation

$$\begin{aligned} g'(t) &= \int_{-\infty}^{+\infty} d\omega\, n(\omega) \cos\omega t, \quad t \geqslant 0, \\ n(\omega) &\in \mathsf{C}^0(\mathbb{R}) \cap L^1(\mathbb{R}), \\ n(\omega) &= n(-\omega) \geqslant 0. \end{aligned} \tag{3.4}$$

Two important examples are provided by the following classical models:

a) The Lorentz damped oscillator model. Here

$$g(t) = 2\rho \exp[-\gamma t] \sin\omega_p t, \ t \geqslant 0, \tag{3.5}$$

with ρ the electron density, $\gamma > 0$ a damping factor and ω_p some resonance frequency. Then

$$\varepsilon_a(\omega) = 1 + \rho\left(\frac{1}{\omega+\omega_p+i\gamma} - \frac{1}{\omega-\omega_p+i\gamma}\right),$$

$$n(\omega) = \frac{\rho\gamma\omega_p\omega^2}{[(\omega-\omega_p)^2+\gamma^2][(\omega+\omega_p)^2+\gamma^2]}. \tag{3.6}$$

Notice that ($\|.\|_1$ is the L^1-norm)

$$\|n\|_1 = 2\rho\omega_p \tag{3.7}$$

does not depend on γ.

b) The Drude free-electron model for metals. Here

$$g(t) = \frac{\omega_p^2}{2\gamma}\{1-\exp[-\gamma t]\},\ t \geqslant 0, \tag{3.8}$$

where ω_p is the so-called plasma frequency. Now

$$\varepsilon_a(\omega) = 1 - \frac{\omega_p^2}{\omega(\omega+i\gamma)},\ n(\omega) = \frac{\gamma\omega_p}{\omega^2+\gamma^2} \tag{3.9}$$

Again

$$\|n\|_1 = \omega_p^2 \tag{3.10}$$

is γ-independent.

The fact that in both cases $\|n\|_1$ does not depend on the damping factor γ will have important consequences for the perturbation theory, developed later on. Note that, according to (3.4),

$$\|n\|_1 = \int_{-\infty}^{+\infty} d\omega\, n(\omega) = g'(0), \tag{3.11}$$

which refers to the system at $t = 0$, at which time no absorption has yet taken place, thus explaining the absence of γ in $\|n\|_1$. Notice also that for $\gamma = 0$ (dispersion but no absorption) Drude's model exhibits a large dielectric contrast in frequency windows close to ω_p. Thus it does not come as a real surprise that gaps open in such windows [**3**]. For related work, see [**7**].

REMARK 3.1. If $\varepsilon_a(\omega)$ is a real function, then $n \equiv 0$, so both $g(0)$ and $g'(0)$ are also identically zero. Hence taking $\gamma = 0$ in the Drude model, for example, is not allowed by our assumptions. In fact the Kramers-Kronig relations break down. However, as we shall discuss in section V, it makes sense to consider this non-absorptive situation as a suitable limit of small absorption.

It will be convenient for the investigation of the limits in (2.13) to add the following assumption:

A_3 : $n(\omega)$ has an analytic continuation in an open neighborhood $\mathcal{O}$ of $\mathbb{R}\backslash\{0\}$.

This condition is satisfied for the Lorentz model with $\mathcal{O}$ a strip along $\mathbb{R}$ and for the Drude model with $\mathcal{O}$ a sector

$$S(\beta) = \{z, |\arg z| < \beta\}, \quad \beta < \frac{\pi}{2}. \tag{3.12}$$

Assumption A_3 implies that $\varepsilon_a(z)$ has an analytic continuation from $\mathbb{C}^+ = \{z, \mathrm{Im} z > 0\}$ into an open set $\mathcal{D}$ containing $\mathbb{R}\backslash\{0\}$. We then have the following striking result:

THEOREM 3.2. *Assume $A_1 - A_3$. Then there are no complete band gaps.*

Below we give some ideas of the proof and refer to [**6**] for details. Below ω is always non-zero. Returning to (2.14) and noting (2.16), we can rewrite it after a Floquet-Bloch decomposition as

$$\tilde{\boldsymbol{E}}(\mathbf{k},\omega) = n(\omega) \lim_{\delta \downarrow 0} \mathsf{R}_e(\mathbf{k},\omega + i\delta)^* \Psi_a(\mathbf{x}) \mathsf{R}_e(\mathbf{k},\omega + i\delta)\varphi_0(\mathbf{k}) \tag{3.13}$$

for all $\mathbf{k} \in \mathcal{B}$, $\mathcal{B}$ being the first Brillouin zone. Here

$$\mathsf{R}_e(\mathbf{k}, z) = [z^2\varepsilon(\mathbf{x}, z) - p^2(\mathbf{k})\mathsf{P}^{\perp}]^{-1}, \tag{3.14}$$

where $p^2(\mathbf{k}) = -\Delta$ with ($\mathbf{k}$-dependent) Bloch boundary conditions, acts in $L^2(\mathcal{C}_0, d\mathbf{x}; \mathbb{C}^3)$ with $\mathcal{C}_0$ the unit cell in coordinate space. Now the proof that $\tilde{\boldsymbol{E}}(\mathbf{k},\omega)$ is non-vanishing for any fixed $\omega \neq 0$ and any $\varphi_0(\mathbf{k})$ requires two steps:

a) Existence of $\mathsf{R}_e(\mathbf{k},\omega + i0) = \lim_{\delta \downarrow 0} \mathsf{R}_e(\mathbf{k},\omega + i\delta)$
b) $\Psi_a(\mathbf{x})\mathsf{R}_e(\mathbf{k},\omega + i0)$ is non-zero.

Let us streamline the arguments for a): First we can show that if $n(\omega) \neq 0$, then the equation

$$[\omega^2\varepsilon(\mathbf{x},\omega) - p^2(\mathbf{k})\mathsf{P}^{\perp}]\mathbf{f} = 0, \quad \mathbf{f} \in \mathcal{D}(p^2(\mathbf{k})) \tag{3.15}$$

has no solution for any $\mathbf{k} \in \mathcal{B}$. Then we are in the position to apply Fredholm analytic perturbation theory using the analyticity of $\varepsilon(\mathbf{x},\omega)$, provided by A_3. However, this requires projecting out the infinite-dimensional null space of $\mathsf{P}^{\perp}$ in order to gain compactness of a modified form of $\mathsf{R}_e(\mathbf{k}, z)$. This can be accomplished using the Feshbach projection method [**6**], resulting in:

THEOREM 3.3. *Assume $A_1 - A_3$. Then $\mathsf{R}_e(\mathbf{k}, z)$ can be continued as a meromorphic function into $\mathcal{D}\backslash\{0\}$, where it can have poles with finite-dimensional residues. In addition it has an infinite-dimensional residue at $z = 0$.*

In addition to being an essential ingredient in the proof of Theorem 3.2, this result will have important implications for the understanding of the non-absorptive limit of such systems.

To conclude this section, notice that, although the non-absorptive Drude model, i.e., the Drude model in the limit $\gamma = 0$, does not satisfy $A_1 - A_3$ as mentioned above, and needed to obtain Theorem 3.2, this theorem can raise some doubts about the interpretation of numerical results for the former, showing large band gaps. However, we shall see later that this non-absorptive Drude model, although unphysical from the point of view of the Kramers-Kronig relations, is nevertheless a reliable limit of the absorptive one. Thus such numerical results are quite relevant for the latter. To demonstrate this we first present a both physically and mathematically efficient way to deal with the non-conservative aspects of our models by extending the Maxwell equation framework to a larger one, possessing a unitary time evolution.

4. Auxiliary fields

As shown in Section II, energy is not conserved in absorptive dielectrics. But, as discussed by one of us [**5**], there is nevertheless a conserved quantity, involving, in addition to $\boldsymbol{E}$ and $\boldsymbol{B}$, two new fields, depending on $\mathbf{x}$, t and a new variable $\lambda \in \mathbb{R}$. The extended set satisfies a set of coupled conservative equations, which are equivalent to the original Maxwell's equations (2.1). As we shall see below these equations lead to a strikingly simple framework for the analysis of the Helmholtz resolvent $\mathsf{R}_e(z)$, given by (2.13). Thus let

$$\boldsymbol{F}_1(\mathbf{x},t) = \boldsymbol{E}(\mathbf{x},t), \quad \boldsymbol{F}_3(\mathbf{x},t) = \boldsymbol{B}(\mathbf{x},t), \tag{4.1}$$

with

$$\boldsymbol{F}_j(\mathbf{x},t) \in L^2(\mathbb{R}^3, d\mathbf{x}; \mathbb{C}^3) = \mathcal{H}_j, \quad j \in \{1,3\} \tag{4.2}$$

and

$$\boldsymbol{F}_j(\mathbf{x},t,\lambda) \in L^2(\mathrm{supp}\Psi_a, d\mathbf{x}; \mathbb{C}^3) \otimes L^2(\mathbb{R}, d\lambda) = \mathcal{H}_j, \quad j \in \{2,4\}. \tag{4.3}$$

Then consider the following set of equations

$$\begin{aligned}
\partial_t \boldsymbol{F}_1(\mathbf{x},t) &= \partial_{\mathbf{x}} \times \boldsymbol{F}_3(\mathbf{x},t) + \int d\lambda\, \sigma(\mathbf{x},\lambda)\boldsymbol{F}_4(\mathbf{x},t,\lambda), \\
\partial_t \boldsymbol{F}_2(\mathbf{x},t,\lambda) &= \lambda \boldsymbol{F}_4(\mathbf{x},t,\lambda), \\
\partial_t \boldsymbol{F}_3(\mathbf{x},t) &= -\partial_{\mathbf{x}} \times \boldsymbol{F}_1(\mathbf{x},t), \\
\partial_t \boldsymbol{F}_4(\mathbf{x},t,\lambda) &= -\sigma(\mathbf{x},\lambda)\boldsymbol{F}_1(\mathbf{x},t) - \lambda \boldsymbol{F}_2(\mathbf{x},t,\lambda),
\end{aligned} \tag{4.4}$$

where $\sigma(\mathbf{x},\lambda) = \sigma(\lambda)\Psi_a(\mathbf{x})$, $\sigma(\lambda)^2 = n(\lambda)$, $\sigma(\lambda) \geqslant 0$. It is easily verified that (4.4) with the initial conditions $\boldsymbol{F}_2(\mathbf{x},t_0,\lambda) = \boldsymbol{F}_4(\mathbf{x},t_0,\lambda) = 0$, leads back to (2.1) and (2.2). In the Hilbert space

$$\mathcal{H} = \oplus_{j=1}^4 \mathcal{H}_j, \tag{4.5}$$

the set (4.4) can be written as an evolution equation

$$-i\partial_t \boldsymbol{F}(t) = \mathsf{K}\boldsymbol{F}(t), \tag{4.6}$$

with K, acting in $\mathcal{H}$, selfadjoint. Thus $\|\mathbf{F}(t)\|^2$, which reduces to $2\mathcal{E}_{em}(t)$ if $\chi(t) \equiv 0$, is conserved. It turns out that K^2 has a rather simple form

$$\mathsf{K}^2 = \begin{pmatrix} \mathsf{H}_e & 0 \\ 0 & \mathsf{H}_m \end{pmatrix}, \tag{4.7}$$

where

$$\mathsf{H}_e = \begin{pmatrix} -\Delta \mathsf{P}^{\perp} + \|n\|_1 \Psi_a & \mathsf{M}_u^* \\ \mathsf{M}_u & \lambda^2 \end{pmatrix}, \tag{4.8}$$

with $\mathsf{M}_u : \mathcal{H}_1 \to \mathcal{H}_2$ and $\mathsf{M}_u^* : \mathcal{H}_2 \to \mathcal{H}_1$ given by

$$\begin{aligned}
(\mathsf{M}_u\varphi_1)(\mathbf{x},\lambda) &= u(\mathbf{x},\lambda)\varphi_1(\mathbf{x},\lambda), \\
(\mathsf{M}_u^*\varphi_2)(\mathbf{x}) &= \int_{\mathbb{R}} d\lambda\, u(\mathbf{x},\lambda)\varphi_2(\mathbf{x},\lambda), \\
u(\mathbf{x},\lambda) &= \lambda n^{1/2}(\lambda)\Psi_a(\mathbf{x}).
\end{aligned} \tag{4.9}$$

REMARK 4.1. In fact $u \notin L^2(\mathbb{R}^3, d\mathbf{x}) \otimes L^2(\mathbb{R}, d\lambda)$ under our general assumptions, so the definitions (4.8) must be handled with care. Let P_1 be the projector upon $\mathcal{H}_1$. Then an application of the Feshbach formula results in

$$\mathsf{P}_1[z^2 - \mathsf{H}_e]^{-1}\mathsf{P}_1 = \mathsf{R}_e(z)\mathsf{P}_1, \quad \mathrm{Im} z > 0. \tag{4.10}$$

REMARK 4.2. According to (4.10) $\mathsf{R}_e(z)$ is analytic in $\mathbb{C}^+$.

REMARK 4.3. The diagonal part of H_e consists of two "good looking" operators, one is

$$\mathsf{h} = -\Delta \mathsf{P}^{\perp} + \|n\|_1 \Psi_a, \tag{4.11}$$

which is a Schrödinger-like operator with a periodic potential and accordingly has band spectrum, whereas the second is multiplication with λ^2 with (absolutely continuous) spectrum $\mathbb{R}^+$. Hence we expect that the spectrum of H_e also contains $\mathbb{R}^+$, thus providing another convincing argument for the non-existence of band gaps. As we shall discuss in the next section, (4.10) provides a way to study $\mathsf{R}_e(z)$ perturbatively, its off-diagonal part being a perturbation of its zero order, the diagonal part.

5. Analytic perturbation theory in the small absorption limit

As shown by (2.14) and (4.10) all information about the behavior of $\tilde{\boldsymbol{E}}(\omega)$ in a given frequency range is contained in the resolvent of the operator H_e. After a Floquet-Bloch decomposition we are then led to investigate the family

$$\{[z^2 - \mathsf{H}_e(\mathbf{k})]^{-1}, \mathbf{k} \in \mathcal{B},\ \mathrm{Im} z > 0\} \tag{5.1}$$

and its behavior as $\delta \downarrow 0$ in $z = \omega \pm i\delta$. Recall that

$$\mathsf{H}_e(\mathbf{k}) = \begin{pmatrix} \mathsf{h}(\mathbf{k}) & \mathsf{M}_u^* \\ \mathsf{M}_u & \lambda^2 \end{pmatrix}, \tag{5.2}$$

where

$$\begin{aligned} \mathsf{h}(\mathbf{k}) &= p^2(\mathbf{k})\mathsf{P}^{\perp} + \|n\|_1 \Psi_a(\mathbf{x}), \\ (\mathsf{M}_u \varphi_1)(\mathbf{x}, \lambda) &= u(\mathbf{x}, \lambda)\varphi_1(\mathbf{x}, \lambda), \\ (\mathsf{M}_u^* \varphi_2)(\mathbf{x}) &= \int_{\mathbb{R}} d\lambda\, u(\mathbf{x}, \lambda)\varphi_2(\mathbf{x}, \lambda), \\ u(\mathbf{x}, \lambda) &= \lambda n^{1/2}(\lambda)\Psi_a(\mathbf{x}). \end{aligned} \tag{5.3}$$

Here we again use the notation Ψ_a for the characteristic function of the absorptive component, restricted to the unit cell $\mathcal{C}_0$ of the Γ-periodic medium. We can split $\mathsf{H}_e(\mathbf{k})$ into its diagonal and off-diagonal parts in an obvious way

$$\mathsf{H}_e(\mathbf{k}) = \mathsf{H}^{(0)}(\mathbf{k}) + \mathsf{V}. \tag{5.4}$$

$\mathsf{H}_0(\mathbf{k})$ has as eigenvalues those of $\mathsf{h}(\mathbf{k})$, which are embedded in the continuum $\{\lambda^2 | \lambda \in \mathbb{R}\}$. In order to study their behavior under the perturbation V it is tempting to use the standard trick of spectral deformation, taking into account the analyticity domain in A_3. In case $\mathcal{O}$ is a sector $S_\beta = \{z | |\arg z| < \beta\}$, as for the Drude model, complex scaling the λ-variable is well adapted. We define for f, analytic in S_β and $|\mathrm{Im}\theta| < \beta$,

$$f_\theta(\lambda) \equiv e^{-\frac{\theta}{2}} f(e^{-\theta}\lambda). \tag{5.5}$$

Then we can consider the family

$$\mathsf{H}_e(\mathbf{k}, \theta) = \mathsf{H}^{(0)}(\mathbf{k}, \theta) + \mathsf{V}_\theta, \tag{5.6}$$

where

$$\mathsf{H}^{(0)}(\mathbf{k}, \theta) = \begin{pmatrix} \mathsf{h}(\mathbf{k}) & 0 \\ 0 & e^{-2\theta}\lambda^2 \end{pmatrix} \tag{5.7}$$

now has isolated eigenvalues for $0 < \mathrm{Im}\,\theta < \beta$ (see Fig. 1)

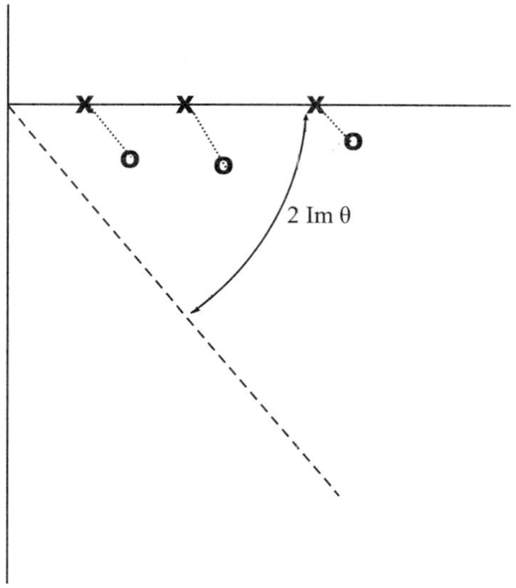

FIGURE 1. The spectrum of $\mathsf{H}_e^{(0)}(\mathbf{k},\theta)$. The crosses are its eigenvalues, the circles perturbed ones, i.e., those of $\mathsf{H}_e(\mathbf{k},\theta)$. The half line going off under an angle $2\mathrm{Im}\theta$ is the continuous spectrum.

and

$$\mathsf{V}_\theta = \begin{pmatrix} 0 & \mathsf{M}^*_{u_\theta} \\ \mathsf{M}_{u_\theta} & 0 \end{pmatrix}. \tag{5.8}$$

This operator is unbounded but is easily seen to be $\mathsf{H}_0(\mathbf{k})$-bounded and analytic in S_β. Thus for sufficiently small $\mathsf{H}^{(0)}(\mathbf{k})$-relative norm of V_θ, non-zero eigenvalues $\mu_m^{(0)}(\mathbf{k})$ of $\mathsf{H}^{(0)}(\mathbf{k})$ will move into the lower half plane, see Fig. 1, and perturbation theory can be applied to study their behavior as long as they stay in S_β, see [**8**]. At this point it is worth noting that, according to the definition (5.3), a small relative norm of V_θ is related to small absorption. However, life is not as simple as it looks. First of all, the functions u_θ, $|\mathrm{Im}\,\theta| < \beta$, are in general not square integrable. This difficulty can be taken care of by standard techniques for the study of relatively bounded perturbations. More seriously, V_θ is, in general, relatively small only in some "frequency windows" of the λ-axis. Thus standard perturbation theory does not apply and we have to develop a local one, again using the Feshbach projection formalism, to project out the bad frequency domains. For the Drude model this is essentially the small frequency region. Summing up, we expect to recover in this way quantitative information about the "resonance spectrum" Σ of $\mathsf{H}_e(\theta)$,

$$\Sigma = \cup_{\mathbf{k}\in\mathcal{B}}\{\mu_m(\mathbf{k}),\ m \in \mathbb{N}\} \tag{5.9}$$

with the $\mu_m(\mathbf{k})$'s the eigenvalues of $\mathsf{H}_e(\mathbf{k},\theta)$, originating from the unperturbed eigenvalues $\mu_m^{(0)}(\mathbf{k})$, $m \in \mathbb{N}$, belonging to the non-zero discrete spectrum of $\mathsf{H}^{(0)}(\mathbf{k})$, i.e., the non-zero spectrum of $\mathsf{h}(\mathbf{k})$. Of course the spectrum of $\mathsf{H}(\theta)$ also has a component arising from the spectrum $e^{-2\theta}\mathbb{R}^+$ of the other diagonal component of $\mathsf{H}_e(\theta)$. But, after projecting out the resolvent of H_e, according to (4.10), this part should play no significant role . In fact, by analytic continuation arguments we have

$$\mathsf{R}_e(z)\mathsf{P}_1 = \mathsf{P}_1[z^2 - \mathsf{H}_e]^{-1}\mathsf{P}_1 = \mathsf{P}_1[z^2 - \mathsf{H}_e(\theta)]^{-1}\mathsf{P}_1, \quad |\mathrm{Im}\,\theta| < \beta, \tag{5.10}$$

and thus only Σ enters, in an intrinsic θ-independent way, in the properties of $\mathsf{R}_e(z)$. In addition (5.10) shows that $\mathsf{R}_e(z)$ has an analytic continuation in z^2 across $\mathbb{R}^+$ up to the upper boundary of Σ, see [**6**]. In this way we obtain another justification of the existence of boundary values in (2.13). On the other hand it is clear from (2.14), that if the imaginary parts of the resonance eigenvalues $\mu_m(\mathbf{k})$ in (5.9) are uniformly small in $\mathbf{k}$ for those eigenvalues close to ω^2, when ω belongs to a given window Δ, then $\tilde{\boldsymbol{E}}(\omega)$ is also small. This can be seen from (4.10) and an expansion of $[z^2 - \mathsf{H}_e(\theta)]^{-1}$ into eigenmodes of $\mathsf{H}_e(\mathbf{k},\theta)$ since the main contribution to the former will come from the eigenvalues in (5.9) closest to ω^2.

Notice that near the band gaps of the non-absorptive limit there should be no contribution from such close resonance eigenvalue when perturbation theory expansions are valid and thus $\tilde{\boldsymbol{E}}(\omega)$ should be negligible in such gaps. This occurs in particular near the complete band gaps of the non-absorptive limit, where perturbation theory works since there should be no such close eigenvalue. Also we can easily convince ourselves that the density of states is also small in such regions. Here we note that the usual definition of the density of states, based upon eigenvalue counting, breaks down. In the original $\mathsf{H}_e(\mathbf{k})$ there are no eigenvalues and in its complex-scaled version they are no longer real. But in physical applications [**5**], it turns out that the role of the density of states is taken over by a function of

$$\mathrm{Im}\{\mathsf{R}_e(\omega + i0) = (2i)^{-1}\{\mathsf{R}_e(\omega + i0) - \mathsf{R}_e(\omega + i0)^*\}, \tag{5.11}$$

which reduces to the usual expression in the absence of absorption.

6. An example: the Drude model

Let us illustrate the perturbative approach for the Drude model (3.9). There

$$u_\theta(\lambda) = \frac{\gamma^{1/2}\omega_p e^{-\theta}\lambda}{\pi^{1/2}[e^{-2\theta}\lambda^2 + \gamma^2]^{1/2}}. \tag{6.1}$$

Obviously u_θ is not uniformly small in the absorption parameter γ; it is even quite large near $\lambda = 0$. However, this "infrared" problem can be taken care of as mentioned above with the result that a convergent perturbation series in γ exists for eigenvalues of $\mathsf{H}_e(\mathbf{k},\theta)$, away from the small frequency region. Consider for example an unperturbed solution

$$\mathsf{h}(\mathbf{k})v_0(\mathbf{k}) = \omega_0^2(\mathbf{k})v_0(\mathbf{k}),\ v_0(\mathbf{k}) \in L^2(\mathcal{C}_0, d\mathbf{x} : \mathbb{C}^3), \tag{6.2}$$

with $0 < \omega_0(\mathbf{k}) < \omega_p$ (recall that ω_p is the plasma frequency and $\|n\|_1 = \omega_p^2$). Note that the discrete spectrum $\sigma_d(\mathsf{h}(\mathbf{k}))$ coincides with the set of real poles of $\mathsf{R}_e(\mathbf{k}, z)$, calculated in the limit $\gamma = 0$, so, in particular $\sigma_d(\mathsf{h}(\mathbf{k})) \cap \Delta) = \emptyset$ for all $\mathbf{k} \in \mathcal{B}$ if Δ is one of the band gaps obtained in this approximation. For sufficiently small γ we then have an expansion of the perturbed, in general complex, eigenvalue $\mu(\mathbf{k})$

$$\mu(\mathbf{k}) = \mu^{(0)}(\mathbf{k}) + \mu^{(2)}(\mathbf{k}) + \mu^{(3)}(\mathbf{k}) + \cdots, \tag{6.3}$$

where $\mu^{(0)}(\mathbf{k}) = \omega_0^2(\mathbf{k})$ and $\mu^{(n)}(\mathbf{k})$is the nthorder correction term. Note that $\mu^{(1)}(\mathbf{k}) = 0$, due to the off-diagonal nature of V. The second order term reads as

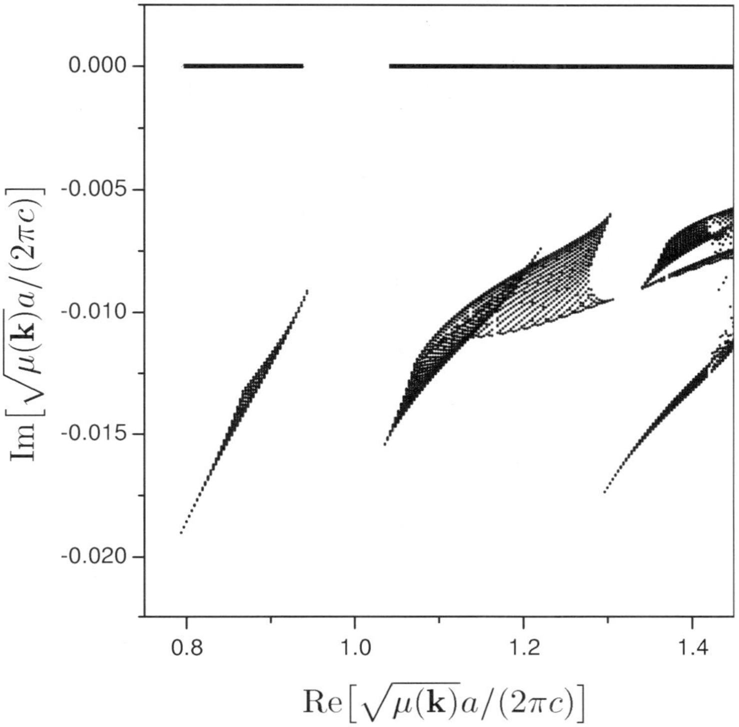

FIGURE 2. Part of the (real) spectrum for $\gamma = 0$ (black line segments on the real axis) and the complex resonance spectrum (black areas in the lower half plane). The considered structure is a square lattice with lattice constant a of Drude metal square rods embedded in vacuum with a filling fraction of 0.65 and a plasma frequency of 1.1 in units $2\pi c/a$, c being the speed of light in vacuum.

follows:

$$\mathrm{Re}\,\mu^{(2)}(\mathbf{k}) = -\|\Psi_a v_0(\mathbf{k})\|^2 \omega_p^2 \frac{\gamma^2}{\omega_0(\mathbf{k})^2 + \gamma^2}, \tag{6.4}$$

$$\mathrm{Im}\,\mu^{(2)}(\mathbf{k}) = -\|\Psi_a v_0(\mathbf{k})\|^2 \omega_p^2 \frac{\gamma\omega_0\,(\mathbf{k}))}{\omega_0(\mathbf{k})^2 + \gamma^2}. \tag{6.5}$$

Such corrections are obviously small for $\gamma << \omega_0(\mathbf{k})$. If, in addition, $\omega_0(\mathbf{k}) << \omega_p$ then the coefficient $\|\Psi_a v_0(\mathbf{k})\|^2$ is also small, the absorptive part of the unit cell $\mathcal{C}_0$ being classically forbidden.

For the particular case of the 2D square rod lattice structure, the mathematical analysis of the non-absorptive limit can be done along the lines developed in [**1**], with the result that large gaps should open. In the small absorption regime our numerical results for the spectrum of $\mathsf{H}_e(\theta)$ are shown in Figs. 2 and 3 for this situation. Even for moderate filling fractions they show that the resonance spectrum Σ remains close to the unperturbed spectrum (note that in these figures ω, with $\omega^2 \in \Sigma$, is plotted in units $2\pi c/a$, c being the speed of light in vacuum and a is the

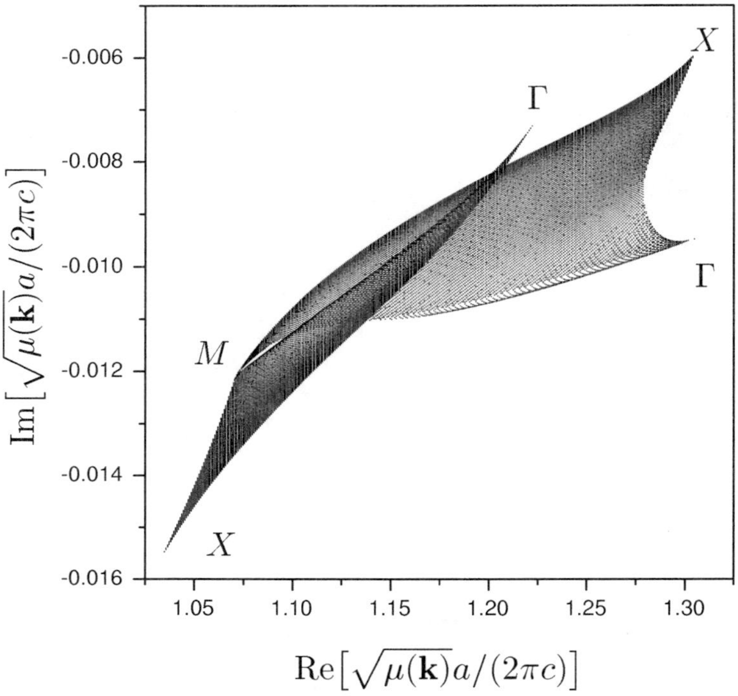

FIGURE 3. Enlarged view of the second area of complex resonances of Fig. 2: $\{\mu_m(\mathbf{k}) \,|\, \mathbf{k} \in \mathcal{B}, m = 2, 3\}$. Γ, X and M correspond to the most degenerated points of the reduced Brillouin zone (triangle ΓXM). Here two contributions with different symmetry partly overlap. Note the small white area to the right of M, where spectrum is absent.

lattice constant). Thus we see that the familiar band structure spectrum of solid state physics and conservative photonic crystals changes into a set of "islands" in the lower half plane. The imaginary part of one of its points is related to the decay in time of the associated solution of Maxwell's equations. Note that stationary solutions no longer exist, the ω in $\exp[-i\omega t]$ now being complex.

7. Concluding remarks and outlook

We have described two types of results which might seem mathematically contradicting. First the non-existence of band gaps for absorptive systems and second the validity of perturbation theory around a certain non-absorptive limit for which band gaps can be shown to exist, both mathematically and numerically. We have seen that nevertheless these results are physically reconciled through the consideration of resonance modes. A notable result is that for the Drude model the eigenvalues $\omega_0^2(\mathbf{k})$ of the selfadjoint operator $\mathsf{h}(\mathbf{k}) = p^2(\mathbf{k})\mathsf{P}^\perp + \|n\|_1 \Psi_a(\mathbf{x})$, cf. (6.2), coincide with the values ω^2 for which

$$(7.1)\quad [\omega^2 \varepsilon^{(0)}(\mathbf{x}, \omega) - p^2(\mathbf{k})\mathsf{P}^\perp]\varphi_\omega = 0, \quad \varepsilon^{(0)}(\mathbf{x}, \omega) = \varepsilon(\mathbf{x}, \omega)|_{\gamma=0} = 1 - \frac{\|n\|_1}{\omega^2}\Psi_a,$$

with frequency-dependent $\varepsilon^{(0)}(\omega)$, has a solution. As stressed earlier, this is due to the fact that $\|n\|_1$ is γ-independent. This is obvious, since we can write (7.1) as

$$[p^2(\mathbf{k})\mathsf{P}^{\perp} + \|n\|_1\Psi_a]\varphi_\omega = \mathsf{h}(\mathbf{k})\varphi_\omega = \omega^2\varphi_\omega, \tag{7.2}$$

but in other cases the situation may be different.

It is an interesting question how to associate the resonance modes with time-decaying states under the effect of absorption, their life-times decreasing with absorption. This is presently under investigation from both mathematical and numerical points of view. The numerical work is concerned with the calculation of Green's functions, associated with the Helmholtz resolvent R_e. The Green's function gives information about the attenuation of an electromagnetic wavepacket travelling through an absorptive device and its "imaginary part" about the radiative decay of embedded atoms. In the first case there is a trade-off between the absorption, occurring once the wavepacket enters an absorptive region and the large reflection from its boundaries that can occur for metallic objects. In an extreme case it may happen that the situation of totally reflecting dielectrics is a useful zero order approximation.

Another interesting issue is what happens with Anderson localization under random perturbations of the lattice. Without absorption localization can occur in the "Lifshits tails" in the gaps of the original unperturbed lattice. Since gaps do not survive if absorption sets in, what happens? This is an important matter, since in experiments absorption and localization give a similar attenuation of a radiation beam crossing the sample under study. A better knowledge of what goes on may help to disentangle the two.

References

[1] A. Figotin and P. Kuchment, SIAM J.Appl.Math **56**, 68, 1561 (1996).
[2] B. Gralak, S. Enoch and G. Tayeb, J. Opt. Soc. Am. A **19**, 1547 (2002).
[3] A. Moroz, Phys. Rev. Lett. **83**, 5274 (1999).
[4] H. van der Lem and A. Moroz, Journal of Optics A: Pure and Applied Optics **2**, 395 (2000).
[5] A. Tip, Phys. Rev. A **57**, 4818 (1998).
[6] A. Tip, A. Moroz and J. M. Combes, J. Phys. A: Math. Gen. **33**, 6223 (2000).
[7] V. Kuzmiak and A. A. Maradudin, Phys. Rev. B **55**, 7427 (1997), V. Yannopapas,A. Modinos and N. Stefanou, Phys. Rev. B **60**, 5359 (1999), W. Y. Zhang, X. Y. Lei, Z. L. Wang, D. G. Zheng, W. Y. Tam, C. T. Chan and Ping Sheng, Phys. Rev. Lett. **84**, 2853 (2000), A. Moroz, A. Tip and J.-M. Combes, Synthetic Metals **116** ,481(2001).
[8] M. Reed and B. Simon, *Methods of Modern Mathematical Physics IV: Analysis of Operators* (Academic Press, New York 1978).

Département de Mathématiques, Université de Toulon et du Var, 83130 La Garde, and, Centre de Physique Théorique, CNRS, Luminy case 907, F13288 Marseille Cedex, France,
E-mail address: combes@cpt.univ-mrs.fr

FOM-Instituut voor Atoom- en Molecuulfysica (AMOLF), Kruislaan 407, Amsterdam, the Netherlands
E-mail address: gralak@amolf.nl

FOM-Instituut voor Atoom- en Molecuulfysica (AMOLF), Kruislaan 407, Amsterdam, the Netherlands
E-mail address: tip@amolf.nl

Contemporary Mathematics
Volume **339**, 2003

Regularity properties for the density of states of random Schrödinger operators

Jean-Michel Combes, Peter D. Hislop, and Frédéric Klopp

ABSTRACT. This paper is a short and introductory review of recent results on the regularity of the density of states of random Schrödinger operators; we mainly review the results from [**8, 4**].

RÉSUMÉ. Cet article présente de façon élémentaire des résultats récents sur les propriétés de régularité de certains opérateurs aléatoires ; l'exposé se concentre sur les résultats obtenus dans [**8, 4**].

0. The random model

On $L^2(\mathbb{R}^d)$, consider the random operator

$$H_\omega = H_0 + V_\omega, \tag{0.1}$$

where $H_0 = (-i\nabla - A_0)^2 + V_0$, and $V_\omega = \sum_{\gamma\in\mathbb{Z}^d} \omega_\gamma V(x-\gamma)$. To avoid technicalities, we assume that the functions $A_0 : \mathbb{R}^d \to \mathbb{R}^d$ and $V_0 : \mathbb{R}^d \to \mathbb{R}$ are smooth. Moreover, the single-site potential V is assumed to be bounded and compactly supported. The parameters $(\omega_\gamma)_\gamma$ are real, bounded, independent and identically distributed (i.i.d.) random variables.

Random Schrödinger operators of the form (0.1) are known as *alloy type* or *Anderson type* models. Their spectral theory has been widely studied during the last 30 years (see [**14, 18**] for an overview). Their success partially stems from their resemblance to the standard discrete Anderson model. This analogy has been widely used to study these models.

1. The integrated density of states

Let us introduce a quantity that proved very useful in the study of the spectral theory of random Schrödinger operators. Define the *integrated density of states* (shortened to *IDS* in the sequel) of H_ω in the following way. Let $\Lambda \subset \mathbb{R}^d$ be a large box and

$$N(E) := \lim_{|\Lambda|\to+\infty} \frac{\#\{\text{eigenvalues of } H_\omega^\Lambda \le E\}}{|\Lambda|}. \tag{1.1}$$

2000 *Mathematics Subject Classification.* Primary 47B80 60H25.

Here, H_ω^Λ is the differential operator H_ω restricted to the box Λ with suitable boundary conditions (Dirichlet or Neumann e.g.). The limit (1.1) exists almost surely under quite general assumptions (e.g. ergodicity); moreover, in many cases, it has been proved to be independent of the boundary conditions one uses to define H_ω^Λ (see e.g. [**14, 13, 10**]). Clearly, $N(E)$ is non-decreasing. Under general assumptions (e.g. ergodicity), it is non-random. Throughout the note, we assume the existence of $N(E)$ for the model we consider.

The question we address in the present note is that of the regularity of the IDS. Not only is this question of interest for its own sake, it is also well known that it plays a crucial role in the spectral theory of random Schrödinger operators, especially in the study of localization through Wegner estimates (see e.g. [**19, 7, 18**])

In complete generality, even for simple models like (0.1), very little is known about the regularity of N. For example, for the model (0.1) with arbitrary randomness, it is not known whether the IDS is a continuous function of energy!

Let us now briefly describe a very general heuristics explaining why the IDS is expected to be regular. Heuristically, the IDS should benefit from the regularity of the IDS of H_0 and of the "IDS" of the potential. Indeed, as $H = H_0 + V_\omega$, if the background operator H_0 and the potential V_ω would commute, one gets $N \sim N_0 * N_V$ where N_0 (resp. N_V) are the density of states of H_0 (resp. V_ω). Actually, as the operators do not commute, one can expect even more regularity for the IDS of the total operator.

2. The Wegner estimate

The main ideas used to obtain regularity results for N were introduced in [**19**] for the simplest discrete model. Except for one dimensional operators, to our knowledge, all known results on the regularity of the IDS use the ideas originating in this paper. We now describe them.

On $\ell^2(\mathbb{Z}^d)$, consider the random operator

$$H_\omega = H_0 + V_\omega$$

where

- $H_0 = ((\delta_{|k-k'|=1}))_{k,k'\in\mathbb{Z}^d}$ is the discrete Laplacian,
- $V_\omega = ((\omega_k \delta_{kk'}))_{k,k'\in\mathbb{Z}^d}$,
- $(\omega_k)_{k\in\mathbb{Z}^d}$ are bounded i.i.d. random variables, distributed according to a bounded density having essential support $[a, b]$.

Let $\Lambda \subset \mathbb{Z}^d$ be a large box. For an open interval $\Delta \subset \mathbb{R}$, let $E_\omega^\Lambda(\Delta)$ denote the spectral projector associated to the energy interval Δ and the operator H_ω^Λ (which, as above, denotes H_ω restricted to Λ). One has

$$\operatorname{tr}(E_\omega^\Lambda(\Delta)) = \operatorname{tr}\left(\int_\Delta \partial_E \mathbf{1}_{(-\infty,0]}(H_\omega^\Lambda - E)dE\right). \tag{2.1}$$

Here, $\mathbf{1}_A$ is the characteristic function of the set A.
Fact 1: The first fact of importance in Wegner's derivation is the following equation

$$-\partial_E(H_\omega^\Lambda - E) = \mathbf{1}_\Lambda = \sum_{\gamma\in\Lambda} \partial_{\omega_\gamma}(H_\omega^\Lambda - E). \tag{2.2}$$

Taking the expectation of (2.1) and using (2.2), one obtains

$$\mathbb{E}[\mathrm{tr}(E_\omega^\Lambda(\Delta))] = -\sum_{\gamma\in\Lambda}\int_\Delta \mathbb{E}[\mathrm{tr}(\partial_{\omega_\gamma}\mathbf{1}_{(-\infty,0]}(H_\omega^\Lambda - E))]dE$$
$$= -\sum_{\gamma\in\Lambda}\int_{\omega'}\mathbb{P}(d\omega')\int_\Delta dE\int_{\omega_\gamma\in[a,b]}\mathbb{P}(d\omega_\gamma)\partial_{\omega_\gamma}\mathbf{1}_{(-\infty,0]}(H_\omega^\Lambda - E).$$

Fact 2: The common probability distribution of the random variables $(\omega_\gamma)_\gamma$ admits a bounded density. Let $[a,b]$ be the essential-support of the random variables $(\omega_\gamma)_\gamma$. Using the boundedness of this density, one computes

(2.3)

$$-\int_{\omega_\gamma\in[a,b]}\mathbb{P}(d\omega_\gamma)\mathrm{tr}(\partial_{\omega_\gamma}\mathbf{1}_{(-\infty,0]}(H_\omega^\Lambda - E)) \le -C\int_{\omega_\gamma\in[a,b]}\mathrm{tr}(\partial_{\omega_\gamma}\mathbf{1}_{(-\infty,0]}(H_\omega^\Lambda - E))d\omega_\gamma$$
$$= C\mathrm{tr}[\mathbf{1}_{(-\infty,0]}(H_{\omega',\omega_\gamma=a}^\Lambda - E) - \mathbf{1}_{(-\infty,0]}(H_{\omega',\omega_\gamma=b}^\Lambda - E)].$$

Here, the operator $H_{\omega',\omega_\gamma=a}^\Lambda$ is the operator H_ω^Λ where ω_γ is fixed $\omega_\gamma = a$ and all the other components $\omega' := (\omega_\beta)_{\beta\neq\gamma}$ are free.

Fact 3: The single-site perturbation $\partial_{\omega_\gamma}V_\omega$ is of rank one. Hence, the trace in the right hand side of (2.3) is bounded by 1. This yields

$$\int_{\omega_\gamma\in[a,b]}\mathbb{P}(d\omega_\gamma)\mathrm{tr}(\partial_{\omega_\gamma}\mathbf{1}_{(-\infty,0]}(H_\omega^\Lambda - E))| \le C. \tag{2.4}$$

Plugging this back into (2.3), and using equations (2.2) and (2.1), one obtains

$$\mathbb{E}(\mathrm{tr}[E_\omega^\Lambda(\Delta)]) \le C\sum_{\gamma\in\Lambda}\int_{\omega'}\mathbb{P}(d\omega')\int_\Delta dE \le C|\Delta||\Lambda|. \tag{2.5}$$

As $\mathrm{tr}[E_\omega^\Lambda(\Delta)]$ is the number of eigenvalues of H_ω^Λ in Δ, estimate (2.5) implies that the IDS is Lipschitz continuous.

Let us now comment on the three crucial facts used during this derivation:

- Fact 1 : Equation (2.2) may be interpreted in the following way : if one wiggles the random parameters, the eigenvalues of the system move. It is easily seen from the Feynman-Hellmann formula that for this property to hold, it is sufficient to have

$$\sum_{\gamma\in\Lambda}V(\cdot-\gamma) \ge C^{-1}\mathbf{1}_\Lambda. \tag{2.6}$$

  Nevertheless, in the case of Hamiltonians of the type (0.1), (2.6) is a strong restriction on the support and sign of single site potential V. We will see in the sequel that this assumption may be weakened quite a bit. It can for example be weakened into

$$\Pi_0(\Lambda)\sum_{\gamma\in\Lambda}V(\cdot-\gamma)\Pi_0(\Lambda) \ge C^{-1}\Pi_0(\Lambda), \tag{2.7}$$

  where $\Pi_0(\Lambda)$ are certain spectral projectors associated to H_0^Λ. As we will see in section 3, the gain is that such assumptions are naturally satisfied by non negative functions V thanks to the Unique Continuation Principle. Under more specific assumptions on H_0, the positivity assumption of V implicit in (2.2) (or (2.6)) can be eliminated altogether. The idea is to read (2.2) as the derivative of $(H_\omega - E)$ with respect to a divergence vector field in probability space. The positivity of this derivative implies that, when the random Hamiltonian is moved along the integral curves of this field, its eigenvalues move. When V does not satisfy (2.6) (or (2.7)), one can look for other vector fields along which H_ω has a "good" behavior. This scheme can be implemented under certain assumptions, as will be explained in section 4.

- Fact 3 : The main consequence of the finiteness of the rank of the single-site perturbation can be interpreted as follows: a change of the potential cannot move too many eigenvalues at a time. In the case of continuous Hamiltonians, the problem is trickier since, obviously, the potential perturbations are not of finite rank. Nevertheless, they are local. It is natural to expect that this should be enough to replace the finite rank assumption. As we will see in section 4, the natural technical tool to put this into practice is Krein's spectral shift function (see e.g. [**1**]). It gives a way of controlling traces similar to the right hand side of (2.3) in terms of certain norms of the potential. Unfortunately, except in dimension 1, the control one gets on these traces is not pointwise as in the finite rank case, but only a control of the L^p-norm for $p \geq 1$; this will induce a loss in the maximal regularity one can prove using this method.
- Fact 2 : The smoothness of the distribution of the $(\omega_\gamma)_\gamma$ is an assumption on the model. This assumption has, up to now, been the most resistant to all the efforts made to get rid of it (except in dimension 1, see e.g. [**2, 16, 6**])! This is quite natural as is seen from the proof shown above. Indeed, the derivation shown above also works if the Hamiltonian is only a pure potential i.e. if $H_\omega = V_\omega$. For this operator, the IDS is nothing but the common probability distribution of the random variables $(\omega_\gamma)_\gamma$. Hence, its regularity alone controls that of the IDS. So, for singular distributions (e.g. if the common distribution of $(\omega_\gamma)_\gamma$ does not admit a density), the contribution of the kinetic energy (i.e. of H_0) will be crucial to obtain some regularity of $N(E)$.

In the sequel, we present results in two settings:

- when the single site potential is non negative but may have arbitrarily small support.
- when the single site potential takes both signs.

As said, the global strategy of the proofs of these results is the one explained above. So we only briefly explain how the two difficulties created by the absence of Facts 1 and 3 are resolved in these settings.

3. For non-negative single site potentials

We now assume that

(H1) the single site potential V is not identically vanishing, non negative, bounded and compactly supported.

(H2) the random variables $(\omega_\gamma)_\gamma$ are i.i.d., bounded with a bounded density.

Then, one proves

THEOREM 3.1 ([**4**]). *Assume hypotheses (H1) and (H2), and that $H_0 = (-i\nabla - A_0)^2 + V_0$ is $\mathbb{Z}^d$-periodic (i.e. A_0 and V_0 are $\mathbb{Z}^d$-periodic).*

Assume, moreover, that H_0 has the Unique Continuation Property.

Then, the IDS for the random operator $H_\omega = H_0 + V_\omega$, is locally Hölder continuous on $\mathbb{R}$, for any Hölder exponent $0 < q < 1$.

If the dimension $d = 1$, and $A_0 = 0$, then the IDS for H_ω is Lipschitz continuous at all energies.

By *locally* Hölder continuous, we mean that the prefactor is locally uniform in the energy. The Hölder exponent q, however, is globally constant. Let us recall the *Unique Continuation Property* (U.C.P. in the sequel). Let H be a differential operator. One says that

$$H \text{ has U.C.P if } H\varphi = E\varphi \text{ and } \varphi \equiv 0 \text{ on open set} \implies \varphi \equiv 0 \text{ everywhere.}$$

It is known that elliptic, second order differential operators with sufficiently regular coefficients have the U.C.P. For example, continuity of the coefficients is sufficient (see e.g. [**20**]).
In the case of Theorem 3.1, Fact 1 is replaced with

PROPOSITION 3.1 ([**4**]). *Assume hypotheses (H1) and (H2), and that H_0 is $\mathbb{Z}^d$-periodic. Let u be a non negative, periodic potential, and $E_0(\cdot)$ be the spectral resolution of H_0.*

If H_0 has the U.C.P. and $u > 0$ on a nonempty open set, then, for $\Delta \subset \mathbb{R}$, a compact interval, there exists a finite constant $C(\Delta, u) > 0$, so that

$$E_0(\Delta) u E_0(\Delta) \geq C(\Delta, u) E_0(\Delta). \tag{3.1}$$

In the proof of Theorem 3.1, equation (3.1) then serves the same role as equation (2.2) in the derivation shown above. Its sufficiency comes from the fact that we only study the local behavior of N; hence, this behavior will be most affected by the local behavior of the integrated density of states of H_0. This fact is also central to the proof of Theorem 3.3

Let us now show how it works a bit more precisely. Let $\Delta \subset \mathbb{R}$ be a bounded energy interval, and $\Lambda \subset \mathbb{R}^d$ be a large cube, as above. Assume Δ is small. Pick $\tilde{\Delta}$ another real interval centered around Δ; the length of $\tilde{\Delta}$ denoted by $|\tilde{\Delta}|$ will be chosen large. The goal is to estimate the trace $\mathbb{E}[\mathrm{tr}(E_\omega^\Lambda(\Delta))]$ by a term of the form $\mathbb{E}[\mathrm{tr}(E_\omega^\Lambda(\Delta)V^\Lambda)]$, where V^Λ is the periodic potential $\sum_{\gamma \in \mathbb{Z}^d} V(\cdot - \gamma)$ restricted to the cube Λ. Therefore, we split it into two pieces

$$\mathrm{tr}(E_\omega^\Lambda(\Delta)) = \mathrm{tr}(E_\omega^\Lambda(\Delta)E_0^\Lambda(\tilde{\Delta})) + \mathrm{tr}(E_\omega^\Lambda(\Delta)E_0^\Lambda({}^c\tilde{\Delta})), \tag{3.2}$$

where ${}^c\tilde{\Delta} \equiv \mathbb{R}\backslash\tilde{\Delta}$. To estimate the second term $\mathrm{tr}(E_\omega^\Lambda(\Delta)E_0^\Lambda({}^c\tilde{\Delta}))$, we pick E to be the center of Δ and write

$$\begin{aligned} \mathrm{tr}(E_\omega^\Lambda(\Delta)E_0^\Lambda({}^c\tilde{\Delta})) = \mathrm{tr}(E_\omega^\Lambda(\Delta)(H_\omega^\Lambda - E)(H_0 - E)^{-1}E_0^\Lambda({}^c\tilde{\Delta})) - \\ - \mathrm{tr}(E_\omega^\Lambda(\Delta)V_\omega^\Lambda(H_0 - E)^{-1}E_0^\Lambda({}^c\tilde{\Delta})) \end{aligned} \tag{3.3}$$

The first term in the right hand side of (3.3) is estimated by $C|\Delta|\mathrm{tr}(E_\omega^\Lambda(\Delta))$. As $0 \leq V_\omega^\Lambda \leq CV^\Lambda$, modulo some additional work, the second term is estimate by $\mathbb{E}[\mathrm{tr}(E_\omega^\Lambda(\Delta)V^\Lambda)]$.
To estimate the first term in the right hand side of (3.2), we use Proposition 3.1. Thanks to (3.1), there exists $C > 0$ (depending only on V and $\tilde{\Delta}$) such that

$$\mathrm{tr}(E_\omega^\Lambda(\Delta)E_0^\Lambda(\tilde{\Delta})) \leq C\mathrm{tr}(E_\omega^\Lambda(\Delta)E_0^\Lambda(\tilde{\Delta})V^\Lambda E_0^\Lambda(\tilde{\Delta})).$$

Replacing $E_0^\Lambda(\tilde\Delta)$ with $\mathbf{1} - E_0^\Lambda({}^c\tilde\Delta)$, we can estimate $\mathrm{tr}(E_\omega^\Lambda(\Delta)E_0^\Lambda(\tilde\Delta))$ with terms of the form $\mathrm{tr}(E_\omega^\Lambda(\Delta)E_0^\Lambda({}^c\tilde\Delta)V^\Lambda E_0^\Lambda(\tilde\Delta)E_\omega^\Lambda(\Delta))$ or $\mathrm{tr}(E_\omega^\Lambda(\Delta)E_0^\Lambda({}^c\tilde\Delta)V^\Lambda E_\omega^\Lambda(\Delta))$. As $\tilde\Delta$ is large, the first type of terms can be incorporated into $\mathrm{tr}(E_\omega^\Lambda(\Delta)E_0^\Lambda(\tilde\Delta))$. Modulo some additional work, the terms of the second type can be estimated by $\mathrm{tr}(E_\omega^\Lambda(\Delta)V^\Lambda)$.
Finally we get that there exists a finite constant $C > 0$ such that, for Δ sufficiently small, one has

$$\mathbb{E}[\mathrm{tr}(E_\omega^\Lambda(\Delta))] \leq C\mathbb{E}[\mathrm{tr}(E_\omega^\Lambda(\Delta)V^\Lambda)]. \tag{3.4}$$

Now, write

$$E_\omega^\Lambda(\Delta)V^\Lambda = \sum_{\gamma\in\Lambda} \partial_{\omega_\gamma}\chi(H_\omega^\Lambda) \tag{3.5}$$

for some finite constant $C > 0$, and some non-negative, increasing function χ. Taking the expectation of (3.5) and performing a partial integration in ω_γ (here, we use the assumption on the regularity of the distribution of ω_γ), we are left with estimating $\mathrm{tr}(\chi(H_b - E) - \chi(H_a - E))$ where $H_b = H^\Lambda_{\omega',\omega_\gamma=b}$.

Let us now explain how we replace Fact 3. Therefore, we follow the ideas and results of [**3**]. Namely, we estimate $\mathrm{tr}(\chi(H_b - E) - \chi(H_a - E))$ using Krein's spectral shift function (SSF) in the following way

$$\mathrm{tr}(\chi(H_b - E) - \chi(H_a - E)) = -\int_{\mathbb{R}} \chi'(\lambda)\zeta(H_b, H_a, \lambda - E)d\lambda.$$

For $1 \leq p < +\infty$ and $2k > pd + 4$, the SSF for a pair of self-adjoint operators $(H_0, H_0 + W)$ satisfies

$$\|\zeta(H_0 + W, H_0, \cdot)\|_{L^p} \leq C(p)\|(i - H_0)^{-k}W\|_{1/p}^{1/p}$$

for some positive constant $C(p)$. Here, $\|A\|_{1/p}^{1/p} = \sum_n \mu_n(A)^{1/p}$ and $\mu_n(A)$ are the singular values of A (see [**3, 9**]). In the application discussed here, $H_0 = H_a$ and $W = H_b - H_a = (b - a)V$, where V is the single site potential. In the argument outlined in section 2, one replaces the $L^1 - L^\infty$-bound performed when going from (2.3) to (2.4) and (2.5) by an $L^p - L^q$-bound. This explains the loss: instead of getting Lipschitz continuity as in the discrete case (section 2), one only recovers Hölder continuity for $N(E)$.

The same kind of ideas can be applied to the Landau model perturbed by an alloy type potential. More precisely, instead of H_0, consider

$$H_L = (-i\nabla - A)^2, \text{ where } A(x_1, x_2) = \frac{B}{2}(-x_2, x_1),$$

and the random model $H_\omega = H_L + V_\omega$. Here, V_ω is the same as above.
One proves

THEOREM 3.2 ([**4**]). *In addition to hypotheses (H1) and (H2), suppose that the magnetic field strength B satisfies a rational flux condition for the unit square i.e.*

$$B \in 2\pi\mathbb{Q}.$$

Then, the IDS for the family $H_\omega = H_L + V_\omega$ is Hölder continuous at all energies with any Hölder exponent $0 < q < 1$.

One cannot apply Theorem 3.1 directly to the magnetic random Hamiltonian H_ω; indeed, the coefficients of this operator are not periodic. Nevertheless, if the rational flux condition is satisfied, they are periodic with respect to the magnetic translations (see e.g. [**17**]). Then, the same arguments as those used to prove Theorem 3.1 yield Theorem 3.2. We believe that the rational flux condition is purely technical (with G. D. Raikov, we recently proved the local Hölder continuity of the IDS at all energies, with no condition of the flux, using a new method, see [**5**]).

We now turn to a different type of result. It is different not so much by the type of property that is proved but rather by the assumptions that are made. Namely, we prove

THEOREM 3.3 ([**4**]). *In addition to hypotheses (H1) and (H2), assume that the operator H_0 admits a density of states $N_0(E)$ that is locally Hölder continuous on $\mathbb{R}$ for some exponent $0 < q_1 \le 1$, i. e. that satisfies*

$$|N_0(E') - N_0(E)| \le C_0 |E' - E|^{q_1}$$

where the finite constant $C_0 > 0$ is locally uniform in energy.
Then, the IDS for H_ω is Hölder continuous on $\mathbb{R}$ for any exponent $0 < q < \dfrac{q_1}{q_1 + 2} \le 1/3$.

Note that H_0 need not be periodic as in Theorem 3.1. This result is certainly not optimal in any way. But, to our knowledge, it is the first to use the regularity of the IDS of the background operator to get some regularity for the total operator. In the proof, we avoid the use of Fact 1 thanks to the regularity of the IDS of the background operator. Let us briefly explain how this is done. Let Δ and Λ be as above. Assume Δ that is small: $|\Delta| < 1$. We now estimate $\mathbb{E}[\mathrm{tr}(E_\omega^\Lambda(\Delta))]$ by terms of the form $\mathbb{E}[\mathrm{tr}(E_\omega^\Lambda(\Delta)V^\Lambda)]$ modulo errors of small size in $|\Delta|$. To estimate the trace $\mathbb{E}[\mathrm{tr}(E_\omega^\Lambda(\Delta))]$, we again start from (3.2). Now, $\tilde{\Delta}$ is chosen small with length $|\Delta|^\alpha$, and $\alpha \in (0,1)$, so that $\Delta \subset \tilde{\Delta}$. The parameter α is fixed later in the argument. Thanks to the regularity of N_0, the expectation of first term in (3.2) can be estimated by $C|\Lambda||\tilde{\Delta}|^{q_1} \sim C|\Lambda||\Delta|^{\alpha q_1}$. To estimate the second term $\mathrm{tr}(E_\omega^\Lambda(\Delta)E_0^\Lambda({}^c\tilde{\Delta}))$, we again split into two pieces as follows

$$\begin{aligned}(3.6)\quad \mathrm{tr}(E_\omega^\Lambda(\Delta)E_0^\Lambda({}^c\tilde{\Delta})) &= \mathrm{tr}(E_\omega^\Lambda(\Delta)(H_\omega^\Lambda - E)(H_0^\Lambda - E)^{-1}E_0^\Lambda({}^c\tilde{\Delta})) + \\ &\quad + \mathrm{tr}(E_\omega^\Lambda(\Delta)V_\omega^\Lambda(H_0^\Lambda - E)^{-1}E_0^\Lambda({}^c\tilde{\Delta}))\end{aligned}$$

where E is the center of Δ. By the definition of $\tilde{\Delta}$ and E, the first term is the right hand side of (3.6) is bounded by $C|\Lambda||\Delta|^{1-\alpha}$. The remaining term can then be compared to $\mathrm{tr}(E_\omega^\Lambda(\Delta)V_\omega^\Lambda)$ and thus dealt with using the same ideas as in the proof of Theorem 3.1.

To end this section, we notice that Theorem 3.3 does not apply to the Landau Hamiltonian (as the IDS for the free Landau Hamiltonian is not Hölder continuous).

4. When the single site potentials change sign

We now assume that

(H1') the single site potential V is not identically vanishing, is bounded and compactly supported;

(H2') the random variables $(\omega_\gamma)_\gamma$ are i.i.d., bounded with an absolutely continuous density.

The lack of positivity of the potential can be partially replaced by some positivity coming from the background operator $H_0 = (-i\nabla - A_0)^2 + V_0$, as above, that need not be periodic.

Let $\Sigma_0 = \inf \sigma(H_0)$ be the bottom of the spectrum of the background operator. Then, we prove

THEOREM 4.1 ([**8**]). *Assume (H1') and (H2'). For any $0 < q < 1$, the IDS is locally Hölder continuous of order q on the interval $(-\infty, \Sigma_0)$.*

Actually the restriction on the energy interval can be relaxed. One can shift the infimum of the spectrum of the background operator

$$H_\omega = \tilde{H}_{t_0} + \tilde{V}_\omega, \text{ where } \tilde{H}_{t_0} = H_0 + t_0 \sum_{\gamma \in \mathbb{Z}^d} V(\cdot - \gamma) \text{ and } \tilde{V}_\omega = \sum_{\gamma \in \mathbb{Z}^d} (\omega_\gamma - t_0) V(\cdot - \gamma).$$

Thus one obtains that the result of Theorem 4.1 is applicable to the energy interval $(-\infty, \Sigma)$, where $\Sigma = \sup_{t_0}[\inf \sigma(\tilde{H}_{t_0})]$.

Let us now briefly describe the proof of Theorem 4.1. Let Λ be a large cube in $\mathbb{Z}^d$ and $\Delta \subset (-\infty, \Sigma_0)$ be a compact interval. Define $\tilde{H}_\omega^\Lambda = H_0 + \sum_{\gamma \in \Lambda} \omega_\gamma V(\cdot - \gamma)$. As above, we want to estimate the expectation of the number of eigenvalues of H_ω^Λ in Δ; as Δ is outside $\sigma(H_0)$, this is roughly equivalent to estimating the eigenvalues of $\tilde{H}_\omega^\Lambda$ in Δ (actually, in a slightly larger interval).

The main idea used to replace Fact 1 dates back to [**12**]: it is to find a vector field in probability space so that, along the flow of this field, the eigenvalues of H_ω^Λ move. Therefore, one introduces the Birman-Schwinger kernel (see e.g. [**15**]) i.e. one writes

$$R_{\omega,\Lambda}(E) = (\tilde{H}_\omega^\Lambda - E_0)^{-1} = (H_0 - E_0)^{-1/2}(1 + \Gamma_\Lambda(E_0;\omega))^{-1}(H_0 - E_0)^{-1/2}$$

where

$$\begin{aligned}\Gamma_\Lambda(E_0;\omega) &= (H_0 - E_0)^{-1/2} V_{\omega,\Lambda} (H_0 - E_0)^{-1/2} \\ &= \sum_{\gamma \in \Lambda} \omega_\gamma (H_0 - E_0)^{-1/2} V(\cdot - \gamma)(H_0 - E_0)^{-1/2}.\end{aligned}$$

The basic fact satisfied by the Birman-Schwinger kernel is

$$(4.1) \qquad E_0 \in (-\infty, \Sigma_0) \cap \sigma(\tilde{H}_\omega^\Lambda) \iff -1 \text{ eigenvalue of } \Gamma_\Lambda(E_0;\omega)$$

and this is valid with multiplicities. The equivalence (4.1) can be extended to

$$(4.2) \quad \begin{aligned}&\#\{\text{eigenvalues of } \tilde{H}_\omega^\Lambda \in [E_0 - \varepsilon, E_0 + \varepsilon]\} \le \\ &\qquad \le \#\{\text{eigenvalues of } \Gamma_\Lambda(E_0;\omega) \in [-1 - C\varepsilon, -1 + C\varepsilon]\}\end{aligned}$$

for some $C > 0$ (independent of ε sufficiently small).

In the probability space, consider now the radial vector field

$$A_\Lambda \equiv \sum_{\gamma \in \Lambda} \omega_\gamma \frac{\partial}{\partial \omega_\gamma}.$$

As $\Gamma_\Lambda(E_0;\omega)$ is homogeneous of degree 1 in the random variables, one has

$$A_\Lambda \Gamma_\Lambda(E_0;\omega) = \Gamma_\Lambda(E_0;\omega).$$

When estimating the right hand side of (4.2), we are looking for eigenvalues of $\Gamma_\Lambda(E_0;\omega)$ close to -1; so, in the region of interest, one has

$$A_\Lambda \lambda_n(E) = \lambda_n(E) \sim -1.$$

Hence, the eigenvalues of $\tilde{H}_\omega^\Lambda$ move when the random variables move along the radial vector field. This is independent of the sign of the single site potential. The main restriction is that we can only study the IDS in region below the spectrum of H_0. Related results on Wegner estimates in dimension 1 were obtained recently in [**11**].

The same ideas also allow to deal with internal gaps of H_0. Therefore, one has to assume that the random perturbation is small. That is, we consider

$$H_\omega(\lambda) = H_0 + \lambda V_\omega,$$

where λ is small. One proves

THEOREM 4.2 ([**8**]). *We assume that H_0 and V_ω satisfy (H1'), (H2') and H_0 has an open gap (B_+, B_-).*

There exists $\lambda_0 > 0$ and $C > 0$ such that, for $|\lambda| < \lambda_0$, for any $q > 1$, on the interval $[B_+ + C\lambda^2, B_- - C\lambda^2]$, the IDS for $H_\omega(\lambda)$ is locally Hölder continuous of order $1/q$.

The ideas used to prove Theorem 4.2 are essentially the ones used to prove Theorem 4.1. But, as we are not studying energies below the spectrum of H_0, we cannot use the Birman-Schwinger kernel directly. We first use the smallness of the random perturbations to decouple the energy region below the gap of H_0, and that above this gap, by the Feshbach projection method. Then, in each of these regions, we are either above or below the spectrum of the free operator and thus, we can use the same technique as for Theorem 4.1.

References

[1] M. Sh. Birman and D. R. Yafaev. The spectral shift function. The papers of M. G. Kreĭn and their further development. *Algebra i Analiz*, 4(5):1–44, 1992.

[2] R. Carmona, A. Klein, and F. Martinelli. Anderson localization for Bernoulli and other random potentials. *Communications in Mathematical Physics*, 108:41–67, 1987.

[3] J.-M. Combes, P. Hislop, and S. Nakamura. The l^p-theory of the spectral shift function, the wegner estimate, and the integrated density of states for some random operators. *Communications in Mathematical Physics*, 218:113–130, 2001.

[4] J.-M. Combes, P. Hislop, and F. Klopp. Hölder continuity of the integrated density of states for some random operators at all energies. *IMRN*, 2003:179–211, 2003.

[5] J.-M. Combes, P. Hislop, F. Klopp, G. Raikov. Global continuity of the integrated density of states for random Landau Hamiltonians. preprint 2003.

[6] D. Damanik, R. Sims, G. Stolz. Localization for one-dimensional, continuum, Bernoulli-Anderson models. *Duke Math. J.*, 114:59–100, 2002.

[7] J. Fröhlich and T. Spencer. Absence of diffusion in the Anderson tight binding model. *Communications in Mathematical Physics*, 88:151–184, 1983.

[8] P. Hislop and F. Klopp. The integrated density of states for some random operators with nonsign definite potentials. *Jour. Func. Anal.*, 195:12–47, 2002.

[9] D. Hundertmark and B. Simon. An optimal l^p-bound on the Krein spectral shift function. *J. d'Analyse Math.*, 87:199–208, 2002.

[10] T. Hupfer, H. Leschke, P. Müller, and S. Warzel. Existence and uniqueness of the integrated density of states for Schrödinger operators with magnetic fields and unbounded random potentials. *Rev. Math. Phys.*, 13(12):1547–1581, 2001.

[11] W. Kirsch and I. Veselic. Existence of the density of states for one-dimensional alloy-type potentials with small support. 2002.

[12] F. Klopp. Localization for some continuous random Schrödinger operators. *Communications in Mathematical Physics*, 167:553–570, 1995.

[13] S. Nakamura. A remark on the Dirichlet-Neumann decoupling and the integrated density of states. *Journal of Functional Analysis*, 179:136–152, 2001.

[14] L. Pastur and A. Figotin. *Spectra of Random and Almost-Periodic Operators*. Springer Verlag, Berlin, 1992.

[15] M. Reed and B. Simon. *Methods of modern mathematical physics. IV. Analysis of operators*. Academic Press [Harcourt Brace Jovanovich Publishers], New York, 1978.

[16] R. Sims and G. Stolz. Localization in one dimensional random media: a scattering theoretic approach. *Communications in Mathematical Physics*, 213(3):575–597, 2000.

[17] J. Sjöstrand. Microlocal analysis for periodic magnetic Schrödinger equation and related questions. In *Microlocal analysis and applications*, volume 1495 of *Lecture Notes in Mathematics*, Berlin, 1991. Springer Verlag.

[18] P. Stollmann. *Caught by disorder*. Birkhäuser, 2001.

[19] F. Wegner. Bounds on the density of states in disordered systems. *Zeitschrift für Physik B*, 44:9–15, 1981.

[20] T. Wolff. Recent work on sharp estimates in second order elliptic unique continuation problems. In *Fourier analysis and partial differential equations (Miraflores de la Sierra, 1992)*, pages 99–128. CRC, Boca Raton, FL, 1995.

(Jean-Michel Combes) Département de Mathématiques, Université de Toulon et du Var, 83130 La Garde, France

E-mail address: Jean-Michel.Combes@cpt.u niv-mrs.fr

(Peter D. Hislop) Department of Mathematics, University of Kentucky, Lexington, KY 40506–0027 USA

E-mail address: hislop@ms.uky.edu

(Frédéric Klopp) LAGA, U.M.R. 7539 C.N.R.S, Institut Galilée, Université de Paris-Nord, 99 Avenue J.-B. Clément, F-93430 Villetaneuse, France

E-mail address: klopp@math.univ-paris13.fr

Contemporary Mathematics
Volume **339**, 2003

Spectral Properties of Schrödinger Operators with a Strongly Attractive δ Interaction Supported by a Surface

Pavel Exner

ABSTRACT. We investigate the operator $-\Delta - \alpha\delta(x-\Gamma)$ in $L^2(\mathbb{R}^3)$, where Γ is a smooth surface which is either compact or periodic and satisfies suitable regularity requirements. We find an asymptotic expansion for the lower part of the spectrum as $\alpha \to \infty$ which involves a "two-dimensional" comparison operator determined by the geometry of the surface Γ. In the compact case the asymptotics concerns negative eigenvalues, in the periodic case Floquet eigenvalues. We also give a bandwidth estimate in the case when a periodic Γ decomposes into compact connected components. Finally, we comment on analogous systems of lower dimension and other aspects of the problem.

1. Introduction

The aim of the present paper is to discuss asymptotic spectral properties of a class of generalized Schrödinger operators in $L^2(\mathbb{R}^3)$. The corresponding potential will be a negative multiple of the Dirac measure supported by a surface $\Gamma \subset \mathbb{R}^3$. In other words, we are going to treat operators corresponding to the formal expression

$$-\Delta - \alpha\delta(x-\Gamma)\,, \tag{1.1}$$

where $\alpha > 0$ is independent of x; properties of Γ will be specified below.

Apart from being an interesting mathematical question in itself, the problem has a natural motivation coming from quantum mechanics. Long time ago, physicists considered a formal "shrinking limit" for a particle localized in the vicinity of a manifold as a natural approach to quantization [**JK, To, dC**]. These considerations inspired studies of spectral and scattering properties of "fat" curved manifolds – see [**DE, DEK**] and references therein. Recently the mentioned limiting argument was reconsidered on a rigorous footing [**FH**], and related results were obtained for other geometric structures such as planar graphs [**RuS, KZ**]. The operator which played role of the Hamiltonian in these models was (a multiple of) the Dirichlet Laplacian in a neighborhood of the manifold. Sometimes other boundary conditions were used: in the papers [**RuS, KZ**] the Dirichlet boundary condition was replaced by the Neumann one.

1991 *Mathematics Subject Classification.* Primary 81Q05; Secondary 35J10, 35P15, 58J05.
Key words and phrases. Schrödinger operators, singular interactions, geometry.
The author obtained a support from NSF for a part of conference expenses and from GAAS Grant #1048101.

Another natural physical application, namely modeling of electron behavior in quantum wires, semiconductor thin films, and similar structures, motivates us to consider a modification of the above scheme in which the particle would be less strictly coupled to the manifold. A way to achieve this goal is to adopt the operator (1.1) as the Hamiltonian of such a system. It is clear that the confinement in this model takes place at negative energies only. Moreover, such a particle can be found at large distances from Γ, although with a small probability, because the exterior of the manifold is a classically forbidden region.

The "shrinking limit" then corresponds to making the δ coupling strong. The main idea is that for large α the eigenfunctions of the operator (1.1) are localized close to Γ. We employ a two-sided estimate of the eigenvalues using minimax principle in combination with a bracketing argument. We take a layer-type neighborhood of Γ and impose Dirichlet and Neumann conditions at its boundary. In view of the strong localization we obtain in this way precise bounds for the negative part of the spectrum. Using the sketched method we have been able to get asymptotic behavior of the eigenvalues as $\alpha \to \infty$ for the operator (1.1) in $L^2(\mathbb{R}^2)$ with Γ belonging to various curve classes – see [**EY1, EY2**] and [**EY3**] for planar loops with a magnetic field. In [**EK1**] the argument was extended to surfaces in $\mathbb{R}^3$ which are diffeomorphic to $\mathbb{R}^2$ and asymptotically planar; the aim was to show that under additional assumptions the operator (1.1) has then a nontrivial discrete spectrum[1].

In the present paper we are going to treat the analogous problem for two other classes of manifolds without a boundary. The first are compact surfaces, the second periodic ones; we will derive an asymptotic expansion for the eigenvalues in the former case and for Floquet eigenvalues in the latter. An important question for periodic systems is the existence of spectral gaps. In case of a nontrivial periodic curve [**EY2**] open gaps always exist for α large enough. This is not true for surfaces. However, we will be able to prove the existence of gaps for a class of non-connected Γ. In conclusion we will comment on extensions of the mentioned two-dimensional results and some other aspects of the problem.

2. Compact surfaces

2.1. Formulation of the problem and the results. Let $\Gamma \subset \mathbb{R}^3$ be a C^4 smooth compact Riemann surface of a finite genus g, i.e. diffeomorphic to a sphere with g handles attached [**Kli**]. As such, it can be parameterized by a finite atlas. The i-th chart $p_i : U_j \to \mathbb{R}^3$ can be expressed in local coordinates $s_\mu^{(i)}$, $\mu = 1, 2$, the particular choice of which will not be important in the following. Usually, we will suppress the chart index. The metric tensor given in the local coordinates by $g_{\mu\nu} = p_{,\mu} \cdot p_{,\nu}$ defines the invariant surface area element $d\Gamma := g^{1/2} d^2 s$, where $g := \det(g_{\mu\nu})$. Furthermore, the tangent vectors $p_{,\mu}$ are linearly independent, and their cross product $p_{,1} \times p_{,2}$ gives, after rescaling, a unit normal field n on Γ. The Weingarten tensor is then obtained by raising the index in the second fundamental form, $h_\mu{}^\nu := -n_{,\mu} \cdot p_{,\sigma} g^{\sigma\nu}$, where $(g^{\mu\nu})$ means conventionally $(g_{\mu\nu})^{-1}$. The eigenvalues $k_\pm$ of $(h_\mu{}^\nu)$ are the principal curvatures. They determine the Gauss curvature K and mean curvature M by

$$K = \det(h_\mu{}^\nu) = k_+ k_- \,, \quad M = \frac{1}{2} \operatorname{Tr}(h_\mu{}^\nu) = \frac{1}{2}(k_+ + k_-) \,. \tag{2.1}$$

[1]The analogous result in the two-dimensional case was proved under weaker assumptions in [**EI**] and for a curve in $\mathbb{R}^3$ in [**EK2**]; in both cases the δ coupling is not required to be strong.

The object of our interest is the generalized Schrödinger operator with an attractive measure-type potential. The latter is a multiple of the Dirac measure μ_Γ defined by $\mu_\Gamma(B) := \mathrm{vol}(B \cap \Gamma)$ for any Borel $B \in \mathbb{R}^3$, where $\mathrm{vol}(\cdot)$ is two-dimensional Hausdorff measure on Γ. Using the trace map $W^{2,1}(\mathbb{R}^3) \to L^2(\mathbb{R}^3, \mu_\Gamma) \cong L^2(\Gamma, d\Gamma)$ which is well defined in view of a standard Sobolev embedding, and abusing slightly the notation, we can define the quadratic form

$$q_\alpha[\psi] = \|\nabla\psi\|^2_{L^2(\mathbb{R}^3)} - \alpha \int_{\mathbb{R}^3} |\psi(x)|^2 d\mu_\Gamma(x), \quad \psi \in W^{2,1}(\mathbb{R}^3). \tag{2.2}$$

By Theorem 4.2 of [**BEKŠ**] this form is bounded from below and closed. Therefore, it is associated with a unique semibounded self-adjoint operator $H_{\alpha,\Gamma}$ which is regarded as the realization of the formal expression (1.1). Let us remark that since Γ is smooth one can define the operator $H_{\alpha,\Gamma}$ alternatively through boundary conditions which involve the jump of the normal derivative across the surface in the same way as in [**EK1**]. This corresponds well to the physicist's concept of the δ interaction.

Since Γ is compact by assumption, the essential spectrum of $H_{\alpha,\Gamma}$ equals $[0, \infty)$; our aim is to investigate the asymptotic behavior of the negative eigenvalues as $\alpha \to \infty$. It will be expressed in terms of the following comparison operator,

$$S = -\Delta_\Gamma + K - M^2 \tag{2.3}$$

on $L^2(\Gamma, d\Gamma)$, where $\Delta_\Gamma = -g^{-1/2}\partial_\mu g^{1/2} g^{\mu\nu}\partial_\nu$ is the Laplace-Beltrami operator on Γ. We denote the j-th eigenvalue of S as μ_j. Notice that it is bounded from above by the j-th eigenvalue of Δ_Γ because the effective potential

$$K - M^2 = -\frac{1}{4}(k_+ - k_-)^2 \le 0;$$

the two coincide when Γ is a sphere. Our first result then reads as follows.

THEOREM 2.1. *(a) $\#\sigma_{\mathrm{d}}(H_{\alpha,\Gamma}) \ge j$ holds for a fixed integer j if α is large enough. The j-th eigenvalue $\lambda_j(\alpha)$ of $H_{\alpha,\Gamma}$ has then an expansion of the form*

$$\lambda_j(\alpha) = -\frac{1}{4}\alpha^2 + \mu_j + \mathcal{O}(\alpha^{-1}\ln\alpha) \quad \text{as} \quad \alpha \to \infty. \tag{2.4}$$

(b) The counting function $\alpha \mapsto \#\sigma_{\mathrm{d}}(H_{\alpha,\Gamma})$ behaves asymptotically as

$$\#\sigma_{\mathrm{d}}(H_{\alpha,\Gamma}) = \frac{|\Gamma|}{16\pi}\alpha^2 + \mathcal{O}(\alpha) \quad \text{for} \quad \alpha \to \infty, \tag{2.5}$$

where $|\Gamma|$ is the Riemann area of the surface Γ.

2.2. Proof of Theorem 2.1. First we construct a family of layer neighborhoods of Γ. Let $\{n(x) : x \in \Gamma\}$ be a field of unit vectors normal to the manifold. Such a field exists globally because Γ is orientable. Define a map $\mathcal{L} : \Gamma \times \mathbb{R} \to \mathbb{R}^3$ by $\mathcal{L}(x, u) = x + un(x)$. Since Γ is smooth by assumption, it is easy to see that there is an $a_1 > 0$ such that for each $a \in (0, a_1)$ the restriction

$$\mathcal{L}_a(x, u) = x + un(x), \quad (x, u) \in \mathcal{N}_a := \Gamma \times (-a, a), \tag{2.6}$$

is a diffeomorphism of $\mathcal{N}_a$ onto its image $\Omega_a = \{x \in \mathbb{R}^3 : \mathrm{dist}(x, \Gamma) < a\}$.

We fix $a \in (0, a_1)$ and estimate (the negative spectrum of) $H_{\alpha,\Gamma}$ using operators acting in the layer Ω_a. To this aim we define the quadratic forms $\eta^{\pm}_{\alpha,\Gamma}[\cdot]$ with the domains $D(\eta^+_{\alpha,\Gamma}) = W^{2,1}_0(\Omega_a)$ and $D(\eta^-_{\alpha,\Gamma}) = W^{2,1}(\Omega_a)$, respectively, which associate with a vector ψ the value

$$\|\nabla\psi(x)\|^2_{L^2(\Omega_a)} - \alpha \int_{\mathbb{R}^3} |\psi(x)|^2 \, d\mu_\Gamma(x) \,.$$

Both the forms are closed and bounded from below; we call the self-adjoint operators in $L^2(\Omega_a)$ associated with them $H^\pm_{\alpha,\Gamma}$. With this notation we can employ the Dirichlet-Neumann bracketing argument [**RS**] which yields the bounds[2]

$$(2.7) \qquad -\Delta^N_{\Sigma_a} \oplus H^-_{\alpha,\Gamma} \le H_{\alpha,\Gamma} \le -\Delta^D_{\Sigma_a} \oplus H^+_{\alpha,\Gamma} \,, \qquad \Sigma_a := \mathbb{R}^3 \setminus \overline{\Omega_a} \,.$$

In the estimation operators the sets Ω_a and Σ_a are decoupled, so $\sigma(H^\pm_{\alpha,\Gamma})$ is the union of the two spectra. As long as we are interested in the negative eigenvalues, we may take into account $H^\pm_{\alpha,\Gamma}$ only, because the "exterior" operators $\Delta^D_{\Sigma_d}$ and $\Delta^N_{\Sigma_d}$ are positive by definition.

In the next step we make use of the natural curvilinear coordinates in Ω_a. More specifically, we transform $H^\pm_{\alpha,\Gamma}$ by means of the unitary operator

$$\hat{U}\psi = \psi \circ \mathcal{L}_a : \; L^2(\Omega_a) \to L^2(\mathcal{N}_a, d\Omega) \,.$$

The measure $d\Omega$ is associated to the pull-back to $\mathcal{N}_a$ of the Euclidean metric tensor in Ω_a. We denote this pull-back metric tensor by G_{ij}; it has the form

$$G_{ij} = \begin{pmatrix} (G_{\mu\nu}) & 0 \\ 0 & 1 \end{pmatrix}, \qquad G_{\mu\nu} = (\delta^\sigma_\mu - uh_\mu{}^\sigma)(\delta^\rho_\sigma - uh_\sigma{}^\rho) g_{\rho\nu} \,,$$

which yields $d\Omega := G^{1/2} d^2 s \, du$ in local coordinates with $G := \det(G_{ij})$ given by

$$G = g\,[(1-uk_+)(1-uk_-)]^2 = g(1-2Mu+Ku^2)^2.$$

Let $(\cdot,\cdot)_G$ denote the inner product in the space $L^2(\mathcal{N}_a, d\Omega)$. Then the operators $\hat{H}^\pm_{\alpha,\Gamma} := \hat{U} H^\pm_{\alpha,\Gamma} \hat{U}^{-1}$ in $L^2(\mathcal{N}_a, d\Omega)$ are associated with the forms $\psi \mapsto \eta^\pm_{\alpha,\Gamma}[\hat{U}^{-1}\psi]$,

$$(2.8) \qquad \eta^\pm_{\alpha,\Gamma}[\hat{U}^{-1}\psi] = (\partial_i\psi, G^{ij}\partial_j\psi)_G - \alpha \int_\Gamma |\psi(s,0)|^2 \, d\Gamma \,,$$

and they differ by their domains, $W^{2,1}_0(\mathcal{N}_a, d\Omega)$ and $W^{2,1}(\mathcal{N}_a, d\Omega)$ for the $\pm$ sign, respectively. As above the expression $\psi(s,0)$ in (2.8) can be given natural meaning using the trace mapping from $W^{2,1}_0(\mathcal{N}_a, d\Omega)$ or $W^{2,1}(\mathcal{N}_a, d\Omega)$ to $L^2(\Gamma, d\Gamma)$.

It is also useful to remove the factor $1 - 2Mu + Ku^2$ from the weight $G^{1/2}$ in the inner product of $L^2(\mathcal{N}_a, d\Omega)$. This is achieved by means of another unitary transformation, namely

$$(2.9) \qquad U\psi = (1-2Mu+Ku^2)^{1/2}\psi : \; L^2(\mathcal{N}_a, d\Omega) \to L^2(\mathcal{N}_a, d\Gamma du) \,.$$

We will denote the inner product in $L^2(\mathcal{N}_a, d\Gamma du)$ by $(\cdot,\cdot)_g$. The operators $B^\pm_{\alpha,\Gamma} := U\hat{H}^\pm_{\alpha,\Gamma}U^{-1}$ acting in $L^2(\mathcal{N}_a, d\Gamma du)$ are associated with the forms $b^\pm_{\alpha,\Gamma}$ given by

[2] This is the conventional way of expressing the argument. A purist might object against inequalities between operators having different domains. However, they make sense in combination with the quadratic form version of the minimax principle which is what we really need.

$b^{\pm}_{\alpha,\Gamma}[\psi] := \eta^{\pm}_{\alpha,\Gamma}[(U\hat{U})^{-1}\psi]$ which again differ by their domains. A straightforward computation, analogous to that performed in [**DEK**], yields

(2.10)
$$\begin{aligned} b^{+}_{\alpha,\Gamma}[\psi] &= (\partial_\mu\psi, G^{\mu\nu}\partial_\nu\psi)_g + (\psi,(V_1+V_2)\psi)_g + \|\partial_u\psi\|^2_g - \alpha\int_\Gamma |\psi(s,0)|^2 d\Gamma\,, \\ b^{-}_{\alpha,\Gamma}[\psi] &= b^{+}_{\alpha,\Gamma}[\psi] + \int_\Gamma M_a(s)|\psi(s,a)|^2 d\Gamma - \int_\Gamma M_{-a}(s)|\psi(s,-a)|^2 d\Gamma \end{aligned}$$

for ψ from $W^{2,1}_0(\Omega_a, d\Gamma du)$ and $W^{2,1}(\Omega_a, d\Gamma du)$, respectively. The quantity $M_u := (M-Ku)(1-2Mu+Ku^2)^{-1}$ here is the mean curvature of the parallel surface characterized by a fixed value of u, and

$$V_1 := g^{-1/2}(g^{1/2}G^{\mu\nu}J_{,\nu})_{,\mu} + J_{,\mu}G^{\mu\nu}J_{,\nu}\,, \quad V_2 := \frac{K-M^2}{(1-2Mu+Ku^2)^2} \tag{2.11}$$

with $J := \frac{1}{2}\ln(1-2Mu+Ku^2)$ is the effective curvature-induced potential [**DEK**].

The operators $B^{\pm}_{\alpha,\Gamma}$ associated with the forms (2.10) are still not easy to handle because the surface and transverse variables are not decoupled. To get a rougher, but still sufficient, estimate we notice that $1-2Mu+Ku^2$ can be squeezed between the numbers $C_\pm(a) := (1\pm a\varrho^{-1})^2$, where $\varrho := \max(\{\|k_+\|_\infty, \|k_-\|_\infty\})^{-1}$. Consequently, the matrix inequality $C_-(a)g_{\mu\nu} \le G_{\mu\nu} \le C_+(a)g_{\mu\nu}$ is valid. Moreover, the first component of the effective potential behaves as $\mathcal{O}(a)$ for $a\to 0$. Hence we have $|V_1| \le va$ for some $v>0$, while V_2 can be squeezed between the functions $C^{-2}_{\mp}(a)(K-M^2)$, both uniformly in the surface variables. These observations motivate us to define the estimation operators in the following way,

$$\tilde{B}^{\pm}_{\alpha,a} := S^{\pm}_a \otimes I + I \otimes T^{\pm}_{\alpha,a} \tag{2.12}$$

with

$$S^{\pm}_a := -C_\pm(a)\Delta_\Gamma + C^{-2}_{\mp}(a)(K-M^2) \pm va$$

in $L^2(\Gamma,d\Gamma)\otimes L^2(-a,a)$, where $T^{\pm}_{\alpha,a}$ are associated with the quadratic forms

(2.13)
$$\begin{aligned} t^{+}_{\alpha,a}[\psi] &:= \int_{-a}^{a} |\partial_u\psi|^2 du - \alpha|\psi(0)|^2\,, \\ t^{-}_{\alpha,a}[\psi] &:= \int_{-a}^{a} |\partial_u\psi|^2 du - \alpha|\psi(0)|^2 - c_a(|\psi(a)|^2 + |\psi(-a)|^2)\,. \end{aligned}$$

In these relations ψ belongs to $W^{2,1}_0(-a,a)$ and $W^{2,1}(-a,a)$, respectively. In distinction to (2.10) the coefficient $c_a := 2(\|M\|_\infty + \|K\|_\infty a)$ in the boundary term of the second expression is independent of the surface variables s. The operators (2.12) provide us with the sought estimate in view of the obvious inequalities

$$\pm B^{\pm}_{\alpha,\Gamma} \le \pm\tilde{B}^{\pm}_{\alpha,a}\,. \tag{2.14}$$

Since $\tilde{B}^{\pm}_{\alpha,a}$ have separated variables their spectra express through those of their constituent operators. To deal with the transverse part, we employ a simple estimate the proof of which can be found in [**EY1**].

LEMMA 2.2. *There are positive numbers c, c_N such that each one of the operators $T^{\pm}_{\alpha,a}$ has a single negative eigenvalue $\kappa^{\pm}_{\alpha,a}$ satisfying the inequalities*

$$-\frac{\alpha^2}{4}\left(1+c_N e^{-\alpha a/2}\right) < \kappa^{-}_{\alpha,a} < -\frac{\alpha^2}{4} < \kappa^{+}_{\alpha,a} < -\frac{\alpha^2}{4}\left(1-8e^{-\alpha a/2}\right)$$

when the attraction is strong enough, $\alpha > c\max\{a^{-1}, c_a\}$.

On the other hand, the surface part requires the following result.

LEMMA 2.3. *The j-th eigenvalues of the operators* $S_a^\pm$ *satisfy the asymptotic bounds* $|\mu_{j,a}^\pm - \mu_j| \le m_j^\pm a$ *with some positive* $m_j^\pm$ *for all a small enough.*

PROOF. We assume $a < \varrho$ so $C_-(a)$ is positive. Using the definitions of $S_a^\pm$ and $C_\pm(a)$ we get easily the asymptotic bound

$$\|S_a^\pm - C_\pm(a)S\| \le \left(v + (\|K\|_\infty + \|M\|_\infty^2)\varrho^{-1}\right) a + \mathcal{O}(a^2) := m(a)\,.$$

Combing this inequality with the minimax principle we find that $|\mu_{j,a}^\pm - C_\pm(a)\mu_j|$ does not exceed $m(a)$. Using once more the definition of $C_\pm(a)$ we conclude that

$$|\mu_{j,a}^\pm - \mu_j| \le m(a) + a\left|(2\varrho^{-1} + \varrho^{-2}a)\mu_j\right|\,,$$

which implies the sought result for small a. □

Armed with these prerequisites we can now prove the asymptotic expansion for eigenvalues of $H_{\alpha,\Gamma}$. By minimax principle they are squeezed between the respective negative eigenvalues of $\tilde{B}_{\alpha,a}^\pm$. Since each of the operators $T_{\alpha,a}^\pm$ has a single negative eigenvalue, the latter are of the form $\kappa_{\alpha,a}^\pm + \mu_{j,a}^\pm$ provided a is small and α is large enough. For definiteness we suppose that these eigenvalues are ordered in the same way as the $\mu_{j,a}^\pm$'s are. Choosing

$$a = a(\alpha) := 6\alpha^{-1}\ln\alpha \tag{2.15}$$

and making use of the above two lemmata we find that

$$\kappa_{\alpha,a}^\pm + \mu_{j,a}^\pm = -\frac{1}{4}\alpha^2 + \mu_j + \mathcal{O}(\alpha^{-1}\ln\alpha) \tag{2.16}$$

holds in this case as $a \to 0$. Since Γ is compact the spectrum of S is purely discrete accumulating at infinity only. Hence, to any positive integer j there is an α_j such that $\kappa_{\alpha,a}^+ + \mu_{j,a}^+ < 0$ holds for $\alpha > \alpha_j$. Consequently, $\tilde{B}_{\alpha,a}^+$ has at least j negative eigenvalues and the same is, of course, true for $H_{\alpha,\Gamma}$. Furthermore, since the upper and lower bound (2.16) differ by the error term only, we arrive at the claim (a).

Using minimax principle again we infer that there is a two-sided estimate

$$\#\sigma_{\rm d}(S_a^+) = \#\sigma_{\rm d}(\tilde{B}_{\alpha,a}^+) \le \#\sigma_{\rm d}(H_{\alpha,\Gamma}) \le \#\sigma_{\rm d}(\tilde{B}_{\alpha,a}^-) = \#\sigma_{\rm d}(S_a^-)\,. \tag{2.17}$$

Using (2.12) together with the definition of $C_\pm(a)$ and the fact that the effective potential is bounded we find that $\#\sigma_{\rm d}(S_a^\pm) = \#\sigma_{\rm d}(S)(1 + \mathcal{O}(a))$. Similarly, the counting function for the operator (2.3) coincides with that of $-\Delta_\Gamma$, up to the same error. Thus it suffices to employ the well-known Weyl formula – see, e.g., [**Ch2**] – to get the claim (b) and to conclude thus the proof.

2.3. Remarks. The assumption that the surface Γ is connected was made mostly for the sake of simplicity. The argument leading to the asymptotic formula (2.4) modifies easily to the case when Γ is a finite disjoint union of C^4 smooth compact Riemann surfaces of finite genera. The situation is, of course, substantially more complicated if the number of compact connected components is infinite; in the next section we will discuss the particular case when such a Γ is periodic.

Furthermore, we have supposed that Γ is a manifold without a boundary. This was important in deriving the asymptotic expansion (2.4) because otherwise the eigenvalues of μ_j of the comparison operator would not be properly defined. On the other hand, the formula (2.5) remains valid even if Γ has a nonempty and smooth boundary. It can be seen by an easy modification of the above argument. If we construct the neighborhood Ω_a in the described way, it will have the boundary consisting of two parts. One of them, $\partial\Omega_a^{(1)}$, contains as before points having normal distance a from Γ. The additional part, $\partial\Omega_a^{(2)}$, is a subset of the normal surface to Γ at $\partial\Gamma$. The form domains of the estimation operators will be again $W_0^{2,1}(\Omega_a)$ and $W^{2,1}(\Omega_a)$, respectively. Consequently, the operators $H_{\alpha,\Gamma}^{\pm}$ will satisfy Dirichlet and Neumann conditions at the whole boundary. In particular, they will satisfy these condition at the additional part $\partial\Omega_a^{(2)}$ of the boundary, and the same will be true for the boundary conditions which $S_a^{\pm}$ must satisfy at $\partial\Gamma$. The eigenvalues of the last named operators no longer differ by an $\mathcal{O}(a)$ term only. However, the Weyl asymptotics entering (2.17) is the same in both cases, because the difference in the number of surface eigenvalues is hidden in the error term.

It is also instructive to compare the formula (2.5) with the known estimates on the number of eigenvalues for generalized Schrödinger operator with measure-type potentials such as the modified Birman-Schwinger bound given in [**BEKŠ**, Sec. 4]. Our result is valid in the asymptotic regime of strong coupling only but by its very nature it has the correct semiclassical behavior. On the other hand, the mentioned bound holds for any $\alpha > 0$ but solvable examples, for instance with Γ being a sphere [**AGS, BEKŠ**], show that it may be rather crude for $\#\sigma_{\mathrm{d}}(H_{\alpha,\Gamma}) > 1$.

3. Periodic surfaces

3.1. Floquet decomposition. Let $\mathcal{T} \equiv \mathcal{T}_r(b)$ be a discrete Abelian group of translations of $\mathbb{R}^3$ generated by an r-tuple $\{b_i\}$ of linearly independent vectors, where $r = 1, 2, 3$. The starting point for the decomposition is a basic period cell $\mathcal{C}$ of $\mathbb{R}^3$, which is a simply connected set such that $\mathcal{C}_n := \mathcal{C} + \sum_i n_i b_i$ is disjoint with $\mathcal{C}$ for any $n = \{n_i\} \in \mathbb{Z}^r$ different from zero and $\bigcup_{n\in\mathbb{Z}^r} \mathcal{C}_n = \mathbb{R}^3$. It is precompact if and only if $r = 3$. The simplest choice of such a period cell is

$$\mathcal{C} = \left\{ \sum_{i=1}^{r} t_i b_i \,:\, 0 \le t_i < 1 \right\} \times \{b_i\}^{\perp}. \tag{3.1}$$

The main geometric object of this section will be a C^4 smooth Riemann surface $\Gamma \subset \mathbb{R}^3$, not necessarily connected, which is supposed to be periodic, i.e. such that $\mathcal{T}$ acts isometrically on Γ and the quotient space $\Gamma/\mathcal{T}$ is compact. A basic period cell of Γ is defined generally in terms of the group $\mathcal{T}$ and its fundamental domain [**Ch1**]. In general the decomposition of Γ into period cells is independent of the above decomposition of the Euclidean space. However, for our purpose it is important that the two are consistent. Hence we choose the period cell of Γ in the form $\Gamma_{\mathcal{C}} := \Gamma \cap \mathcal{C}$. It is clear that $\partial\Gamma_{\mathcal{C}} = \Gamma \cap \partial\mathcal{C}$ is generally nonempty, in particular, if Γ is connected. The boundary is piecewise smooth if $\partial\mathcal{C}$ has the same property.

We are interested again in the generalized Schrödinger operator $H_{\alpha,\Gamma}$ with a δ interaction supported now by the periodic surface. It is defined as above by means of the quadratic form (2.2); recall that Theorem 4.2 of [**BEKŠ**] used there does not require the compactness of Γ. As usual in a periodic situation our main tool will be the Floquet analysis. We introduce the family of quadratic forms

$$
\begin{aligned}
q_{\alpha,\theta}\,[\psi] &= \|\nabla\psi\|^2_{L^2(\mathcal{C})} - \alpha \int_{\mathcal{C}} |\psi(x)|^2 d\mu_\Gamma(x)\,, \\
\mathrm{Dom}(q_{\alpha,\theta}) &= \{\, \psi \in W^{2,1}(\mathcal{C}) \,:\, \psi(x+b_i) = \mathrm{e}^{i\theta_i}\psi(x) \,\}\,,
\end{aligned}
\tag{3.2}
$$

where $\theta = \{\theta_i\} \in [0, 2\pi)^r$, and denote by $H_{\alpha,\theta}$ the self-adjoint operators associated with them. For simplicity we will not indicate the dependence of these forms and operators on Γ. Modifying the standard reasoning [**RS, EY2**] to the present situation we get the sought decomposition.

LEMMA 3.1. *There is a unitary map* $\mathcal{U} : L^2(\mathbb{R}^3) \to \int^{\oplus}_{[0,2\pi)^r} L^2(\mathcal{C})\, d\theta$ *such that*

$$
\mathcal{U} H_{\alpha,\Gamma} \mathcal{U}^{-1} = \int^{\oplus}_{[0,2\pi)^r} H_{\alpha,\theta}\, d\theta
\tag{3.3}
$$

and

$$
\sigma(H_{\alpha,\Gamma}) = \bigcup_{[0,2\pi)^r} \sigma(H_{\alpha,\theta})\,.
\tag{3.4}
$$

The spectrum of $H_{\alpha,\theta}$ *is purely discrete if* $r = 3$ *while* $\sigma_{\mathrm{ess}}(H_{\alpha,\theta}) = [0, \infty)$ *if* $r = 1, 2$*; the eigenvalues (conventionally arranged in the ascending order, with their multiplicity taken into account) are continuous functions of the quasimomenta* θ_i.

Consequently, behavior of the spectral bands of $H_{\alpha,\Gamma}$ can be found through properties of eigenvalues of the fiber operators. The difference between the situation with $r = 3$ and the "partially periodic" cases, $r = 1, 2$, is that in the former we will get the asymptotic behavior for all bands. Of course, the error term will not be uniform in the band index. Another difference is that in the case $r = 3$ the spectrum is known to be absolutely continuous [**SŠ**], even under weaker assumptions than used here, while for $r = 1, 2$ this remains to be an open problem.

3.2. Fiber operator eigenvalues asymptotics. As before we need a comparison operator. In the present case it is defined on $L^2(\mathcal{C}, \mu_\Gamma) \cong L^2(\Gamma_{\mathcal{C}}, d\Gamma)$ by

$$
S_\theta = -\Delta_\Gamma + K - M^2
\tag{3.5}
$$

with the domain consisting of those $\phi \in W^{2,1}(\Gamma_{\mathcal{C}})$ with $\Delta_\Gamma \phi \in L^2(\Gamma_{\mathcal{C}}, d\Gamma)$. If $\Gamma_{\mathcal{C}}$ has a nontrivial boundary we have to require in addition that ϕ satisfies the Floquet conditions at the points of $\partial\Gamma_{\mathcal{C}}$. One can always choose the atlas in such a way that the local charts are periodic with respect to the group $\mathcal{T}$. In that case the conditions read $\phi(x + b_i) = \mathrm{e}^{i\theta_i}\phi(x)$ and

$$
\frac{\partial\phi(x+b_i)}{\partial s_\mu} = \mathrm{e}^{i\theta_i}\frac{\partial\phi(x)}{\partial s_\mu}\,, \quad 1 \le i \le r\,, \quad \mu = 1, 2\,,
\tag{3.6}
$$

for derivatives with respect to the surface coordinates[3]. Since $\Gamma_{\mathcal{C}}$ is precompact and the curvatures involved are bounded, the spectrum of S_θ is purely discrete for each $\theta \in [0, 2\pi)^r$; we denote the j-th eigenvalue of S_θ as $\mu_j(\theta)$.

---

[3]One can also use a coordinate-free way, for instance, by bending the elementary cell into a torus and moving the quasimomentum from the boundary conditions into the operator.

THEOREM 3.2. *Under the stated assumptions the following claims are valid:*
(a) Fix λ as an arbitrary number if $r = 3$ and a non-positive one for $r = 1, 2$. To any $j \in \mathbb{N}$ there is $\alpha_j > 0$ such that $H_{\alpha,\theta}$ has at least j eigenvalues below λ for any $\alpha > \alpha_j$ and $\theta \in [0, 2\pi)^r$. The j-th eigenvalue $\lambda_j(\alpha, \theta)$ has then the expansion

$$\lambda_j(\alpha, \theta) = -\frac{1}{4}\alpha^2 + \mu_j(\theta) + \mathcal{O}(\alpha^{-1}\ln\alpha) \quad as \quad \alpha \to \infty, \tag{3.7}$$

where the error term is uniform with respect to θ.
(b) If the set $\sigma(S) := \bigcup_{\theta\in[0,2\pi)^r} \sigma(S_\theta)$ has a gap separating a pair of bands, then the same is true for $\sigma(H_{\alpha,\Gamma})$ if α is large enough.

PROOF. The argument is the same as in Section 2.2, one has just to modify the domains of the quadratic forms involved and to check that the used estimates are uniform in θ which follows from the continuity of the Floquet eigenvalues. □

3.3. Compactly disconnected periodic surfaces. By the second part of Theorem 3.2 the operator $H_{\alpha,\Gamma}$ has open spectral gaps in the asymptotic regime if the comparison operator has the same property. The latter may or may not be true depending on the geometry of Γ. The situation is different, however, if Γ is not connected and each one of its connected component is compact and contained in (an interior of) a translate of the period cell $\mathcal{C}$. Let us stress that the last named property is a nontrivial assumption; to see that this is the case imagine a family of annular surfaces interlaced neighborwise into an infinite periodic "chain"[4]. An equally important observation is that while in Section 3.2 the choice of the basic period cell $\mathcal{C}$ was mostly irrelevant and one could settle for the simplest one represented by (3.1), it clearly matters here. To give an example[5] consider an infinite array of a boomerang-shaped surfaces: they cannot be stacked in individual rectangular boxes if one wants them to be close enough to each other.

If the assumptions of this section are valid then the domain of the comparison operator is independent of the Floquet conditions (3.6). In that case, S_θ does not depend on θ, and it has the form of a finite direct sum of operators of the type (2.3) for finite-genae surfaces to which the basic period cell $\Gamma_\mathcal{C}$, now automatically closed, can be decomposed (we have noted that in the first remark of Section 2.3). On the other hand, the Floquet decomposition (3.3) of the operator $H_{\alpha,\Gamma}$ is nontrivial and its spectral bands have generically nonzero widths. By the absolute continuity result mentioned above, we know that this is always true in the "fully periodic" case, $r = 3$, while for $r = 1, 2$ the analogous claim is presently just a conjecture.

An easy way to estimate the spectral band widths is to employ a bracketing argument again. Inspecting the domains of the quadratic forms (3.2) we see that the Floquet eigenvalues can be bound from above and below if the boundary conditions for the fiber operator $H_{\alpha,\theta}$ are changed to Dirichlet and Neumann, respectively. Since $\mathrm{dist}(\partial\mathcal{C}, \Gamma_\mathcal{C}) > 0$ holds by assumption, the neighborhood Ω_a of $\Gamma_\mathcal{C}$ is contained in the interior of $\mathcal{C}$ for all $a > 0$ small enough. The negative part of the spectrum of the two estimation operators can be then treated in the exactly the same way as in the case of a compact surface, singly or finitely connected, because the "exterior"

[4]In fact the proof of Theorem 3.3 can be modified to this case too. We use this assumption to avoid a cumbersome formulation needed in more general situations.

[5]A proper name would be an "Australian gift shop" example.

region $\Gamma_C \setminus \mathcal{N}(a)$ contributes to the positive part only (more exactly, from the first eigenvalue up if $r=3$). We arrive thus at the following result.

THEOREM 3.3. *Let Γ be a C^4 smooth periodic surface such that each one of its connected component is compact and contained in (an interior of) a translate of a fixed period cell C of the group $\mathcal{T}$. Denote by μ_j the j-th eigenvalue of the comparison operator (3.5). Then the j-th Floquet eigenvalue $\lambda_j(\alpha,\theta)$ from the decomposition (3.3) of the operator $H_{\alpha,\Gamma}$ behaves asymptotically as*

$$\lambda_j(\alpha,\theta) = -\frac{1}{4}\alpha^2 + \mu_j + \mathcal{O}(\alpha^{-1}\ln\alpha) \quad \textit{for} \quad \alpha\to\infty\,, \tag{3.8}$$

with the error term uniform with respect to the quasimomenta θ. Consequently, the number of open gaps in $\sigma(H_{\alpha,\Gamma})$ exceeds any fixed integer if α is large enough.

4. Concluding remarks

4.1. Curves in the plane. A two-dimensional analogue of the present results was discussed in [**EY1, EY2**], with Γ being a smooth loop or an infinite smooth connected periodic curve. We have derived asymptotic expansions of the form (2.4) and (3.7) where μ_j and $\mu_j(\theta)$, respectively, are eigenvalues of the operator

$$S = -\partial_s^2 - \frac{1}{4}k(s)^2\,. \tag{4.1}$$

Here s is the arc length variable and k is the signed curvature of Γ. The boundary conditions were periodic for the loop and the Floquet ones over the period in the other case. As in the three-dimensional situation, it is easy to extend these results to Hamiltonians with the δ interaction supported by a family of curves, be it a finite number of nonintersecting loops or a periodic system with multiple curves[6].

The latter includes families of curves periodic in two directions, i.e. $r=2$ in the terminology of the preceding section. In that case we know from [**BSŠ**] that the spectrum of $H_{\alpha,\Gamma}$ is purely absolutely continuous so none of the spectral bands is degenerate; for $r=1$ this is an open problem again. From the viewpoint of open gaps, it is the absence of noncompact connected components of Γ which is important. If Γ can be broken into finite families of loops confined within the interior of the period cells, the analogue of Theorem 3.3 is valid and the system has many gaps for large α. By [**EY2**] a single periodic connected curve which is not a straight line gives rise to an open gap for large α, because the comparison operator (4.1) has the same property. It is not a priori clear whether the same is true for two or more such curves, because then we compare with a union of band spectra in which the gaps in one component may overlap with bands in the other one. It is not excluded, of course, that some gaps may survive since the curves have the same periodicity group; the problem deserves a deeper investigation.

4.2. Semiclassical interpretation. The results discussed here and in the earlier work mentioned in the introduction can be viewed also from a different perspective. Recall that the deviation of the spectrum of $H_{\alpha,\Gamma}$ from the one corresponding to the ideal manifold described by the comparison operator (2.3) are due to quantum tunneling. Hence they must be sensitive to the appropriate parameter,

[6] A model similar to the last named case was treated by a different technique in [**KK**] and earlier in [**FK**]. The setting used in these papers differs slightly from the present one. It concerns the roles of the coupling and spectral parameters which are switched there.

i.e. the Planck's constant if we reintroduce it into the picture. However, the operator $-h^2\Delta - v\delta(x-\Gamma)$ is the h^2 multiple of (1.1) if we denote $\alpha := vh^{-2}$. In this sense the obtained asymptotic formulae represent a semiclassical approximation.

4.3. Open problems. One can ask whether the "wide" gaps which one has if α is large and Γ is decomposed into compact components will persist when the assumption about non-connected character of Γ is weakened. It is natural to conjecture that the answer depends on properties of the corresponding operator (2.3). For the Laplace-Beltrami operator a construction of connected periodic manifolds exhibiting gaps has been presented recently [**Po1**]. It is based on connecting compact components of a non-connected surface by thin cylinders. It is worth examining what will be the effect of the curvature-induced potential $K - M^2$ which represents additional "potential wells" at the cylinders and connecting necks.

Spectral properties of the Laplace-Beltrami operator were studied also for other surface classes such as locally perturbed periodic ones for which eigenvalues in the gaps may appear [**Po2**]. A similar behavior may be expected for the operator (2.3) and one can ask whether the same will be true asymptotically for the corresponding $H_{\alpha,\Gamma}$. Our present method can yield information only on eigenvalues below the threshold of $\sigma_{\mathrm{ess}}(H_{\alpha,\Gamma})$ because it employs the minimax principle in a substantial way, so another approach is needed.

An extension to higher dimensions, to an m-dimensional Γ in $\mathbb{R}^n$, is also interesting. One conjectures that a similar asymptotic formula will be valid with the effective potential replaced by the function of the principal curvatures derived in [**To**], see also [**FH**]. It is needed, however, that the operator corresponding to the symbol (1.1) makes sense. If the $\operatorname{codim}\Gamma = n-m$ equals one, it is defined as here in terms of quadratic forms. If $n-m = 2, 3$ one can proceed as in [**EK2**] using generalized boundary conditions (and $-\alpha^2/4$ will be replaced by the point-interaction eigenvalue in dimension $n-m$); for $\operatorname{codim}\Gamma > 3$ there is no meaningful operator $H_{\alpha,\Gamma}$, at least as long as we stay within the Hilbert-space theory.

Useful comments by S. Kondej, D. Krejčiřík, and P. Kuchment are gratefully acknowledged. I also appreciate the referee who checked every sentence twice at least.

References

[AGS] J.-P. Antoine, F. Gesztesy, J. Shabani: *Exactly solvable models of spere interaction in quantum mechanics*, J. Phys. **A20** (1987), 3627–3712.

[BSŠ] M.S. Birman, T.A. Suslina, R.G. Shterenberg: *Absolute continuity of the two-dimensional Schrödinger operator with delta potential concentrated on a periodic system of curves*, Algebra i Analiz **12** (2000), 140–177; translated in St. Petersburg Math. J. **12** (2001), 535–567.

[BEKŠ] J.F. Brasche, P. Exner, Yu.A. Kuperin, P. Šeba: *Schrödinger operators with singular interactions*, J. Math. Anal. Appl. **184** (1994), 112–139.

[Ch1] I. Chavel, *Riemannian Geometry; A Modern Perspective*, Cambridge University Press 1993.

[Ch2] I. Chavel, *The Laplacian on Riemannian manifolds*, Proc. ICMS Instructional Conference "Spectral Theory and geometry", Cambridge University Press 1999, pp. 30–75.

[dC] R.C.T. da Costa: *Quantum mechanics of a constrained particle*, Phys. Rev. **A23** (1981), 1982–1987.

[DE] P. Duclos, P. Exner: *Curvature-induced bound states in quantum waveguides in two and three dimensions*, Rev. Math. Phys. **7** (1995), 73–102.

[DEK] P. Duclos, P. Exner, D. Krejčiřik: *Bound states in curved quantum layers*, Commun. Math. Phys. **223** (2001), 13–28.

[EI] P. Exner, T. Ichinose: *Geometrically induced spectrum in curved leaky wires*, J. Phys. **A34** (2001), 1439–1450.

[EK1] P. Exner, S. Kondej, *Bound states due to a strong δ interaction supported by a curved surface*, J. Phys. **A36** (2003), 443–457.
[EK2] P. Exner, S. Kondej: *Curvature-induced bound states for a δ interaction supported by a curve in $\mathbb{R}^3$*, Ann. H. Poincaré **3** (2002), 967–981.
[EY1] P. Exner, K. Yoshitomi, *Asymptotics of eigenvalues of the Schrödinger operator with a strong δ-interaction on a loop*, J. Geom. Phys. **41** (2002), 344–358.
[EY2] P. Exner, K. Yoshitomi, *Band gap of the Schrödinger operator with a strong δ-interaction on a periodic curve*, Ann. H. Poincaré **2** (2001), 1139–1158.
[EY3] P. Exner, K. Yoshitomi, *Persistent currents for 2D Schrödinger operator with a strong δ-interaction on a loop*, J. Phys. **A35** (2002), 3479–3487.
[FK] A. Figotin, P. Kuchment: *Band-gap structure of spectra of periodic dielectric and acoustic media I, II*, SIAM J. Appl. Math. **56** (1996), 68–88, 1561–1620.
[FH] R. Froese, I. Herbst: *Realizing holonomic constraints in classical and quantum mechanics*, Commun. Math. Phys. **220** (2001), 489–535.
[JK] H. Jensen, H. Koppe: *Quantum mechanics with constraints*, Ann. Phys. **63** (1971), 586–591.
[Kli] W. Kligenberg, *A Course in Differential Geometry*, Springer Verlag, New York 1978.
[KK] P. Kuchment, L. Kunyansky: *Spectral properties of high-contrast band-gap materials and operators on graphs*, Experimental Math. **8** (1998), 1–28.
[KZ] P. Kuchment, Hongbiao Zeng: *Convergence of spectra of mesoscopic systems collapsing onto a graph*, J. Math. Anal. Appl. **258** (2001), 671–700.
[Po1] O. Post: *Periodic manifolds with spectral gaps*, J. Diff. Eq., to appear; `math-ph/0207017`.
[Po2] O. Post: *Eigenvalues in spectral gaps of a perturbed periodic manifold*, `math-ph/0207018`.
[RS] M. Reed, B. Simon, *Methods of Modern Mathematical Physics, IV. Analysis of Operators*, Academic Press, New York 1978.
[RuS] J. Rubinstein, M. Schatzman, *Variational problems on multiply connected thin strips, I. Basic estimates and convergence of the Laplacian spectrum*, Arch. Rat. Mech. Anal. **160** (2001), 271–308.
[SŠ] T.A. Suslina, R.G. Shterenberg: *Absolute continuity of the spectrum of the Schrödinger operator with the potential concentrated on a periodic system of hypersurfaces*, Algebra i Analiz **13** (2001), 197–240.
[To] J. Tolar: *On a quantum mechanical d'Alembert principle*, in "Group Theoretical Methods in Physics", Lecture Notes in Physics, vol. 313, Springer, Berlin 1988; pp. 268-274.

Department of Theoretical Physics, Nuclear Physics Institute, Academy of Sciences, 25068 Řež near Prague, Czech Republic
E-mail address: `exner@ujf.cas.cz`

Contemporary Mathematics
Volume **339**, 2003

Absolute Continuity of the Spectra of Periodic Waveguides

Leonid Friedlander

ABSTRACT. Let G be a domain in $\mathbb{R}^n$, $n \geq 2$, that is periodic in x_1-direction and bounded in the complementary directions. Under an additional symmetry assumption, we prove that the spectrum of an x_1-periodic self-adjoint second order differential operator in G is purely absolutely continuous.

Let G be a connected domain in $\mathbb{R}^n$ with smooth boundary such that
1. it is invariant under the shift $x_1 \mapsto x_1 + 2\pi$;
2. the set $\{x \in G : |x_1| \leq \pi\}$ is bounded;
3. it is invariant under the reflection $x_1 \mapsto -x_1$.

In the domain G, we consider a second order self-adjoint elliptic differential operator

$$L(x, D) = -\sum_{p,l=1}^{n} \frac{\partial}{\partial x_p} g_{pl}(x) \frac{\partial}{\partial x_l} + \frac{1}{i} \sum_{l=1}^{n} \left(a_l(x) \frac{\partial}{\partial x_l} + \frac{\partial}{\partial x_l} a_l(x) \right) + V(x), \tag{1}$$

together with either the Dirichlet boundary condition

$$u(x) = 0, \quad x \in \partial G, \tag{2}$$

or the condition

$$\frac{\partial u(x)}{\partial \nu} + c(x) u(x) = 0 \tag{3}$$

on the boundary of G, where

$$\frac{\partial}{\partial \nu} = \sum_{p,l=1}^{n} g_{pi}(x) n_p \frac{\partial}{\partial x_l},$$

with $n_p(x)$ being the components of the outward unit normal vector $n(x)$ to ∂G. The coefficients $g_{pl}(x)$, $a_l(x)$, and $V(x)$ are assumed to be real-valued, smooth up to the boundary, and they are 2π-periodic in x_1. The function $c(x)$ in (3) is assumed to be smooth on ∂G, 2π-periodic in x_1, and

$$\operatorname{Im} c(x) = \sum_{p=1}^{n} a_p(x) n_p(x). \tag{4}$$

2000 *Mathematics Subject Classification.* Primary 35P05; Secondary 47F05.

By L we will denote the operator $L(x, D)$, together with either the Dirichlet boundary conditions or the boundary condition (3). This is a self-adjoint operator in $L^2(G)$.

Our additional assumption will be that the operator L is invariant under the reflection $x_1 \mapsto -x_1$. We will use the following notations. The indices that take values from 1 to n will be denoted by Roman letters; the indices that take values from 2 to n will be denoted by Greek letters. If $x = (x_1, x_2, \dots x_n)$ then $x' = (x_2, \dots, x_n)$, so $x = (x_1, x')$. In terms of the coefficients of L, our symmetry assumption means that

$$
\begin{aligned}
g_{11}(-x_1, x') &= g_{11}(x_1, x'), \quad g_{\alpha\beta}(-x_1, x') = g_{\alpha\beta}(x_1, x'), \\
a_\alpha(-x_1, x') &= a_\alpha(x_1, x'), \quad V(-x_1, x') = V(x_1, x'), \\
g_{1\alpha}(-x_1, x') &= -g_{1\alpha}(x_1, x'), \quad a_1(-x_1, x') = -a_1(x_1, x'), \\
c(-x_1, x') &= c(x_1, x').
\end{aligned} \tag{5}
$$

In this paper, we prove the following theorem.

THEOREM. *Let the operator L be invariant under the shift $x_1 \mapsto x_1 + 2\pi$ and under the reflection $x_1 \mapsto -x_1$. Then its spectrum is purely absolutely continuous.*

Absolute continuity of the spectrum for different periodic differential operators has been studied by many authors in the last 30 years. We are not going to give a historical review here. One can find an extensive list of the literature on the subject in [**KL**]. Absolute continuity in periodic waveguides was studied in [**D**], [**K**], [**SW**], [**SS**]. It is still an open problem, whether the spectrum of any self-adjoint elliptic second order differential operator in $\mathbb{R}^n$ is purely absolutely continuous. We proved in [**F**] that this is the case if one imposes an additional symmetry assumption on the operator. With some small adjustments, the proof from [**F**] is extendible to operators in waveguides. Here, we are making this extension.

Let us recall some facts from Floquet's theory (e.g., see [**K**], [**BS**].) Let $L(k)$, $k \in \mathbb{R}$, be the operator L acting on functions that satisfy the condition

$$
u(x_1 + 2\pi, x') = e^{2\pi i k} u(x_1, x'). \tag{6}
$$

Then the spectrum of the operator L is the union of the spectra of operators $L(k)$. One can interpret $L(k)$ as an operator L acting on functions on a compact set $C = \{x \in G : |x_1| \le \pi\}$ that satisfy the condition (2) or (3) on the lateral boundary $\partial G \cap (-\pi, \pi)$, together with

$$
u(\pi, x') = e^{2\pi i k} u(-\pi, x'), \quad \frac{\partial u}{\partial x_1}(\pi, x') = e^{2\pi i k} \frac{\partial u}{\partial x_1}(-\pi, x'). \tag{7}
$$

By making the substitution $u(x) = \exp(ikx_1)v(x)$, one can see that $L(k)$ is unitarily equivalent to the operator

$$
\begin{aligned}
L'(k) = &- \sum_{p,l=1}^{n} \Big(\frac{\partial}{\partial x_p} + ik\delta_{1p} \Big) g_{pl}(x) \Big(\frac{\partial}{\partial x_l} + ik\delta_{1l} \Big) \\
&+ \frac{1}{i} \sum_{l=1}^{n} \Big(a_l(x) \Big(\frac{\partial}{\partial x_l} + ik\delta_{1l} \Big) + \Big(\frac{\partial}{\partial x_l} + ik\delta_{1l} \Big) a_l(x) \Big) + V(x)
\end{aligned} \tag{8}
$$

acting on x_1-periodic functions. In the case when L is taken with the Dirichlet boundary conditions (2), the operators $L'(k)$ should be taken with the same conditions, so $L'(k)$ becomes an analytic family of operators. The case of boundary conditions (3) is a little more complicated. Namely, if $u(x)$ satisfies (3) then the boundary condition for $v(x)$ is of the same type, but $c(x)$ should be replaced by

$$c(x,k) = c(x) + ik \sum_{p=1}^{n} g_{p1}(x) n_p(x).$$

Note that $L'(k)$ is an operator of type (1), with $a_l(x)$ replaced by $a_l(x)+kg_{1l}(x)$, so the condition (4) holds for $L'(k)$. The problem is that the domain of the operator $L'(k)$ depends on k. To handle this problem, chose a real-valued, smooth in $\bar{G}$, 2π-periodic in x_1 function $\alpha(x)$ such that

$$\frac{\partial \alpha}{\partial \nu}(x) = \sum_{p=1}^{n} g_{1p}(x) n_p(x)$$

on ∂G. One can easily construct such a function by defining it in a small neighborhood of ∂G first, and then, say, multiplying it by a function supported in a smaller neighborhood of ∂G and equal to 1 on ∂G. Then the operator

$$\tilde{L}(k) = e^{ik\alpha(x)} L'(k) e^{-ik\alpha(x)}$$

is of the form (1), with $\partial/\partial x_j$ replaced by $\partial/\partial x_j + ik(\delta_{j1} - \partial\alpha/\partial x_j)$, and this operator should be taken with the boundary conditions (3) on ∂G, together with the periodic conditions in x_1.

In both cases, of the Dirichlet boundary condition and the boundary condition (3), we have found a model for $L(k)$ ($L'(k)$ in the first case and $\tilde{L}(k)$ in the second case) that forms a k-analytic family of operators. One can apply the argument of Thomas, [**Th**], to conclude that absolute continuity of the spectrum of L is equivalent to the following statement: no value of λ is an eigenvalue of $L(k)$ for all *complex* values of k. Here, we understand $L(k)$ as an operator L acting on functions in C that satisfy (2) or (3), and (7). One can replace L by $L - \lambda$, and to prove the theorem, it is sufficient to show that, when its assumptions are met, *there exists a complex number k such that the equation $Lu = 0$ has only a trivial solution in C that satisfies the boundary conditions (2) or (3) and (7).* We assume that it is not the case and, finally, we will get a contradiction.

Let $\Gamma_{\pm} = \{x \in G : x_1 = \pm\pi\}$. Notice that the boundary ∂G is transversal to the planes $\{x_1 = \pm\pi\}$. Indeed, if that were not the case then the tangent plane to ∂G at some point x^0 with $x_1^0 = \pm\pi$ would be perpendicular to the x_1-axis, and, by the implicit function theorem, the boundary ∂G would be given by an equation $x_1 = f(x')$ in a neighborhood of x^0, which contradicts to the fact that G is invariant under reflections in the planes $\{x_1 = \pm\pi\}$ (a consequence of the assumptions about G.) This transversality implies that $\Gamma_{\pm}$ are smooth manifolds with boundary. The symmetry assumptions (5) imply that $g_{1\alpha}(\pm\pi, x') = 0$, so the x_1 direction is normal to both Γ_+ and Γ_-. Then $\partial_\nu = \partial/\partial\nu = \pm g_{11}(\pm\pi, x')\partial/\partial x_1$ is the outward normal derivative to $\Gamma_{\pm}$ at $(\pm\pi, x')$. With this convention, one has the standard Green formula

$$(9) \qquad (Lu, v)_C = (u, Lv)_C - (\partial_\nu u, v)_\Gamma + (u, \partial_\nu v)_\Gamma$$

where $\Gamma = \Gamma_+ \cup \Gamma_-$, and $(\cdot,\cdot)$ is the usual L^2 scalar product. Notice that $a_1(\pm\pi, x') = 0$ (see (5)), so there are no boundary terms that come from first order terms in (1). Both functions u and v in (9) must satisfy the appropriate boundary condition, either (2) or (3). Let $\zeta = \exp(2\pi i k)$. Then the assumption, which will eventually lead to a contradiction, is that the problem

$$(10)\qquad Lu = 0 \quad \text{in } C, \quad u(\pi, x') = \zeta u(-\pi, x'), \quad \frac{\partial u}{\partial \nu}(\pi, x') + \zeta \frac{\partial u}{\partial \nu}(-\pi, x') = 0,$$

together with either the Dirichlet boundary condition or the boundary condition (3), has a non-trivial solution for every $\zeta \neq 0$. In what follows, we will always assume that all functions in question satisfy either (2) or (3).

We will make a reduction of the problem (10) to a problem on Γ_-. Let $\mathcal{M}$ be the space of solutions of the problem $Lu = 0$ in C, $u(\pm\pi, x') = 0$. This space is finite dimensional. We introduce a space

$$(11)\qquad \mathcal{L} = \{\phi(x') \in L^2(\Gamma_-) : \phi(x') = \partial_\nu u(-\pi, x') \quad \text{for some } u \in \mathcal{M}\}.$$

Notice that, in this definition, one can replace $-\pi$ by π because the operator L is invariant under the reflection $x_1 \mapsto -x_1$. It is a standard fact from the elliptic theory that the boundary value problem

$$(12)\qquad Lu = 0 \quad \text{in } C, \quad u(-\pi, x') = \psi(x'),\ u(\pi, x') = 0$$

is solvable if and only if $\psi \perp \mathcal{L}$. If the problem (12) is solvable then its solution is not unique, but one can find the unique solution that satisfies an additional constraint $\partial_\nu u(-\pi, x') \perp \mathcal{L}$. Such a solution will be denoted by $P\psi$. (P stands for "Poisson operator.") For a function $u(x)$ in C, we define $j_\pm u$ to be its normal derivatives on $\Gamma_\pm$. Finally we define the Dirichlet-to-Neumann operators

$$(13)\qquad N_0\psi = j_- P\psi, \quad N_1\psi = j_+ P\psi, \quad \psi \in \mathcal{L}^\perp.$$

In words, one takes the solution $u(x)$ of (12) that satisfies the additional condition $\partial_\nu u(-\pi, x') \perp \mathcal{L}$; then $N_0\psi = \partial_\nu u(-\pi, x')$ and $N_1\psi = \partial_\nu u(\pi, x')$. Clearly, N_0 maps $\mathcal{L}^\perp$ into itself. Because the operator L is invariant under the symmetry $x_1 \mapsto -x_1$, one can interchange Γ_+ and Γ_-. This means that if $u(x)$ is the solution of $Lu = 0$ such that $u(-\pi, x') = 0$, $u(\pi, x') = \psi$, and $\partial_\nu u(\pi, x') \in \mathcal{L}^\perp$ then $\partial_\nu u(\pi, x') = N_0\psi$ and $\partial_\nu u(-\pi, x') = N_1\psi$. This is actually the main reason why the symmetry assumption is helpful.

The situation here is somewhat more complicated than in [**F**], where C was $\mathbb{T}^{n-1} \times [-\pi, \pi]$, and $\Gamma_\pm \cong \mathbb{T}^{n-1}$. In the case when Γ_- is a manifold without boundary, N_0 is an elliptic pseudodifferential operator of order 1, and its principal symbol is the same as the principal symbol of $\sqrt{\Delta}$ where Δ is the Laplace–Beltrami operator that corresponds to the Riemannian metric, with $g_{pl}(x)$ being its contravariant metric tensor components. The calculus of pseudodifferential operators on manifolds with boundary is more complicated than when there is no boundary. One can show that, in this case, N_0 is a sum of a pseudodifferential operator and a singular Green operator (e.g., see [**G**],) and its principal part is also the same as of $\sqrt{\Delta}$ where the Laplacian on Γ_- is taken with the Diriclet boundary conditions if one treats the boundary conditions (2) for L, or it is taken with the Neumann boundary conditions if one treats the problem (3). Notice that the principal symbol of the boundary conditions (3) is the same as for the Neumann boundary conditions. The only information about N_0 that we will need is that N_0 *is a self-adjoint, bounded* *from below operator with discrete spectrum.* N_1 is an integral operator with smooth kernel. In particular, it is a *bounded* operator.

To make the reduction of the problem (10) to an equation on Γ_-, we first disregard the last condition in (10). Let us state couple of facts that are proved in [**F**], and the proof of which is word-by-word the same in the case of waveguides. Here, we are keeping almost the same notations.

1. Let $\zeta \neq \pm 1$. Then the problem

$$Lu = 0 \quad \text{in } C, \quad u(-\pi, x') = \psi(x'), \quad u(\pi, x') = \zeta\psi(x') \tag{14}$$

is solvable if and only if $\psi(x') \in \mathcal{L}^\perp$.

2. Let $\zeta \neq \pm 1, 0$. For $\psi \in \mathcal{L}^\perp$, the general solution of the problem (14) is

$$u(x) = (P\psi)(x_1, x') + \zeta(P\psi)(-x_1, x') + v(x), \quad v \in \mathcal{M}.$$

The last boundary condition from (10) can be written down as

$$2\zeta N_0\psi + (1 + \zeta^2)N_1\psi + \frac{\partial v}{\partial \nu}(\pi, x') + \zeta\frac{\partial v}{\partial \nu}(-\pi, x') = 0.$$

In particular,

$$2\zeta N_0\psi + (1 + \zeta^2)N_1\psi \in \mathcal{L}. \tag{15}$$

Then, for $\zeta \neq \pm 1, 0$, the problem (10) has a non-trivial solution if and only if there exists a non-zero function $\psi \in \mathcal{L}^\perp$ for which (15) holds.

Let Q be the orthogonal projection onto the space $\mathcal{L}^\perp$, and let $z = (\zeta^2 + 1)/2\zeta$. Then (15) can be rewritten as an equation

$$N_0\psi + z\tilde{N}_1\psi = 0, \quad \psi \in \mathcal{L}^\perp, \tag{16}$$

where $\tilde{N}_1 = QN_1$. The assumption of the spectrum of L not being purely absolutely continuous leads to the conclusion that the equation (16) has a non-trivial solution for every $z \neq \pm 1$. As in [**F**], the operators N_0 and $\tilde{N}_1$ are self-adjoint, and the kernel of $\tilde{N}_1$ is trivial. The proof is absolutely the same. Selfadjointness is a consequence of the Green formula (9), and triviality of $\ker\tilde{N}_1$ follows from the unique continuation theorem for solutions of second order differential equations. Now, the Theorem follows from the following

LEMMA. *Let A and B be self-adjoint operators in a Hilbert space H. Suppose that A is bounded from below and its spectrum is discrete; B is a bounded operator. Assume that there exists an open domain $\Omega \subset \mathbb{C}$ such that the operator $A(z) = A + zB$ has a non-trivial kernel for every $z \in \Omega$. Then* $\ker B \neq 0$.

PROOF. The proof of the lemma is basically the same as in [**F**], and it is a simple exercise in elementary Functional Analysis. We present it here for the sake of completeness. It follows from the theory of analytic families of operators (e.g., see [**Ka**]) that the dimension of $\ker A(z)$ is constant for all complex numbers z, outside a discrete set $E \subset \mathbb{C}$. The assumptions of the Lemma imply that this constant is positive. Moreover, the Riesz projections onto $\ker A(z)$ depend analytically on $z \in \mathbb{C} \setminus E$. In particular, one can find a continuous family of unit vectors $\psi(x) \in \ker A(x)$, $x \in \mathbb{R} \setminus E$. We will show that

$$(B\psi(x_1), \psi(x_2)) = 0 \tag{17}$$

for every $x_1, x_2 \in \mathbb{R} \setminus E$. If $x_1 \neq x_2$, then

$$(B\psi(x_1), \psi(x_2)) = \frac{(A(x_1)\psi(x_1), \psi(x_2)) - (\psi(x_1), A(x_2)\psi(x_2))}{x_1 - x_2} = 0.$$

If $x_1 = x_2$, then we take the limit $x_2 \to x_1$ in (17). Here we use the assumption that the operator B is bounded.

Equation (17) implies that

$$(A\psi(x_1), \psi(x_2)) = 0 \tag{18}$$

for every $x_1, x_2 \in \mathbb{R} \setminus E$. Let $\mathcal{F}$ be the linear span of all vectors $\psi(x)$, $x \in \mathbb{R} \setminus E$. Then the quadratic form associated to the operator A vanishes on $\mathcal{F}$. Because the operator A is bounded from below, and it has discrete spectrum, one concludes that $\mathcal{F}$ is finite-dimensional.

Now, let $x_j \in \mathbb{R} \setminus E$, and $x_j \to \infty$. The vectors $\psi_j = \psi(x_j)$ lie on a finite-dimensional sphere, so one can assume that $\psi_j \to \psi$, $\|\psi\| = 1$. The restriction of the operator A to $\mathcal{F}$ is bounded, so the norms of $A\psi_j$ are bounded. Therefore,

$$B\psi = \lim_{j\to\infty} B\psi_j = -\lim_{j\to\infty} \frac{A\psi_j}{x_j} = 0.$$

□

References

[BS] M. Sh. Birman, M. Solomyak, *On the negative discrete spectrum of a periodic elliptic operator in a waveguide-type domain, perturbed by decaying potential*, J. D'Analyse Math. **83** (2001), 337–391.

[D] V. I. Derguzov, *On the discreteness of the spectrum of the periodic boundary-value problem associated with the study of periodic wave guides*, Siberian Math. J. **21** (1980), 664–672.

[F] L. Friedlander, *On the Spectrum of a Class of Second Order Periodic Elliptic Differential Operators*, Commun. Math. Phys. **229** (2002), 49–55.

[G] G. Grubb, *Functional Calculus of Pseudodifferential Boundary Problems*, Birkhäuser, Boston, 1986.

[Ka] T. Kato, *Perturbation theory for linear operators*, Springer Verlag, Berlin–Heidelberg–New York, 1966.

[K] P. Kuchment, *Floquet Theory for Partial Differential Equations*, Birkhäuser Verlag, Basel, 1993.

[KL] P. Kuchment, S. Lenemdorskiî, *On the structure of spectra of periodic elliptic operators*, Trans. Amer. Math. Soc. **354** (2002), 537–569.

[SS] R. G. Shterenberg, T. A. Suslina, *Absolute continuity of the spectrum of a magnetic Schrödinger operator in a two-dimensional periodic waveguide*, Algebra i Analiz **14** (2002), 159–206; English transl., St. Petersburg Math. J. **14** (2003).

[SW] A. V. Sobolev, J. Walthoe, *Absolute Continuity in Periodic Waveguides*, Proc. London Math. Soc. **(3)85** (2002), 717–741.

[Th] L. E. Thomas, *Time Dependent Approach to Scattering from Impurities in a Crystal*, Commun. Math. Phys **33** (1973), 335–343.

Department of Mathematics, University of Arizona, Tucson, Arizona 85721
E-mail address: friedlan@math.arizona.edu

Contemporary Mathematics
Volume **339**, 2003

The Anderson metal-insulator transport transition

François Germinet and Abel Klein

ABSTRACT. We discuss a new approach to the metal-insulator transition for random operators, based on transport instead of spectral properties. It applies to random Schrödinger operators, acoustic operators in random media, and Maxwell operators in random media. We define a local transport exponent $\beta(E)$, and set the *metallic transport region* to be the part of the spectrum with nontrivial transport (i.e., $\beta(E) > 0$). The *strong insulator region* is taken to be the part of the spectrum where the random operator exhibits strong dynamical localization in the Hilbert-Schmidt norm, and hence no transport. For the standard random operators satisfying a Wegner estimate, these metallic and insulator regions are shown to be complementary sets in the spectrum of the random operator, and the local transport exponent $\beta(E)$ provides a characterization of the *metal-insulator transport transition.* If such a transition occurs, then $\beta(E)$ has to be discontinuous at a *transport mobility edge*: if the transport is nontrivial then $\beta(E) \geq \frac{1}{2bd}$, where d is the space dimension and $b \geq 1$ is the power of the volume in Wegner's estimate. We also examine the transport transition for random polymer models, where the random dimer models provide explicit examples of the transport transition and of a transport mobility edge.

1. Introduction

In this survey we discuss a new approach to the metal-insulator transition for random operators based on transport instead of spectral properties, introduced in [**GK3**]. This new point of view, in addition to being closer to the physical meaning of a "metal-insulator" transition, is shown to give a better understanding of the transition.

By a random operator we always mean a $\mathbb{Z}^d$-ergodic random self-adjoint operator H_ω on either $\mathrm{L}^2(\mathbb{R}^d, \mathrm{d}x; \mathbb{C}^n)$ or $\ell^2(\mathbb{Z}^d; \mathbb{C}^n)$, where ω belongs to a probability set Ω with a probability measure $\mathbb{P}$ and expectation $\mathbb{E}$. Note that ergodicity implies the existence of a nonrandom set Σ, such that $\sigma(H_\omega) = \Sigma$ with probability one, where $\sigma(A)$ denotes the spectrum of the operator A. In addition, there are nonrandom sets Σ_{pp}, Σ_{ac}, and Σ_{sc}, which are the pure point spectrum, absolutely continuous spectrum, and singular continuous spectrum of H_ω, respectively, with probability one (see [**KM**]).

2000 *Mathematics Subject Classification.* Primary 82B44; Secondary 47B80, 60H25.
A.K. was supported in part by NSF Grant DMS-0200710.

For a Schrödinger operator with a random potential and spectrum of the form $[E_0, \infty)$, the following picture of a metal-insulator transition has emerged (e.g., [**LGP**, Section 4.2]): Near the bottom of the spectrum E_0 the spectrum is controlled by large fluctuations of the potential, with the corresponding states localized primarily in the regions of such fluctuations. But in three or more dimensions, at very large energies the kinetic term should dominate the fluctuations of the potential to produce extended states. Thus a transition must occur from an *insulator regime*, characterized by localized states, to a very different *metallic regime* characterized by extended states. The energy E_{me} at which this *metal-insulator transition* occurs is called the *mobility edge*. The medium should have zero conductivity in the *insulator region* $[E_0, E_{\mathrm{me}}]$ and nonzero conductivity in the *metallic region* $[E_{\mathrm{me}}, \infty)$.

The standard mathematical interpretation of the metal-insulator transition is as a spectral transition: the random Schrödinger operator should have pure point spectrum with exponentially decaying eigenstates in the interval $[E_0, E_{\mathrm{me}}]$ and absolutely continuous spectrum on the interval $[E_{\mathrm{me}}, \infty)$.

The existence of exponential localization (i.e., pure point spectrum with exponentially decaying eigenstates) is by now well established (e.g., [**GMP, KS, FS, HM, FMSS, CKM, vDK, AM, Ai, CH1, Klo2, KSS, FLM, ASFH, Wa2, GK1, Klo3, Klo4, U, GK4, GK5**]). But there are no mathematical results on the existence of continuous spectrum and a metal-insulator transition. (Except for the special case of the Anderson model on the Bethe lattice, where one of us has proved that for small disorder the random operator has purely absolutely continuous spectrum in a nontrivial interval [**Kle1**] and nontrivial transport [**Kle2**].)

In [**GK3**] we proposed a new approach to the metal-insulator transition based on transport instead of spectral properties, which we describe in this survey. It relies on a local transport exponent $\beta(E)$; the *metallic transport region* is set to be the part of the spectrum with nontrivial transport (i.e., $\beta(E) > 0$). The *strong insulator region* is defined as the part of the spectrum where the random operator exhibits strong dynamical localization in the Hilbert-Schmidt norm, and hence no transport. There is a natural definition of a *transport mobility edge* between the strong insulator and the metallic transport regions. (See Section 2.)

For the standard random operators satisfying a Wegner estimate, these metallic and insulator regions are shown to be complementary sets in the spectrum of the random operator. (This rules out the possibility of trivial transport, i.e., transport with $\beta(E) = 0$.) Thus the local transport exponent $\beta(E)$ provides a characterization of the *metal-insulator transport transition*. Moreover, $\beta(E)$ is discontinuous at a transport mobility edge: if the transport is nontrivial then $\beta(E) \geq \frac{1}{2bd}$, where d is the space dimension and $b \geq 1$ is the power of the volume in Wegner's estimate. (See Section 3.)

We illustrate these concepts in the case of random polymer models. They may have critical energies, and Jitomirskaya, Schulz-Baldes and Stolz [**JSBS**] showed that critical energies creates transport. We extend their results to local tranport exponents, proving that critical energies must be in the metallic transport region. For the special case of the random dimer model, we combine this fact with the results of De Bièvre and Germinet [**DBG**] to give a complete description of the transport transition and exhibit tranport mobility edges. (See Section 4.)

2. The transport transition

Let H_ω be a random operator on $\mathrm{L}^2(\mathbb{R}^d, \mathrm{d}x; \mathbb{C}^n)$ (or $\ell^2(\mathbb{Z}^d; \mathbb{C}^n)$). By χ_x we denote the characteristic function of the the cube of side 1 centered at $x \in \mathbb{R}^d$. If $x \in \mathbb{R}^d$ we write $\langle x \rangle = \sqrt{1+|x|^2}$, and let $\langle X \rangle$ denote the operator given by multiplication by the function $\langle x \rangle$. Given an open interval $I \subset \mathbb{R}$, we denote by $C_c^\infty(I)$ the class of real valued infinitely differentiable functions on $\mathbb{R}$ with compact support contained in I, with $C_{c,+}^\infty(I)$ being the subclass of nonnegative functions. The Hilbert-Schmidt norm of an operator A is written as $\|A\|_2$, i.e., $\|A\|_2^2 = \operatorname{tr} A^*A$.

The (random) moment of order $n \geq 0$ at time t for the time evolution in the Hilbert-Schmidt norm, initially spatially localized in the cube of side one around the origin, and "localized" in energy by the function $\mathcal{X} \in C_{c,+}^\infty(\mathbb{R})$, is given by

$$M_\omega(n, \mathcal{X}, t) = \left\| \langle X \rangle^{\frac{n}{2}} \mathrm{e}^{-itH_\omega} \mathcal{X}(H_\omega) \chi_0 \right\|_2^2 , \tag{1}$$

its expectation by

$$\mathbf{M}(n, \mathcal{X}, t) = \mathbb{E}\left\{ M_\omega(n, \mathcal{X}, t) \right\} , \tag{2}$$

and its time averaged expectation by

$$\mathcal{M}(n, \mathcal{X}, T) = \frac{2}{T} \int_0^\infty \mathrm{e}^{-\frac{2t}{T}} \mathbf{M}(n, \mathcal{X}, t)\, \mathrm{d}t. \tag{3}$$

These quantities are finite for the usual random operators, including random Schrödinger operators, random Landau Hamiltonians, and acoustic and Maxwell operators in random media. (The required properties are a trace estimate and a kernel polynomial decay estimate as stated in [**GK3**, Lemmas A.4 and A.5]; they hold in great generality.) If $\mathcal{X}(H_\omega) \neq 0$ ($\mathcal{X}(H_\omega)$ is either $= 0$ or $\neq 0$ with probability one) we have [**GK3**, Proposition 3.1]:

$$0 \leq M_\omega(0, \mathcal{X}, 0) \leq M_\omega(n, \mathcal{X}, t) \leq C_{d,\mathcal{X},n} \langle t \rangle^{\left[n + \frac{3d}{2}\right]+3} \quad \text{for } \mathbb{P}\text{-a.e. } \omega , \tag{4}$$

$$0 < \mathbf{M}(0, \mathcal{X}, 0) \leq \mathbf{M}(n, \mathcal{X}, t) \leq C_{d,\mathcal{X},n} \langle t \rangle^{\left[n + \frac{3d}{2}\right]+3} , \tag{5}$$

$$0 < \mathbf{M}(0, \mathcal{X}, 0) \leq \mathcal{M}(n, \mathcal{X}, T) \leq C'_{d,\mathcal{X},n} \langle T \rangle^{\left[n + \frac{3d}{2}\right]+3} , \tag{6}$$

where $[u]$ denotes the largest integer $\leq u$.

To measure the rate of growth of moments of initially spatially localized wave packets under the time evolution, "localized" in energy by $\mathcal{X} \in C_{c,+}^\infty(\mathbb{R})$ with $\mathcal{X}(H_\omega) \neq 0$, we compute the (lower) transport exponent

$$\beta(n, \mathcal{X}) = \liminf_{T \to \infty} \frac{\log \mathcal{M}(n, \mathcal{X}, T)}{n \log T} . \tag{7}$$

If $\mathcal{X}(H_\omega) = 0$ we set $\beta(n, \mathcal{X}) = 0$. We define the *n-th transport exponent* in an open interval I by

$$\beta(n, I) = \sup_{\mathcal{X} \in C_{c,+}^\infty(I)} \beta(n, \mathcal{X}) , \tag{8}$$

and the *n-th local transport exponent* at the energy E by

$$\beta(n, E) = \inf_{I \ni E} \beta(n, I) . \tag{9}$$

($\beta(n, E)$ provides a measure of the rate of transport for which E is responsible.) The exponent $\beta(n, E)$ is increasing in n and satisfy the ballistic bound

$$0 \leq \beta(n, \mathcal{X}),\ \beta(n, I),\ \beta(n, E) \leq 1\,, \tag{10}$$

as shown in [**GK3**, Proposition 3.2]. Note that $\beta(n, E) = 0$ if $E \notin \Sigma$.

The *local (lower) transport exponent* may be now defined by

$$\beta(E) = \lim_{n\to\infty} \beta(n, E) = \sup_n \beta(n, E)\,, \tag{11}$$

and we have $0 \leq \beta(E) \leq 1$, with $\beta(E) = 0$ if $E \notin \Sigma$. Note that $\beta(E) > 0$ if and only if $\beta(n, E) > 0$ for some $n > 0$.

This motivates the following definition.

DEFINITION 2.1. *The metallic transport region Σ_{MT} for H_ω is defined as the set of energies with nontrivial transport:*

$$\Sigma_{\mathrm{MT}} = \{E \in \mathbb{R},\ \beta(E) > 0\} = \{E \in \Sigma,\ \beta(E) > 0\}\,. \tag{12}$$

Its complementary set in the spectrum will be called the trivial transport region Σ_{TT} (note that logarithmic transport is not excluded a priori):

$$\Sigma_{\mathrm{TT}} = \Sigma \backslash \Sigma_{\mathrm{MT}} = \{E \in \Sigma,\ \beta(E) = 0\}\,. \tag{13}$$

To make the connection with the absolutely continuous spectrum, we recall that the Guarneri-Combes-Last bound [**Gu, Co, La**] says that $\beta(E) \geq \frac{1}{d}$ if $E \in \Sigma_{\mathrm{ac}}$. (While the Guarneri-Combes-Last bound is stated for a fixed self-adjoint operator, the same bound follows for random operators using Fatou's Lemma and Jensen's inequality.) Thus

$$\Sigma_{\mathrm{ac}} \subset \left\{E \in \Sigma,\ \beta(E) \geq \frac{1}{d}\right\} \subset \Sigma_{\mathrm{MT}}\,. \tag{14}$$

But the converse to the Guarneri-Combes-Last bound is not true, a lower bound on the local transport exponent does not specify the spectrum (e.g., [**DRJLS, La, DBF, BGT, CM, GKT**]).

The trivial transport region should be connected to localization, and indeed it is, but to the right kind of localization. Pure point spectrum does not imply trivial transport, there are counterexamples [**DRJLS, JSBS, GKT**]. The right notion of localization is dynamical localization (which implies pure point spectrum).

DEFINITION 2.2. *The random operator H_ω exhibits strong HS-dynamical localization in the open interval I if for all $\mathcal{X} \in C^\infty_{c,+}(I)$ we have*

$$\mathbb{E}\left\{\sup_{t\in\mathbb{R}} M_\omega(n, \mathcal{X}, t)\right\} < \infty \quad \textit{for all } n \geq 0\,. \tag{15}$$

The random operator H_ω exhibits strong HS-dynamical localization at the energy $E \in \mathbb{R}$ if there exists an open interval I, with $E \in I$, such that there is strong HS-dynamical localization in the open interval I.

The intuitive idea behind the last definition is that the moments of an initially localized wave packet remain uniformly bounded under time evolution "localized" in an open interval around the energy E. The Hilbert-Schmidt norm takes into account all possible wave packets localized in a given bounded region.

DEFINITION 2.3. *The strong insulator region Σ_{SI} for H_ω is defined as*

$$\Sigma_{\mathrm{SI}} = \{E \in \Sigma\,;\ H_\omega \textit{ exhibits strong HS-dynamical localization at } E\}\,.$$

We clearly have

$$\Sigma_{\text{SI}} \subset \Sigma_{\text{TT}}\,. \tag{16}$$

DEFINITION 2.4. *An energy* $E \in \Sigma_{\text{MT}} \cap \overline{\Sigma}_{\text{SI}}$ *will be called a transport mobility edge.*

Since the strong insulator region is a relatively open subset of the spectrum Σ, we have $\Sigma_{\text{SI}} = \left\{\bigcup_{j=1}^{N} I_j\right\} \cap \Sigma$, where the I_j's are disjoint open intervals; N may be either finite or infinite. A transport mobility edge must be an edge of one of the intervals I_j.

3. A characterization of the transport transition

Instead of giving a technical definition of the relevant class of random operators, let us simply say that in this section by a standard random operator we mean one of the following:

The Anderson model:

$$H_\omega = -\Delta + V_\omega \quad \text{on} \quad \ell^2(\mathbb{Z}^d)\,, \tag{17}$$

where Δ is the finite difference Laplacian and $\{V_\omega(x);\ x \in \mathbb{Z}^d\}$ are independent identically distributed bounded random variables. (E.g., [**FS, FMSS, CKM, vDK, AM, Ai, ASFH, Wa2**].)

Random Schrödinger operators (Anderson Hamiltonians):

$$H_\omega = -\Delta + V_{\text{per}} + V_\omega \quad \text{on} \quad \mathrm{L}^2(\mathbb{R}^d, \mathrm{d}x), \tag{18}$$

where Δ is the Laplacian operator, V_{per} is a periodic potential (by rescaling we take the period to be one) of the form $V_{\text{per}} = V_{\text{per}}^{(1)} + V_{\text{per}}^{(2)}$, with $V_{\text{per}}^{(i)}$, $i = 1, 2$, periodic with period one, $0 \le V_{\text{per}}^{(1)} \in \mathrm{L}^1_{\text{loc}}(\mathbb{R}^d, \mathrm{d}x)$, $V_{\text{per}}^{(2)}$ relatively form-bounded with respect to $-\Delta$ with relative bound < 1, and V_ω a random potential of the form

$$V_\omega(x) = \sum_{i \in \frac{1}{q}\mathbb{Z}^d} \omega_i\, u(x - i), \tag{19}$$

where $q \in \mathbb{N}$, $\omega = \{\omega_i;\ i \in \frac{1}{q}\mathbb{Z}^d\}$ are independent identically distributed bounded random variables, u is a real valued measurable function with compact support, $u \in \mathrm{L}^p(\mathbb{R}^d, \mathrm{d}x)$ with $p > \frac{d}{2}$ if $d \ge 2$ and $p = 2$ if $d = 1$. (See [**CH1, Klo2, KSS, GK3, GK4, GK5**].)

Random Landau Hamiltonians:

$$H_\omega = H_0 + V_\omega \quad \text{on} \quad \mathrm{L}^2(\mathbb{R}^2, \mathrm{d}x), \tag{20}$$

where $H_0 = (-i\nabla - A)^2$, $A = \frac{B}{2}(x_2, -x_1)$, and the random potential V_ω is as in (19) with $q = 1$ and $u(x)$ bounded. (See [**CH2, Wa1, GK4**].)

Maxwell operators in random media:

$$H_\omega = \frac{1}{\sqrt{\mu_\omega(x)}} \nabla\times \frac{1}{\varepsilon_\omega(x)} \nabla\times \frac{1}{\sqrt{\mu_\omega(x)}} \quad \text{on} \quad \mathrm{L}^2\left(\mathbb{R}^3, dx; \mathbb{C}^3\right) \tag{21}$$

where $\nabla\times$ is the operator given by the curl, $\varepsilon_\omega(x)$ is the random dielectric constant and $\mu_\omega(x)$ is the random magnetic permeability. We take

$$\varepsilon_\omega(x) = \varepsilon_0(x)\gamma_\omega(x) \text{ , with } \gamma_\omega(x) = 1 + \sum_{i\in\frac{1}{q}\mathbb{Z}^3} \omega_i u(x-i), \tag{22}$$

$$\mu_\omega(x) = \mu_0(x)\beta_\omega(x) \text{ , with } \beta_\omega(x) = 1 + \sum_{i\in\frac{1}{q}\mathbb{Z}^3} \omega_i v(x-i), \tag{23}$$

where $q \in \mathbb{N}$, $\omega = \{\omega_i;\ i \in \frac{1}{q}\mathbb{Z}^d\}$ are independent identically distributed bounded random variables taking values in the interval $[-1,1]$, $\varepsilon_0(x)$ and $\mu_0(x)$ are periodic measurable functions (by rescaling we take the period to be one), such that $0 < \varepsilon_- \le \varepsilon(x) \le \varepsilon_+ < \infty$ and $0 < \mu_- \le \mu(x) \le \mu_+ < \infty$ for some constants $\varepsilon_\pm$ and $\mu_\pm$, $u(x)$ and $v(x)$ are nonnegative measurable real valued functions with compact support, such that

$$0 \le U_- \le U(x) \equiv \sum_{i\in\frac{1}{q}\mathbb{Z}^3} u_i(x) \le U_+ < \infty, \tag{24}$$

$$0 \le V_- \le V(x) \equiv \sum_{i\in\frac{1}{q}\mathbb{Z}^3} v_i(x) \le V_+ < \infty, \tag{25}$$

for some constants $U_\pm$ and $V_\pm$, with $U_- + V_- > 0$ and $\max\{U_+, V_+\} < 1$. (See [**FK2, Kle3, KK2**].)

Acoustic operators in random media:

$$H_\omega = -\frac{1}{\sqrt{\kappa_\omega(x)}}\nabla\cdot\frac{1}{\rho_\omega(x)}\nabla\frac{1}{\sqrt{\kappa_\omega(x)}} \quad \text{on} \quad \mathrm{L}^2(\mathbb{R}^d, \mathrm{d}x), \tag{26}$$

where ∇ is the gradient operator, and, the random compressibility $\kappa_\omega(x)$ and the random mass density $\varrho_\omega(x)$ are of the same form as $\varepsilon_\omega(x)$ and $\mu_\omega(x)$ in (22) and (23). (See [**FK1, KK2**]).

These random operators satisfy all the requirements for the bootstrap multiscale analysis [**GK1**] in appropriate intervals, with the possible exception of a Wegner estimate. They also satisfy an appropriate interior estimate [**GK3**, Lemma A.2], and the kernel polynomial decay estimate of [**GK2**, Theorem 2] with nonrandom constants. (To be precise, a standard random operator is a random operator satisfying Assumptions SLI, EDI, IAD, NE, and SGEE in [**GK1**], the interior estimate of [**GK3**, Lemma A.2], and the Assumptions of [**GK4**] with nonrandom constants; these properties are routinely verified for the usual random operators.)

Although we frame our discussion for random operators on the continuum, our results apply also to random operators on the lattice.

In our results a Wegner estimate in an open interval (Assumption W in [**GK1**]) will be an explicit hypothesis. To state it we need to consider the restriction of a random operator H_ω to a finite box. By $\Lambda_L(x)$ we denote the open box (or cube) of side $L > 0$:

$$\Lambda_L(x) = \{y \in \mathbb{R}^d;\ \|y - x\| < L/2\}, \tag{27}$$

and by $\overline{\Lambda}_L(x)$ the closed box, where $\|x\| = \max\{|x_i|,\ i = 1, \dots, d\}$. (We will use $|x|$ to denote the usual Euclidean norm.) We will also use the notation

$$\chi_{x,L} = \chi_{\Lambda_L(x)}\,, \quad \text{in particular } \chi_x = \chi_{x,1} = \chi_{\Lambda_1(x)}\,. \tag{28}$$

The operator $H_{\omega,x,L}$ is defined as an appropriate restriction of H_ω to the box $\Lambda_L(x)$ (e.g., to the open box $\Lambda_L(x)$ with Dirichlet boundary condition, or to the closed box $\overline{\Lambda}_L(x)$ with periodic boundary condition). We write $R_{\omega,x,L}(z) = (H_{\omega,x,L} - z)^{-1}$ for the finite volume resolvent.

We say that the random operator H_ω satisfies a Wegner estimate in an open interval $\mathcal{I}$ if there exist constants $b \geq 1$, $0 < \eta_{\mathcal{I}} \leq 1$, and $Q_{\mathcal{I}}$, such that

$$\mathbb{P}\left\{\operatorname{dist}(\sigma(H_{\omega,x,L}), E) \leq \eta\right\} \leq Q_{\mathcal{I}} \eta L^{bd}\ , \tag{29}$$

for all $E \in \mathcal{I}$, $0 < \eta \leq \eta_{\mathcal{I}}$, $x \in \mathbb{Z}^d$, and $L \in 2\mathbb{N}$.

Wegner estimates have been proven for a large variety of random operators [**Weg, HM, CKM, CH1, Klo2, CH2, CHM, Ki, FK1, FK2, Wa1, KSS, St, CHN, HK, CHKN, KK2**], under certain assumptions. (E.g., for random Schrödinger operators the probability distribution of the random variable ω_i is assumed to have a bounded density, and $u(x) \geq 0$.) Usually $b = 1$ or 2.

If we have a Wegner estimate, we can prove a lower bound on the local transport exponent in the metallic transport region [**GK3**, Theorem 2.10], which may be compared to the Guarneri-Combes-Last bound (see the paragraph containing (14)). We use the notation $B^{\mathcal{I}} = B \cap \mathcal{I}$ for a subset B of $\mathbb{R}$.

THEOREM 3.1. *Let H_ω be a standard random operator satisfying a Wegner estimate in an open interval $\mathcal{I}$. If $\beta(E) > 0$ for some $E \in \mathcal{I}$ then $\beta(E) \geq \frac{1}{2bd}$, i.e. the metallic transport region in $\mathcal{I}$ is given by*

$$\Sigma_{\mathrm{MT}}^{\mathcal{I}} = \left\{E \in \mathcal{I},\ \beta(E) \geq \frac{1}{2bd}\right\}. \tag{30}$$

In fact, if $E \in \Sigma_{\mathrm{MT}}^{\mathcal{I}}$, then $\beta(n, E) \geq \frac{1}{2bd} - \frac{9b+2}{2bn}$ for all $n \geq 0$.

If the standard random operator satisfies a Wegner estimate we have equality in (16) [**GK3**, Theorem 2.8]:

THEOREM 3.2. *Let H_ω be a standard random operator satisfying a Wegner estimate in an open interval $\mathcal{I}$. Then*

$$\Sigma_{\mathrm{SI}}^{\mathcal{I}} = \Sigma_{\mathrm{TT}}^{\mathcal{I}}. \tag{31}$$

In particular, the strong insulator region and the metallic transport region are complementary sets in the spectrum $\Sigma^{\mathcal{I}}$ of H_ω in $\mathcal{I}$, i.e.,

$$\Sigma_{\mathrm{SI}}^{\mathcal{I}} \cap \Sigma_{\mathrm{MT}}^{\mathcal{I}} = \emptyset \quad \text{and} \quad \Sigma_{\mathrm{SI}}^{\mathcal{I}} \cup \Sigma_{\mathrm{MT}}^{\mathcal{I}} = \Sigma^{\mathcal{I}}. \tag{32}$$

Theorem 3.2 shows that the local transport exponent $\beta(E)$ provides a characterization of the metal-insulator transport transition. Theorem 3.1 says that if this transition occurs, $\beta(E)$ has to be discontinuous at a transport mobility edge.

The existence of a nontrivial strong insulator region is now proven for standard random operators. It is a consequence of well established results on Anderson localization and of the bootstrap multiscale analysis [**GK1**, Theorem 3.4] that yields strong HS-dynamical localization [**GK1**, Corollary 3.10]. (See also [**GDB, DS**]. On the lattice strong HS-dynamical localization turns out to be the same as strong dynamical localization (of wave packets) and was originally proven by Aizenman [**Ai, ASFH**].) The relevant results, adapted for this article, will now be stated.

Given $x \in \mathbb{Z}^d$, $L \in \mathbb{N}$, we set

$$\Gamma_{x,L} = \sum_{y \in \Upsilon_L(x)} \chi_y\,, \text{ where } \Upsilon_L(x) = \left\{y \in \mathbb{Z}^d;\ \|y - x\| = \tfrac{L}{2} - 1\right\}. \tag{33}$$

Given $\theta > 0$, $E \in \mathbb{R}$, $x \in \mathbb{Z}^d$, and $L \in 6\mathbb{N}$, we say that the box $\Lambda_L(x)$ is (θ, E)-*suitable* for H_ω if $E \notin \sigma(H_{\omega,x,L})$ and

$$\|\Gamma_{x,L} R_{\omega,x,L}(E) \chi_{x,L/3}\| \leq \frac{1}{L^\theta}\,. \tag{34}$$

THEOREM 3.3. *Let H_ω be a standard random operator satisfying a Wegner estimate in an open interval $\mathcal{I}$. Given $\theta > d$, for each $E \in \mathcal{I}$ there exists a finite scale $\overline{\mathcal{L}}_\theta(E)$ (depending on θ, E, $Q_{\mathcal{I}}$, d), bounded in compact subintervals of $\mathcal{I}$, such that, if for some $E \in \Sigma \cap \mathcal{I}$ we can verify at some finite scale $\mathcal{L} > \overline{\mathcal{L}}_\theta(E)$ that*

$$\mathbb{P}\{\Lambda_{\mathcal{L}}(0) \text{ is } (\theta, E)\text{-suitable}\} > 1 - \frac{1}{841^d}\,, \tag{35}$$

then there exists an open interval $I \ni E$, such that H_ω has pure point spectrum in I, with exponentially decaying eigenfunctions, and exhibits strong HS-dynamical localization in I. In particular, $E \in \Sigma_{\mathrm{SI}}$.

Theorem 3.3 suggests the following definition.

DEFINITION 3.4. *The multiscale analysis region Σ_{MSA} is defined as the set of energies where we can perform the bootstrap multiscale analysis:*

$$\Sigma_{\mathrm{MSA}} = \{E \in \Sigma;\ \text{ the hypotheses of Theorem 3.3 hold at } E\}\,.$$

(Note that if Σ_{AM} denotes the set of energies in the spectrum satisfying the starting hypothesis of the Aizenman-Molchanov method [**AM, Ai, ASFH**], we have $\Sigma_{\mathrm{AM}} = \Sigma_{\mathrm{MSA}}$.)

It follows that

$$\Sigma_{\mathrm{MSA}} \subset \Sigma_{\mathrm{SI}} \subset \Sigma_{\mathrm{TT}}\,. \tag{36}$$

With a Wegner estimate we have equality in (36) [**GK3**, Theorem 2.8]:

THEOREM 3.5. *Let H_ω be a standard random operator satisfying a Wegner estimate in an open interval $\mathcal{I}$. Then $\Sigma_{\mathrm{TT}}^{\mathcal{I}} \subset \Sigma_{\mathrm{MSA}}^{\mathcal{I}}$ and hence*

$$\Sigma_{\mathrm{MSA}}^{\mathcal{I}} = \Sigma_{\mathrm{SI}}^{\mathcal{I}} = \Sigma_{\mathrm{TT}}^{\mathcal{I}}\,. \tag{37}$$

The equality (37) shows that the strong insulator region is canonical in the sense that it may be defined by three equivalent conditions or properties, all very natural. In fact the number of such conditions/properties is actually much larger [**GK3**, Theorem 4.2]. In the analogy with classical statistical mechanics the strong insulator region corresponds to the region of complete analyticity [**DoSh1, DoSh2**].

Theorem 3.2 is a corollary of Theorem 3.5. Theorems 3.1 and 3.5 are consequences of the fact that slow transport cannot take place for random operators satisfying our assumptions [**GK3**, Theorem 2.11]. (A weaker form of this result has been discussed by Martinelli and Scoppola [**MS**, Section 8] for the discrete Anderson model.)

THEOREM 3.6. *Let H_ω be a standard random operator satisfying a Wegner estimate in an open interval $\mathcal{I}$. Let $\mathcal{X} \in C_{c,+}^\infty(\mathbb{R})$, with $\mathcal{X} \equiv 1$ on some open interval $J \subset \mathcal{I}$, $\alpha \geq 0$, and $n > 2bd\alpha + (9b+2)d$. If*

$$\liminf_{T \to \infty} \frac{1}{T^\alpha} \mathcal{M}(n, \mathcal{X}, T) < \infty\,, \tag{38}$$

then $J \cap \Sigma \subset \Sigma_{\mathrm{MSA}}$, and hence $J \cap \Sigma \subset \Sigma_{\mathrm{SI}}$.

Theorem 3.6 has the following immediate corollary, which can be read as follows: if the transport at an energy E is too slow (i.e., $\beta(n, E) < \frac{1}{2bd} - \frac{9b+2}{2bn}$ for some $n > (9b+2)d$), then strong HS-dynamical localization has to hold at E.

COROLLARY 3.7. *Let H_ω be a standard random operator satisfying a Wegner estimate in an open interval $\mathcal{I}$. If $E \in \mathcal{I} \cap \Sigma$ and $\beta(n, E) < \frac{1}{2bd} - \frac{9b+2}{2bn}$ for some $n > (9b+2)d$, then $E \in \Sigma_{\mathrm{MSA}} \subset \Sigma_{\mathrm{SI}}$.*

Theorem 3.5 follows immediately from Corollary 3.7, since $\beta(E) = 0 \Rightarrow \beta(n, E) = 0$ for all $n \geq 0$. The same is true for Theorem 3.1, since if $\beta(n, E) < \frac{1}{2bd} - \frac{9b+2}{2bn}$ for some $n > (9b+2)d$, it follows from Corollary 3.7 that $E \in \Sigma_{\mathrm{SI}}$ and hence $\beta(E) = 0$.

Some comments on the proof of Theorem 3.6 are in order. Its main hypothesis, condition (38), is formulated in terms of the dynamics, but the starting hypothesis of the bootstrap multiscale analysis, condition (35), is stated in terms of resolvents. The first step is to reformulate condition (38) in terms of resolvents. To do so, we used the Hilbert-Schmidt norm when we defined the moments in (1), so we can use Plancherel Theorem to get

$$\mathcal{M}(n, \mathcal{X}, T) = \frac{1}{\pi T} \int_{\mathbb{R}} \mathbb{E} \left\| \langle X \rangle^{\frac{n}{2}} R_\omega(E + i\tfrac{1}{T}) \mathcal{X}(H_\omega) \chi_0 \right\|_2^2 \mathrm{d}E \,, \tag{39}$$

for any $n > 0$, $T > 0$, and $\mathcal{X} \in C^\infty_{c,+}(\mathbb{R})$, where $R_\omega(z) = (H_\omega - z)^{-1}$. Thus, if we set

$$\Omega_\varepsilon(n, \mathcal{X}, E) = \mathbb{E} \left(\left\| \langle X \rangle^{\frac{n}{2}} R_\omega(E + i\varepsilon) \mathcal{X}(H_\omega) \chi_0 \right\|_2^2 \right) \,, \tag{40}$$

condition (38) in Theorem 3.6 is the same as

$$\Omega \equiv \liminf_{\varepsilon \to 0^+} \varepsilon^{1+\alpha} \int_{\mathbb{R}} \Omega_\varepsilon(n, \mathcal{X}, E) \, \mathrm{d}E < \infty \,. \tag{41}$$

The proof of Theorem 3.6 requires that we obtain the finite volume condition (35) for the bootstrap multiscale analysis out of the infinite volume condition (41). The kernel polynomial decay estimate from [**GK2**] plays an important role in the proof, as well as the Wegner estimate. We refer to [**GK3**] for the details.

4. Transport transition in random polymer models

Following Jitomirskaya, Schulz-Baldes and Stolz [**JSBS**], we define a random polymer model by

$$H_\omega = -\Delta + V_\omega \quad \text{on} \quad \ell^2(\mathbb{Z}) \,, \tag{42}$$

where Δ is the centered discrete Laplacian and the random potential V_ω is defined as follows: Let $\hat{v}_\pm = (\hat{v}_\pm(1), \cdots, \hat{v}_\pm(L_\pm))$ be two finite sequences of real numbers, where $L_\pm \in \mathbb{N}$. A discrete one dimensional random polymer is constructed by randomly juxtaposing blocks of sign $+$ and sign $-$: let $\omega = \{\omega_i\,;\ i \in \mathbb{Z}\}$ be a sequence of independent identically distributed random variables taking the value $+$ with probability $p \in (0, 1)$ and $-$ with probability $1 - p$; $\{V_\omega(x);\ x \in \mathbb{Z}\}$ is the sequence of blocks $(\ldots, \hat{v}_{\omega_0}, \hat{v}_{\omega_1}, \ldots)$. Note that H_ω is a $\mathbb{Z}$-ergodic self-adjoint random operator [**JSBS**].

The spectrum of a random polymer model is always pure point with exponentially decaying eigenfunctions, so there is no spectral transition. Moreover, they exhibit strong HS-dynamical localization on energy intervals not containing a discrete set of special energies. (See [**DBG**] for the random dimer model, for random polymer models these results can be proven by the methods of [**DSS**].) Nevertheless, random polymer models may exhibit a transport transition, as we shall see.

Random polymer models are similar to Bernoulli models (i.e., $L_\pm = 1$); they do not satisfy a Wegner estimate as stated in Section 2, only a subexponential Wegner estimate (i.e., with $\eta \leq \mathrm{e}^{-L^\beta}$ with $0 < \beta < 1$ in (29), see [**CKM**, Theorem 4.1]) on appropriate intervals. The bootstrap multiscale analysis can still be performed with this subexponential Wegner estimate [**GK3**, Remark 3.13], and it gives pure point spectrum with exponentially decaying eigenfunctions, and strong HS-dynamical localization. But Theorem 3.6 (and hence Theorems 3.1 and 3.2) is not established for these models, since the proof relies on the polynomial decay estimates of [**GK2**], which can only beat polynomial bounds obtained by using the Wegner estimate (i.e., $\eta = L^{-s}$, $s > 0$ in (29)). Nevertheless, we can say a lot about the metal-insulator tranport transition for random polymer models.

Random polymer models may have *critical energies*, which we now define following [**JSBS**]. The transfer matrices associated to the blocks $\pm$ at the energy $E \in \mathbb{R}$ are given by:

$$T_\pm^E = \prod_{L_\pm}^{i=1} T(\hat{v}_\pm(i) - E)\,, \quad \text{where} \quad T(v) = \begin{pmatrix} v & -1 \\ 1 & 0 \end{pmatrix}. \tag{43}$$

An energy $E_c \in \mathbb{R}$ is said to be critical for the random operator H_ω if the matrices $T_\pm^{E_c}$ are either elliptic (i.e., $\left|\mathrm{Tr} T_\pm^{E_c}\right| < 2$) or equal to $\pm\mathrm{I}$, and they commute (i.e., $\left[T_+^{E_c}, T_-^{E_c}\right] = 0$).

Critical energies may or may not exist. An example is given by the random dimer model: $L_\pm = 2$, $\hat{v}_\pm(i) = \pm\lambda$ for $i = 1, 2$, with $0 < \lambda$. De Bièvre and Germinet [**DBG**] proved:

(i): If $0 < \lambda \leq 1$ there are exactly two critical energies: $\pm\lambda$.
(ii): If $1 < \lambda$ there are no critical energies.

Dunlap, Philipps and Wu [**DWP**] found numerically an interesting phenomenon of delocalization for the random dimer model at the critical energies, due to the absence of reflection at these energies (and thus perfect transmission). Such a phenomenon has been contested by physicists (e.g., see [**LGP**]). De Bièvre and Germinet [**DBG**] showed that energies that are not critical cannot contribute to this delocalization, and strong dynamical localization holds outside the critical energies (if $\lambda \neq \frac{1}{\sqrt{2}}, \sqrt{2}$). Jitomirskaya, Schulz-Baldes and Stolz [**JSBS**] confirmed by a rigorous proof the numerical computations of [**DWP**].

The transport exponents for random polymer models are defined as in Section 2. Note that in the lattice χ_0 is simply the orthogonal projection on the vector δ_0, the element of $\ell^2(\mathbb{Z})$ taking the value 1 at site 0 and the value 0 everywhere else. Thus (1) may be rewritten as

$$M_\omega(n, \mathcal{X}, t) = \left\| \langle X \rangle^{\frac{n}{2}} \mathrm{e}^{-itH_\omega} \mathcal{X}(H_\omega)\delta_0 \right\|^2 . \tag{44}$$

Since the random operator H_ω is bounded, uniformly in ω, we may remove the restriction of $\mathcal{X}$ having compact support; in particular we may calculate moments and exponents with $\mathcal{X} \equiv 1$. It follows from [**JSBS**, Main Theorem], using Fatou's

Lemma and Jensen's inequality, that if H_ω has a critical energy E_c we have

$$\beta(n, \mathcal{X} \equiv 1) \geq 1 - \frac{1}{n} \quad \text{for all} \quad n \geq 0 . \tag{45}$$

We can extend this result to local transport exponents, proving that critical energies lie in the metallic transport region, and hence the existence of the transport transition.

THEOREM 4.1. *Let H_ω be a random polymer Hamiltonian as in* (42). *If E_c is a critical energy, then*

$$\beta(n, E_c) \geq 1 - \frac{1}{n} \quad \textit{for all} \quad n \geq 0 . \tag{46}$$

In particular, $\beta(E_c) = 1$ and $E_c \in \Sigma_{\mathrm{MT}}$.

The proof of Theorem 4.1 is given at the end of this section.

Under an extra hypothesis on the transfer matrices $T_\pm^{E_c}$, the bound (45) can be improved to $\beta(n, \mathcal{X} \equiv 1) \geq 1 - \frac{1}{2n}$ [**JSBS**, Theorem 4]; such a derivation requires a large deviation estimate for products of transfer matrices. In this case the proof of Theorem 4.1 yields $\beta(n, E_c) \geq 1 - \frac{1}{2n}$.

We may summarize the results on a transport transition for random polymer models in the following theorem

THEOREM 4.2. *Let H_ω be a random polymer Hamiltonian as in* (42). *Then $\Sigma \backslash \Sigma_{\mathrm{SI}}$ is a discrete set containing the critical energies (and hence $\Sigma_{\mathrm{SI}} \neq \emptyset$). Moreover, the set of critical energies is contained in Σ_{MT}, so if there are critical energies we have $\Sigma_{\mathrm{MT}} \neq \emptyset$ and there is a transport transition.*

For the special case of the random dimer model, we may use the results of [**DBG**] to get more detailed information.

THEOREM 4.3. *Let Let H_ω be a random dimer Hamiltonian i.e., H_ω is as in* (42) *with $L_\pm = 2$, $\hat{v}_\pm(i) = \pm\lambda$ for $i = 1, 2$, with $0 < \lambda$. Then*

(i): *If $0 < \lambda \leq 1$, we have a transport transition and $\pm\lambda$ are transport mobility edges. Moreover,*

$$\Sigma_{\mathrm{MT}} = \{-\lambda, \lambda\} , \quad \Sigma_{\mathrm{SI}} = \Sigma \backslash \Sigma_{\mathrm{MT}} \quad \textit{if} \quad \lambda \neq \frac{1}{\sqrt{2}} , \tag{47}$$

$$\Sigma_{\mathrm{MT}} \supset \left\{ -\tfrac{1}{\sqrt{2}}, \tfrac{1}{\sqrt{2}} \right\} , \quad \Sigma_{\mathrm{SI}} \supset \Sigma \backslash \left\{ -\tfrac{3}{\sqrt{2}}, -\tfrac{1}{\sqrt{2}}, \tfrac{1}{\sqrt{2}}, \tfrac{3}{\sqrt{2}} \right\} \; \textit{if} \; \lambda = \frac{1}{\sqrt{2}} . \tag{48}$$

(ii): *If $1 < \lambda$, $\lambda \neq \sqrt{2}$, we do not have a transport transition:*

$$\Sigma_{\mathrm{MT}} = \emptyset \quad \textit{and} \quad \Sigma_{\mathrm{SI}} = \Sigma . \tag{49}$$

(iii): *If $\lambda = \sqrt{2}$ we have*

$$\Sigma_{\mathrm{SI}} \supset \Sigma \backslash \{0\} . \tag{50}$$

PROOF OF THEOREM 4.1. Let E_c be a critical energy, and consider open bounded intervals I, I', with $E_c \in I \subset I'$, $\delta = \mathrm{dist}(I, \mathbb{R} \backslash I') > 0$, and $\mathcal{X} \in C_{c,+}^\infty(\mathbb{R})$ with $\mathcal{X} \equiv 1$ on I'.

As in [**JSBS**, Proposition 1], it is an immediate consequence of the definition of a critical energy that $T_\pm^{E_c}$ can be represented as two rotations in some appropriate basis, and thus, there exists a constant $C > 0$ such that, for $z \in \mathbb{C}$ with $|z - E_c|$

small enough, we have $\| \prod_{i=k}^{l} T_{\omega_i}^{z} \| \le C \exp(C|z - E_c|\,|k - l|)$. Following [**JSBS**, Section 6, Proof of Theorem 1], or [**DT**, Section 2, Proof of Theorem 1.2], it implies that there exists a constant $C > 0$ such that, for all $\omega \in \Omega$,

$$\frac{1}{2\pi T} \int_I \left\| \langle X \rangle^{n/2} R_\omega \left(E + iT^{-1}\right) \delta_0 \right\|^2 \mathrm{d}E \ge C T^{n-1} . \tag{51}$$

We exploit this lower bound together with the kernel decay estimate of [**GK2**, Theorem 2] to get a lower bound on $\mathcal{M}(n, \mathcal{X}, t)$, which is defined as in (3) but without taking the expectation. Using Plancherel's Theorem as in [**GK3**, Lemma 6.3], writing $\mathcal{X} = 1 - (1 - \mathcal{X})$, and using the inequality $(a-b)^2 \ge \frac{1}{2}a^2 - b^2$, for a, b real, we get

$$\frac{2}{T} \int_0^\infty e^{-2t/T} M_\omega(n, \mathcal{X}, t)\, \mathrm{d}t \tag{52}$$

$$= \frac{1}{\pi T} \int_{\mathbb{R}} \left\| \langle X \rangle^{n/2} R_\omega \left(E + iT^{-1}\right) \mathcal{X}(H_\omega) \delta_0 \right\|^2 \mathrm{d}E, \tag{53}$$

$$\ge \frac{1}{2\pi T} \int_I \left\| \langle X \rangle^{n/2} R_\omega \left(E + iT^{-1}\right) \delta_0 \right\|^2 \mathrm{d}E \tag{54}$$

$$- \frac{1}{\pi T} \int_I \left\| \langle X \rangle^{n/2} R_\omega \left(E + iT^{-1}\right) \left(1 - \mathcal{X}(H_\omega)\right) \delta_0 \right\|^2 \mathrm{d}E . \tag{55}$$

Proceeding as in [**GK3**, Eq. (6.34)], using [**GK2**, Theorem 2], we get

$$\sup_{E \in I} \sup_{T > 0} \left\| \langle X \rangle^{n/2} R_\omega \left(E + iT^{-1}\right) \left(1 - \mathcal{X}(H_\omega)\right) \delta_0 \right\|^2 \le C(n, \delta). \tag{56}$$

where $\delta = \operatorname{dist}(I, \mathbb{R} \backslash I') > 0$, and the finite constant $C(n, \delta)$ does not depend on ω. Combining (51) and (56), we get that there exists a constant $C' > 0$, depending only on the constants of the polymer, I, I', and $\mathcal{X}$, such that, for all $\omega \in \Omega$,

$$\frac{2}{T} \int_0^\infty e^{-2t/T} M_\omega(n, \mathcal{X}, t)\, \mathrm{d}t \ge C' T^{n-1} \tag{57}$$

for all $T > 0$. Taking the expectation, we obtain

$$\mathcal{M}(n, \mathcal{X}, T) \ge C' T^{n-1} \tag{58}$$

for all $T > 0$. It follows that $\beta(n, I) \ge 1 - \frac{1}{n}$, and since $I \ni E_c$ is arbitrary, $\beta(n, E_c) \ge 1 - \frac{1}{n}$. The theorem follows. □

References

[Ai] Aizenman, M.: Localization at weak disorder: some elementary bounds. Rev. Math. Phys. **6**, 1163-1182 (1994)

[AM] Aizenman, M., Molchanov, S.: Localization at large disorder and extreme energies: an elementary derivation. Commun. Math. Phys. **157**, 245-278 (1993)

[ASFH] Aizenman, M., Schenker, J., Friedrich, R., Hundertmark, D.: Finite volume fractional-moment criteria for Anderson localization. Commun. Math. Phys. **224**, 219-253 (2001)

[An] Anderson, P.: Absence of diffusion in certain random lattices. Phys. Rev. **109**, 1492-1505 (1958)

[BGT] Barbaroux, J-M., Germinet, F., Tcheremchantsev, S. : Fractal dimensions and the phenomenon of intermittency in quantum dynamics. Duke Math. J. **110**, 161-193 (2001)

[CKM] Carmona, R., Klein, A., Martinelli, F.: Anderson localization for Bernoulli and other singular potentials. Commun. Math. Phys. **108**, 41-66 (1987)

[Co] Combes, J.M.: Connection between quantum dynamics and spectral properties of time evolution operators in *Differential Equations and Applications in Mathematical Physics*, Eds. W.F. Ames, E.M. Harrel, J.V. Herod, Academic Press, 59-69 (1993)

[CH1] Combes, J.M., Hislop, P.D.: Localization for some continuous, random Hamiltonian in d-dimension. J. Funct. Anal. **124**, 149-180 (1994)

[CH2] Combes, J.M., Hislop, P.D.: Landau Hamiltonians with random potentials: localization and the density of states. Commun. Math. Phys. **177**, 603-629 (1996)

[CHKN] Combes, J.M., Hislop, P.D., Klopp, F. Nakamura, S.: The Wegner estimate and the integrated density of states for some random operators. Spectral and inverse spectral theory (Goa, 2000), Proc. Indian Acad. Sci. Math. Sci. **112** , 31-53 (2002)

[CHM] Combes, J.M., Hislop, P.D., Mourre, E.: Spectral averaging, perturbation of singular spectra, and localization. Trans. Amer. Math. Soc. **348**, 4883-4894 (1996)

[CHN] Combes, J.M., Hislop, P.D., Nakamura, S.: The L^p-theory of the spectral shift function, the Wegner estimate and the integrated density of states for some random operators. Commun. Math. Phys. **218**, 113-130 (2001)

[CM] Combes, J.M., Mantica, G.: Fractal dimensions and quantum evolution associated with sparse potential Jacobi matrices. Long time behaviour of classical and quantum systems (Bologna, 1999), 107–123, Ser. Concr. Appl. Math., 1, World Sci. Publishing, River Edge, NJ, 2001

[CFKS] Cycon, H.L., Froese, R.G., Kirsch, W., Simon, B.: *Schrödinger operators*. Heidelberg: Springer-Verlag, 1987

[DSS] Damanik, D., Sims, R., Stolz, G.: Localization for one dimensional, continuum, Bernoulli-Anderson models. Duke Math. J. **114**, 59-100 (2002)

[DS] Damanik, D., Stollmann, P.: Multi-scale analysis implies strong dynamical localization. Geom. Funct. Anal. **11**, 11-29 (2001)

[DT] Damanik, D., Tcheremchantsev, T.: Power-law bounds on transfer matrices and quantum dynamics in one dimension. Preprint

[DBF] De Bièvre, S., Forni, G.: Transport properties of kicked and quasiperiodic Hamiltonians. J. Statist. Phys. **90**, 1201-1223 (1998)

[DBG] De Bièvre, S., Germinet, F.: Dynamical localization for random dimer Schrödinger operator. J. Stat. Phys. **98**, 1135-1147 (2000)

[DRJLS] Del Rio, R., Jitomirskaya, S., Last, Y., Simon, B.: Operators with singular continuous spectrum IV: Hausdorff dimensions, rank one pertubations and localization. J. d'Analyse Math. **69**, 153-200 (1996)

[DoSh1] Dobrushin, R., Shlosman, S.: Completely analytical Gibbs fields. Prog in Phys. **10**, 347-370 (1985)

[DoSh2] Dobrushin, R., Shlosman, S.: Completely analytical interactions. J. Stat. Phys. **46**, 983-1014 (1987)

[vDK] von Dreifus, H., Klein, A.: A new proof of localization in the Anderson tight binding model. Commun. Math. Phys. **124**, 285-299 (1989)

[DWP] Dunlap, D.H., Wu, H.-L., Phillips, P.W.: Absence of localization in random-dimer model. Phys. Rev. Lett. **65**, 88-91 (1990)

[FK1] Figotin, A., Klein, A.: Localization of classical waves I: Acoustic waves. Commun. Math. Phys. **180**, 439-482 (1996)

[FK2] Figotin, A., Klein, A.: Localization of classical waves II: Electromagnetic waves. Commun. Math. Phys. **184**, 411-441 (1997)

[FLM] Fischer, W., Leschke, H., Müller, P.: Spectral localization by Gaussian random potentials in multi-dimensional continuous space. J. Statist. Phys. **101** 935–985 (2000)

[FMSS] Fröhlich, J., Martinelli, F., Scoppola, E., Spencer, T.: Constructive proof of localization in the Anderson tight binding model. Commun. Math. Phys. **101**, 21-46 (1985)

[FS] Fröhlich, J., Spencer, T.: Absence of diffusion with Anderson tight binding model for large disorder or low energy. Commun. Math. Phys. **88**, 151-184 (1983)

[GDB] Germinet, F., De Bièvre, S.: Dynamical localization for discrete and continuous random Schrödinger operators. Commun. Math. Phys. **194**, 323-341 (1998)

[GKT] Germinet, F., Kiselev, A., Tcheremchantsev, S.: Transfer matrices and transport for 1D Schrödinger operator with singular spectrum. In preparation

[GK1] Germinet, F., Klein, A.: Bootstrap multiscale analysis and localization in random media. Commun. Math. Phys. **222**, 415-448 (2001)

[GK2] Germinet, F, Klein, A.: Decay of operator-valued kernels of functions of Schrodinger and other operators. Proc. Amer. Math. Soc. 131, 911-920 (2003)

[GK3] Germinet, F., Klein, A.: A characterization of the Anderson metal-insulator transport transition. Preprint

[GK4] Germinet, F., Klein, A.: Explicit finite volume criteria for localization in random media and applications. Preprint

[GK5] Germinet, F., Klein, A.: Higher order localization for random Schrödinger operators through explicit finite volume criteria. Markov Process. Related Fields. To appear

[GMP] Gol'dsheid, Ya., Molchanov, S., Pastur, L.: Pure point spectrum of stochastic one dimensional Schrödinger operators. Funct. Anal. Appl. **11**, 1-10 (1977)

[Gu] Guarneri, I., Spectral properties of quantum diffusion on discrete lattices, Europhys. Lett. 10 (1989), 95-100; On an estimate concerning quantum diffusion in the presence of a fractal spectrum, Europhys. Lett. **21**, 729-733 (1993)

[HK] Hislop, P.D., Klopp, F.: The integrated density of states for some random operators with nonsign definite potentials. J. Funct. Anal. **195**, 12-47 (2002)

[HM] Holden, H., Martinelli, F.: On absence of diffusion near the bottom of the spectrum for a random Schrödinger operator. Commun. Math. Phys. **93**, 197-217 (1984)

[JSBS] Jitomirskaya, S., Schulz-Baldes, H., Stolz, G.: Delocalization in random polymer models. Commun. Math. Phys., to appear

[JiL] Jitomirskaya, S., Last, Y.: Anderson localization for the almost Mathieu equation, III. Semi-uniform localization, continuity of gaps, and measure of the spectrum. Commun. Math. Phys. **195**, 1-14 (1998)

[Ki] Kirsch, W.: Wegner estimates and Anderson localization for alloy-type potentials. Math. Z. **221**, 507–512 (1996)

[KM] Kirsch, W., Martinelli, F. : On the ergodic properties of the spectrum of general random operators. J. Reine Angew. Math. **334**, 141-156 (1982)

[KSS] Kirsch, W., Stolz, G., Stollmann, P.: Localization for random perturbations of periodic Schrödinger operators. Random Oper. Stochastic Equations **6**, 241-268 (1998)

[Kle1] Klein, A.: Extended states in the Anderson model on the Bethe lattice. Adv. Math. **133**, 163–184 (1998)

[Kle2] Klein, A.: Spreading of wave packets in the Anderson model on the Bethe lattice. Commun. Math. Phys. **177**, 755–773 (1996)

[Kle3] Klein, A.: Localization of light in randomized periodic media. In *Diffuse Waves in Complex Media*, J.-P. Fouque, ed., pp. 73-92, Kluwer, The Netherlands, 1999

[KK1] Klein, A., Koines, A.: A general framework for localization of classical waves: I. Inhomogeneous media and defect eigenmodes. Math. Phys. Anal. Geom. **4**, 97-130 (2001)

[KK2] Klein, A., Koines, A.: A general framework for localization of classical waves: II. Random media. Preprint

[Klo1] Klopp, F.: Localization for semiclassical continuous random Schrödinger operators. II. The random displacement model. Helv. Phys. Acta **66**, 810–841 (1993)

[Klo2] Klopp, F.: Localization for continuous random Schrödinger operators. Commun. Math. Phys. **167**, 553-569 (1995)

[Klo3] Klopp, F.:Weak disorder localization and Lifshitz tails. Commun. Math. Phys. **232**,125-155 (2002)

[Klo4] Klopp, F.: Weak disorder localization and Lifshitz tails: continuous Hamiltonians. Ann. I.H.P. **3**, 711-737 (2002)

[KS] Kunz, H., Souillard, B.: Sur le spectre des operateurs aux differences finies aleatoires. Commun. Math. Phys. **78**, 201-246 (1980)

[La] Last, Y: Quantum dynamics and decompositions of singular continuous spectra. J. Funct. Anal. **142**, 406-445 (1996)

[LGP] Lifshits, I.M., Gredeskul, A.G., Pastur, L.A.: *Introduction to the Theory of Disordered Systems*, Wiley-Interscience, New York, 1988

[MS] Martinelli, F., Scoppola, E.: Introduction to the mathematical theory of Anderson localization. Riv. Nuovo Cimento **10**, 1-90 (1987)

[St] Stollmann. P.: Wegner estimates and localization for continuum Anderson models with some singular distributions. Arch. Math. (Basel) **75**, 307-311 (2000)

[U] Ueki, N.: Wegner estimates and localization for Gaussian random potentials. Preprint

[Wa1] Wang, W.-M.: Microlocalization, percolation, and Anderson localization for the magnetic Schrödinger operator with a random potential. J. Funct. Anal. **146**, 1-26 (1997)

[Wa2] Wang, W.-M.: Localization and universality of Poisson statistics for the multidimensional Anderson model at weak disorder. Invent. Math. **146**, 365-398 (2001)

[Weg] Wegner, F.: Bounds on the density of states in disordered systems. Z. Phys. B **44**, 9-15 (1981)

Université de Lille 1, UMR 8524 CNRS, UFR de Mathématiques, F-59655 Villeneuve d'Ascq Cédex, France

E-mail address: germinet@agat.univ-lille1.fr

University of California, Irvine, Department of Mathematics, Irvine, CA 92697-3875, USA

E-mail address: aklein@uci.edu

Contemporary Mathematics
Volume **339**, 2003

Theory and Computation of Scattering Resonances of Photonic Microstructures

S.E. Golowich and M.I. Weinstein

ABSTRACT. The optical properties of microstructured photonic waveguides are governed, in the scalar approximation, by a Schrödinger equation with a high contrast and oscillatory potential. We outline a scattering resonance perturbation theory for this non-selfadjoint eigenvalue problem. The imaginary parts of the scattering resonance "eigenvalues" correspond to the radiative leakage rates from these photonic waveguides. We implement a multiple scale approach to the study of such waveguides in the scalar approximation. The leading order theory agrees with classical homogenization theory, describing an effective homogeneous medium with dielectric properties given by an appropriate averaging of the refractive index profile. We compute the first non-trivial correction, which takes into account the microstructure, and find that the higher order homogenization expansion gives very good agreement with direct numerical simulations by Fourier and multipole methods. The higher order expansion is crucial for estimation of the leakage rates; in various examples of physical interest the leading order (homogenization) gives a substantial underestimation. We rigorously derive our 2nd order homogenization expansion for the case of $N-$ fold symmetric structures.

CONTENTS

2000 *Mathematics Subject Classification.* Primary 78A40; Secondary 78A48 78M40 78A45.
Key words and phrases. Microstructure, nanostructure, photonic band gap, photonic crystal, scattering frequency, scattering resonance, quasi-mode, radiation, leaky mode, tunneling loss.

1. Introduction and outline

The propagation of light in inhomogeneous media is a rich source of physical phenomena and is of huge technological importance. Recent advances in fabrication technology have made possible the fabrication of micro- and nanostructure media. Such media are often called *photonic* micro- or nanostructures, because by variation of the distribution of micrometer, respectively nanometer, scale structures, their individual geometry and material contrasts, one can influence the light propagation properties, linear and nonlinear, by finely tuning *optical* properties such as group velocity, dispersion and loss. Regardless of the characteristic length scale of the individual features, we shall loosely refer to such structures as *photonic microstructures*. Potential applications range from devices for telecommunication networks [**10**] to the exploration of quantum electrodynamics (QED) in microcavities [**19**].

In this paper, we focus on a class of photonic microstructure waveguides which have attracted a great deal of recent interest. Photonic crystal waveguides or "holey" optical fibers are waveguides with novel cross-sections consisting of holes surrounded by glass. The holes may be empty or filled with a material chosen to influence the propagation. The ability to vary the transverse geometry, combined with the large index contrasts possible with such structures, give multiple new degrees of freedom that potentially enable designs with radically different properties than is possible with standard fiber. See [**7**] for a discussion of interesting physical phenomena observed in such waveguides.

Efficient and accurate mathematical modeling of light propagation in microstructured waveguides is clearly necessary for their design and analysis. A feature of most such structures is that they are inherently leaky due to the existence of paths leading from the waveguide "core" to the outer region that avoid the holes and pass only through the background glass. Such structures support no true guided modes (the L^2- spectrum is continuous, corresponding to extended radiation states), but they will have leaky modes characterized by complex-valued propagation constants or effective indexes (scattering resonances); the leakage rate is given by the imaginary part. Physically, this leakage is due to a combination of tunneling through the holes and propagation through the glass surrounding them. The ability to calculate such rates is clearly of fundamental importance. Other quantities of interest include the real parts of the effective indexes, which determine the response of the structure to longitudinal variations such as gratings, and the dispersion relations of the various leaky modes, which may be quite unusual compared to standard waveguides due to the presence of large index contrasts and interference effects.

In the scalar approximation, the leaky modes and effective indexes are obtainable from the solution of the eigenvalue problem for the Schrödinger equation

$$\begin{aligned} H\psi \;=\; (\,-\Delta \;+\; V\,)\,\psi \;=\; E\psi \\ \psi \text{ outgoing as } r = |\mathbf{x}| \to \infty. \end{aligned} \tag{1.1}$$

Formulations of this *scattering resonance problem* are in section 3 and section 4. Here, $\Delta = \Delta_\perp$ denotes the two-dimensional Laplace operator in the transverse plane. (In the following, we shall omit the subscript $\perp$.) Since the outgoing radiation condition at infinity is not self-adjoint, we expect the eigenvalues, E, to be complex. The imaginary parts of the scattering resonance energies, E, determine the leakage rates; see section 2.

We view the potential as consisting of two parts:

$$V(x) = V_0(x) + \delta V(x),$$

where in the above waveguide context, V_0 is a slowly varying "average" and $\delta V(x)$ is oscillatory but, possibly, pointwise large, corresponding to a microstructure with high contrast.

We now give a brief description of the results presented in this article.

I. Theory of scattering resonances of the Schrödinger equation for high contrast and oscillatory potentials: In section 4 we formulate the scattering resonance problem with potential V, $SRP(V)$, as an appropriate Lippman-Schwinger equation. We prove that if $SRP(V_0)$ has a scattering resonance pair $(\psi_0, E_0) \in H^1(\chi(x)dx) \times \mathbb{C}$, then $SRP(V_0 + \delta V)$ has a "nearby" scattering resonance pair, which varies smoothly with δV for small δV in appropriate norm $|||\delta V|||$. This norm of the perturbation is small if δV is sufficiently oscillatory. In particular, for a potential $\delta V_N = \delta V(r, N\theta)$, where $\delta V(r, \Theta)$ is 2π periodic in Θ, corresponding to an N-fold symmetric potential, $|||\delta V_N||| = \mathcal{O}(N^{-1})$. In the microstructure waveguide context, smallness of $|||\delta V_N|||$ is implied the smallness of $\lambda^{-2}\, N^{-1}$, where λ denotes the wavelength of light.

II. Multiple scales, higher order homogenization expansion and approximate numerical methods: We develop a multiple scale perturbation theory of a large class of "sufficiently oscillatory" structures suitable for the analytical study and efficient numerical computation of such quantities as leakage rates, group velocities and dispersion. The procedure becomes more accurate as the transverse structure becomes more oscillatory, in a sense which we make precise below. If the transverse structure is invariant under rotation by $2\pi/N$ (*e.g.* in the case of a single ring of N holes uniformly distributed in an annulus) we expand the leaky modes and the effective indexes (scattering resonances) in powers of N^{-1}. The leading order, $\mathcal{O}(N^0)$, term in the expansion is derived from a suitably averaged or homogenized refractive index profile [**4, 12**]. We derive and numerically compute the leading order corrections due to microstructure (which enter at order N^{-2}) of both the real and imaginary parts of the effective indexes of the leaky modes. Since the leading order behavior is given by a homogenized (in angle) effective medium, we refer to this expansion as a homogenization expansion. Results on first-order corrections to homogenized eigenvalues of periodic composite media were obtained in [**16, 14**].

A few of the consequences of the homogenization expansion and numerical implementation that we wish to highlight are:

(1) Computational algorithm: Our analytic theory leads to a natural efficient computational algorithm for the modes and spectral characteristics, *e.g.* leakage rates, dispersion. Multiscale analysis enables us to eliminate the "stiff" aspects of the computation due to the rapidly varying structure. Therefore, one can expect significant reduction in computational effort. This is especially important for increasingly microstructured media, for which the approximation improves, as well as for simulation based optimization of light-guiding characteristics.

(2) Arbitrary geometry and index contrasts: Although we require N-fold symmetry of the structure, the individual microfeatures may have arbitrary geometry and large index contrasts. In our current implementation, we approximate an arbitrary

microfeature by a *simple layered structure*, defined in section 7.2, for which many of the calculations required can be done explicitly.

(3) Very good agreement with full numerical simulation: In section 6 we present results of numerical simulations based on our theory for several waveguides with transverse microstructure. A more extensive discussion of and comparison with other numerical approaches is presented in [**7**]. We focus on the complex effective indexes as they vary with the geometry of the microstructure. We are particularly interested in the imaginary parts of the effective indexes, which correspond to the leakage rates. We present comparisons of our predicted loss rates with results obtained by solving the Helmholtz equation directly by Fourier decomposition algorithm [**15**] and the vector Maxwell equation by multipole methods [**20**]. One expects homogenization theory to be valid when the wavelength of light is long compared with the individual microfeatures. We find very good agreement of our theory with full simulations even in regimes where $\lambda_{\rm fs}/d$, the ratio of wavelength to length scale of microfeatures, is as small as $3/2$. Although, the expansion is derived for N large, with the $N = \infty$ limit being the leading order term, examples of section 6 show agreement in cases where $N = 3$ and $N = 6$. We also observe the expected departure of our approximate methods from the results of full simulations for sufficiently small $\lambda_{\rm fs}/d$.

(4) Sensitivity of leakage rates to microstructure: In section 6 we have used our theory to compute the first two nontrivial terms of the effective indexes (scattering resonances) of the leaky modes. The first term corresponds to an average theory and the second term is a correction due to microstructure. The imaginary parts of the effective indexes, corresponding to the leakage rates, are very sensitive to the introduction of microstructure and their accurate approximation requires both terms. In contrast, as noted above, the real parts of the effective indexes are relatively insensitive to microstructure, and are well captured by the leading order term.

(5) Corrected fields: We find that corrections, due to microstructure, of the mode fields predicted by the averaged structure are compactly supported in space if the microstructure perturbation is compactly supported in space. This is not possible for the true solution of the mathematical model. An additional observation indicating a limitation of the homogenization expansion is seen for a family of structures consisting of two rings of holes, parametrized by the relative phase angle of the arrangement in one ring to another. Since the variation in refractive index in one ring occurs on a set disjoint from that for the second ring, we find the homogenization expansion predicts complex effective indexes that are *independent* of the phase angle. This suggests that in the regime where certain interference effects are important, the expansion of this paper may have limited use. Nevertheless, we have proved in [**6**] (see also section 4) that two-term approximate expansion of modes and effective indexes is valid for sufficiently large N. This rigorous theory gives field corrections which, at any finite N, are not compactly supported.

Acknowledgments: The authors wish to thank B. Eggleton, R. Windeler, R. Bise, D. DiGiovanni and J. Jaspara for introducing them to the subject of microstructured optical fiber and for discussions on physical aspects. The authors also wish to thank R.V. Kohn, G.C. Papanicolaou, T. Salamon and M. Vogelius for stimulating and informative discussions during the preparation of this work.

2. Maxwell's equations in a wave guide and the scalar approximation

Maxwell's equations in a three dimensional dielectric medium with spatial coordinates, $\mathbf{x} = (x_1, x_2, x_3)$, and time coordinate, t are:

$$\begin{aligned} \nabla \cdot \vec{\mathcal{D}} &= 0 \quad \nabla \wedge \vec{\mathcal{E}} = -\partial_t \vec{\mathcal{B}} \\ \nabla \cdot \vec{\mathcal{B}} &= 0, \ \nabla \wedge \vec{\mathcal{H}} = \partial_t \vec{\mathcal{D}} \end{aligned} \tag{2.1}$$

$\vec{\mathcal{E}}$ denotes the electric field, $\vec{\mathcal{D}}$ the electric displacement, $\vec{\mathcal{B}}$, the magnetic induction and $\vec{\mathcal{H}}$ the magnetic field. We have assumed no sources of free charge or current. A linear nonmagnetic dielectric medium is characterized by the linear constitutive laws:

$$\vec{\mathcal{D}} = \epsilon \vec{\mathcal{E}}, \quad \vec{\mathcal{H}} = \mu_0^{-1} \vec{\mathcal{B}}.$$

The free space (vacuum) values of ϵ and μ are denoted by ϵ_0 and μ_0. The *speed of light* in a vacuum, c, is given by $(\epsilon_0 \mu_0)^{-\frac{1}{2}}$, while that in a general nonmagnetic medium $(\epsilon \mu_0)^{-\frac{1}{2}}$. The *index of refraction* of the medium, $n(\mathbf{x})$, is the ratio of c to the speed of light in the medium:

$$n(\mathbf{x}) = \frac{\sqrt{\epsilon_0 \mu_0}}{\sqrt{\epsilon \mu_0}} = \sqrt{\frac{\epsilon_0}{\epsilon}}$$

and therefore

$$\epsilon(\mathbf{x}) = n^2(\mathbf{x}) \, \epsilon_0 \tag{2.2}$$

We fix a frequency, ω, and derive the equation of the corresponding modes of the waveguide. Thus, we let

$$\vec{\mathcal{E}} = \vec{\mathcal{E}}(\mathbf{x}, \omega) e^{-i\omega t}, \ \vec{\mathcal{H}} = \vec{\mathcal{H}}(\mathbf{x}, \omega) e^{-i\omega t} \tag{2.3}$$

The relationship between frequency and wavelength in free space is given by

$$\omega = ck = \frac{2\pi c}{\lambda_{\text{fs}}},$$

where λ_{fs} denotes the free space wavelength.

Substitution of (2.3) into (2.1) yields:

$$\vec{\nabla} \wedge \vec{\mathcal{E}} = i \sqrt{\frac{\mu_0}{\epsilon_0}} \, k \, \vec{\mathcal{H}} \tag{2.4}$$

$$\vec{\nabla} \wedge \vec{\mathcal{H}} = -i \sqrt{\frac{\epsilon_0}{\mu_0}} \, k \, n^2 \, \vec{\mathcal{E}} \tag{2.5}$$

Apply the curl operator, $\vec{\nabla} \wedge$ to (2.4) and using (2.5) yields the vector Helmholtz wave equation:

$$\vec{\nabla} \left(\vec{\nabla} \cdot \vec{\mathcal{E}} \right) - \Delta \vec{\mathcal{E}} = k^2 n^2 \vec{\mathcal{E}}. \tag{2.6}$$

The condition $\vec{\nabla} \cdot \vec{\mathcal{D}} = 0$ becomes

$$\vec{\nabla} \cdot \left(n^2 \, \vec{\mathcal{E}} \right) = 0$$

which can be used in (2.6) to give

$$\Delta\vec{\mathcal{E}} + k^2 n^2 \vec{\mathcal{E}} = -\vec{\nabla}\left[\vec{\mathcal{E}} \cdot \vec{\nabla} \ln n^2 \right] \tag{2.7}$$

We consider propagation in a waveguide, with x_3 taken to be the longitudinal coordinate and $\mathbf{x}_\perp = (x_1, x_2)$ the transverse coordinates and assume the waveguide has a uniform transverse cross-section (translation invariance in x_3):

$$n(\mathbf{x}) = n(\mathbf{x}_\perp) = n(x_1, x_2).$$

It is then natural to seek elementary modes which propagate in the x_3 direction

$$\vec{\mathcal{E}}(\mathbf{x}) = \vec{e}(\mathbf{x}_\perp) e^{i\beta x_3} = \left[\vec{e}_\perp(\mathbf{x}_\perp) + e_{\text{lon}}(\mathbf{x}_\perp)\hat{k} \right] e^{i\beta x_3}$$

where $\hat{k} = (0,0,1)$, $\vec{e}_\perp \cdot \hat{k} = 0$. The vectors $\vec{e}_\perp$ and e_{lon} denote transverse and longitudinal field amplitudes. Substituting into (2.7), we obtain the following equations [**18**]

$$\left(\Delta_\perp + k^2 n^2(\mathbf{x}_\perp) - \beta^2 \right) \vec{e}_\perp = \mathcal{V}_1 \vec{e}_\perp, \tag{2.8}$$

$$\left(\Delta_\perp + k^2 n^2(\mathbf{x}_\perp) - \beta^2 \right) e_{\text{lon}} = -i\beta\, \vec{e}_\perp \cdot \nabla_\perp \ln n^2, \tag{2.9}$$

where

$$\mathcal{V}_1 \vec{e}_\perp \equiv -\vec{\nabla}_\perp \left(\vec{e}_\perp \cdot \vec{\nabla}_\perp \ln n^2 \right)$$

Note that (2.8) is a closed equation for $\vec{e}_\perp$. Therefore, we can first solve for $\vec{e}_\perp$, and then obtain e_{lon} solving (2.9).

We shall work in the *scalar approximation*, which entails neglecting the term $\mathcal{V}_1\, \vec{e}_\perp$ in (2.8) that arises due to the vector nature of the fields. For modes that are localized away from the microstructure, which includes most of the examples in section 6, the vector corrections are expected to be small because the field $\vec{e}_\perp$ is small at the interfaces, which are the only places at which it contributes. However, this will not be true in other examples of interest, and we expect that our methods can be extended to the full vector case. In any case, we will not discuss vector effects further in this work.

In the scalar approximation the components of $\vec{e}_\perp$ satisfy independent scalar Helmholtz equations. We let φ denote either of these transverse components. Then,

$$\left(\Delta_\perp + k^2 n^2 \right) \varphi = \beta^2 \varphi \tag{2.10}$$

where $\Delta_\perp$ denotes the Laplace operator in the transverse variables, $x_\perp = (x_1, x_2)$.

Introducing the notation

$$V = k^2(n_g^2 - n^2), \quad E = k^2 n_g^2 - \beta^2, \tag{2.11}$$

the equation for the φ and propagation constant β can be viewed as a Schrödinger equation with potential $V(\mathbf{x}_\perp)$ and energy parameter E:

$$H\varphi \equiv (-\Delta_\perp + V)\varphi = E\,\varphi. \tag{2.12}$$

Corresponding to the physical problem described in section 3 we assume that for some $r_* > 0$,

$$\begin{aligned} V(\mathbf{x}) &\geq 0, \text{for all } \mathbf{x} \text{ and} \\ V(\mathbf{x}) &= 0, \text{for } r \geq r_* \end{aligned} \tag{2.13}$$

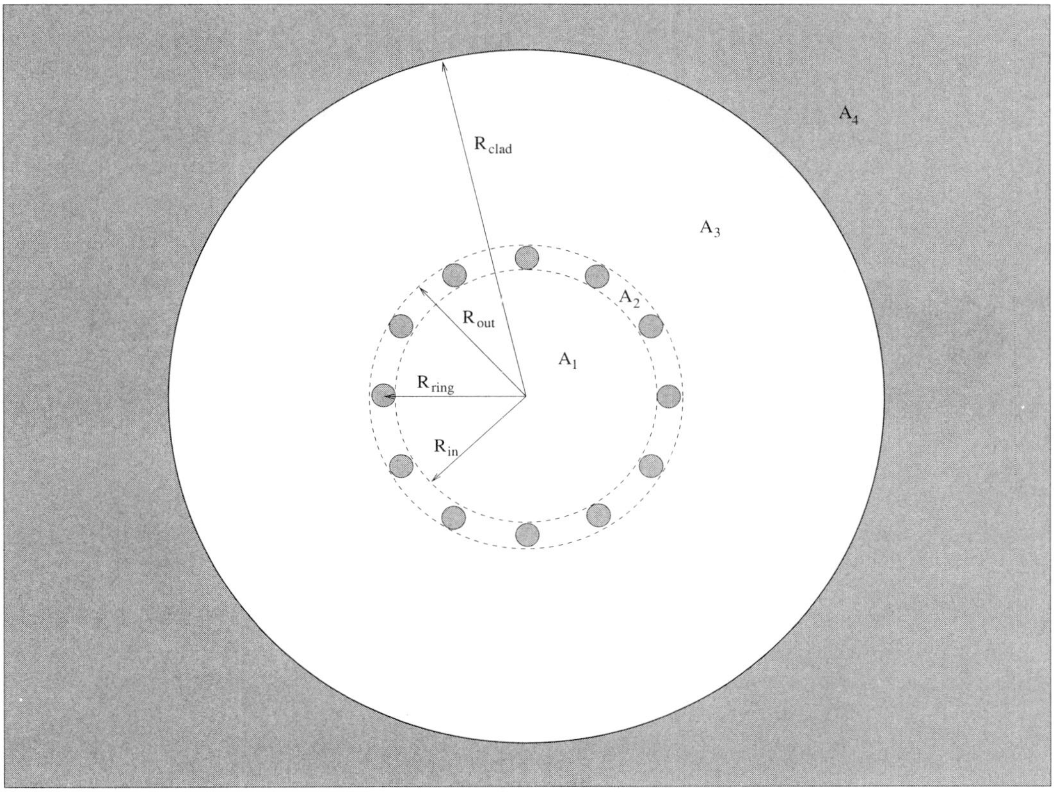

FIGURE 1. A cross section of a microstructured waveguide.

The potential, V, therefore does not support bound states (guided modes) and only has scattering states (radiation modes), along with scattering resonances (leaky modes).

3. The resonance problem for microstructure waveguides

To fix ideas, consider a cylindrical waveguide composed of two materials "g" and "h", with corresponding refractive indexes n_g and n_h. We specify a transverse cross-section of the waveguide in the 2-dimensional plane, with coordinates $\mathbf{x} = (x, y) = (r, \theta)$. In figure 1 we show a schematic picture of a glass waveguide with a single ring of 12 air holes in a solid glass cylinder. Four regions are illustrated:

A_1, the region $|\mathbf{x}| \leq R_{\text{in}}$, a disc of glass with refractive index n_g;

A_2, the region $R_{\text{in}} \leq |\mathbf{x}| \leq R_{\text{out}}$, an annular region with 12 equally spaced air-holes. In A_2, n takes on two values, n_g in the bulk and $n_h \sim 1$ in the holes;

A_3, the region $R_{\text{out}} \leq |\mathbf{x}| \leq R_{\text{clad}}$, consisting of a solid glass, $n = n_g$, and finally;

A_4, the region $|\mathbf{x}| \geq R_{\text{clad}}$ consisting of air, $n = n_h$.

We are interested in how well light is confined to the core (A_1) region in such microstructured fiber. This structure has actual guided, in addition to leaky, modes, due to the lower refractive index of region A_4. However, we view any propagation of light from A_1 to A_3 as loss. Therefore, the appropriate mathematical problem for our purposes is the mode equation (2.12) posed on the spatial domain: $\{|\mathbf{x}| \leq R_{\rm in}\} \cup \{R_{\rm in} \leq |\mathbf{x}| \leq R_{\rm out}\} \cup \{R_{\rm out} \leq |\mathbf{x}|\}$, subject to an *outward going radiation condition as* $|\mathbf{x}| \to \infty$. In practice, the glass optical fiber is usually coated with a high index and high loss polymer, so any light that does leak into A_3 is rapidly attenuated, further justifying the use of this boundary condition.

We now present a more general mathematical formulation of the problem.

The scattering resonance problem: *Given a potential (index profile)* $V = V(r,\theta)$ *of compact support, we seek* E *and nonzero* φ *for which*

$$H\varphi \equiv (-\Delta + V)\varphi = E\varphi \tag{3.1}$$

subject to the outward going radiation condition, as $r \to \infty$. *We call a pair* (φ, E) *a scattering resonance pair.*

A scattering resonance pair is sometimes called a *quasiresonance* [**9**].

In our setting, the radiation condition can be simply described. Because $V(r)$ is identically zero for $r \geq r_*$, equation (2.12) becomes:

$$-(\Delta_\perp + E)\varphi = 0, \quad r \geq r_* \tag{3.2}$$

Solutions of (3.2) can be expanded in a series

$$\varphi = \sum_{l=-\infty}^{+\infty} \left(c_l^{(1)} e^{il\theta} H_l^{(1)}(\sqrt{E}\, r) + c_l^{(2)} e^{il\theta} H_l^{(2)}(\sqrt{E}\, r) \right). \tag{3.3}$$

The functions $H_l^{(1)}$ and $H_l^{(2)}$ are first and second Hankel functions of order l and satisfy the following asymptotic formulae for large arguments [**1**] (page 364, equations 9.2.3 and 9.2.4):

$$\begin{aligned} H_l^{(1)}(\zeta) &\sim \sqrt{\frac{2}{\pi\zeta}}\, e^{i(\zeta - \frac{1}{2}l\pi - \frac{\pi}{4})}, \quad -\pi < \arg\zeta < 2\pi \\ H_l^{(2)}(\zeta) &\sim \sqrt{\frac{2}{\pi\zeta}}\, e^{-i(\zeta - \frac{1}{2}l\pi - \frac{\pi}{4})}, \quad -2\pi < \arg\zeta < \pi \end{aligned} \tag{3.4}$$

The functions $H_l^{(1)}$ are interpreted as *outgoing radiation* because, in the time dependent setting, they have outward moving phase fronts for increasing t:

$$e^{-i\omega t} H_l^{(1)}(\zeta) \sim \frac{h_l}{\sqrt{\zeta}} e^{i(\zeta - \omega t)},$$

where h_l is a complex constant. Therefore the requirement that φ be outgoing is equivalent to the vanishing of all coefficients $c_l^{(2)}$ in (3.3):

$$\textit{Outward going radiation} \Longleftrightarrow c_l^{(2)} = 0 \text{ for all } l. \tag{3.5}$$

Therefore, the scattering resonance problem can be formulated as the eigenvalue problem (3.1), (3.5).

Since (3.5) is a non-self adjoint boundary condition at infinity E can be expected to be complex. However, there are constraints on the location of E in the complex plane. For the φ to be oscillatory at infinity, we must have $\Re E > 0$ and, for φ to be outgoing $\Re\sqrt{E} > 0$. Furthermore, $\Im E \leq 0$. For, if $\Im E > 0$, then $\Im\sqrt{E} > 0$.

From the series expansion (3.3) with $c_l^{(2)} = 0$ we conclude that φ is exponentially decaying as $r \to \infty$, and therefore a eigenstate of the self-adjoint operator H with non-real eigenvalue. This yields a contradiction. To summarize:

If E is an energy for which there is a solution of the resonance problem, then $\Re E > 0$ and $\Im E \leq 0$ Correspondingly, by (2.11), we have $\Re\beta > 0$ and $\Im\beta > 0$. Since the electric field $\vec{\mathcal{E}}(\mathbf{x}, \omega) = \vec{e}_\perp(\mathbf{x}_\perp)e^{i\beta x_3}$, the complex resonances E correspond to the radiative decay or leakage rates of the structure.

In the following section, section 4, we discuss an alternative formulation of the scattering resonance problem, more amenable to rigorous theory. In section 5 we embark on a multiple scale / homogenization approach, which can be justified using the results of section 4.

4. Scattering resonances of the Schrödinger equation for high contrast and oscillatory potentials

Motivated by a class of problems arising in wave propagation through high contrast microstructures, we now study the scattering resonances for the Schrödinger equation with high contrast and oscillatory potentials. We present an outline of results, specialized to the case of $N-$ fold symmetric potentials. The detailed proofs can be found in [**6**].

Let $H = -\Delta + V$, where V is sufficiently smooth and of compact support. The essential spectrum of H is the postive half-line, $[0, \infty)$. For $E \notin [0, \infty)$, the resolvent $(H - E)^{-1}$ is a well-defined bounded operator on L^2. Restricted to the dense subspace of smooth compactly supported functions, it can be analytically continued across the essential spectrum to the lower half plane, where it defines a meromorphic function, with the poles corresponding to scattering resonances; see, for example, [**5, 17, 3**]. We denote this analytical continuation by $(H - E)_+^{-1}$.

Consider $H = -\Delta$ and dimension $n = 2$. Then, for compactly supported f

$$(4.1) \qquad (-\Delta - E)_+^{-1} f = \frac{i}{4} \int H_0^{(1)}(\sqrt{E}|x - y|)\, f(y)\, dy;$$

see [**9**], page 165, and [**13**].

In this section we shall work with the following formulation of the scattering resonance problem for the potential V, $SRP(V)$:

Given V a bounded function of compact support, find (E, ψ) such that $\psi \in L^2_{\rm loc}$ and

$$(4.2) \qquad \psi = -(-\Delta - E)_+^{-1}\, V\psi.$$

Equation (4.2) is often called a Lippman-Schwinger equation.

Suppose (E_0, ψ_0) is a solution to $SRP(V_0)$. E_0 is a complex number in the fourth quadrant, and $|\psi_0(x)| \leq e^{\alpha_0|x|}$. Consider the scattering resonance problem $SRP(V_0 + \delta V)$, where $\delta V(x)$ is pointwise large (high contrast) and oscillatory. We next reformulate (4.2), in a manner appropriate for class of perturbations we consider.

We introduce a smooth function, $\chi(x)$, with $\chi(x) \sim e^{-\alpha|x|}$ for large $|x|$, for $\alpha > \alpha_0$ sufficiently large. Define the operators

$$(4.3) \qquad T_V = \langle D\rangle^{-1}\, \chi^{-1} V \chi^{-1}\, \langle D\rangle^{-1}$$

$$(4.4) \qquad T_+(E) = \langle D\rangle\, \chi\, (-\Delta - E)_+^{-1}\, \chi\, \langle D\rangle$$

Finally, define

$$\Psi \;=\; \langle D\rangle\;\chi(x)\;\psi(x), \tag{4.5}$$

where

$$\langle D\rangle \;=\; (\;I-\Delta\;)^{\frac{1}{2}}\;.$$

Instead of studying (4.2), we shall consider the reformulation

$$(\;I\;+\;T_+(E)T_V\;)\;\Psi\;=\;0 \tag{4.6}$$

We shall refer to this as a *preconditioned* Lippman-Schwinger equation. In terms of this formulation, given a nontrivial solution (Ψ_0, E_0) of (4.6), we want to study the solutions $(\Psi_0+\delta\Psi, E_0+\delta E)$ of

$$(\;I\;+\;T_+(E_0+\delta E)T_{V_0+\delta V}\;)\;(\Psi_0+\delta\Psi)\;=\;0 \tag{4.7}$$

for δV sufficiently "small".

Concerning the definition of Ψ in (4.5), the purpose of multiplication by $\chi(x)$ is to give a formulation in L^2, while the operator multiplier, $\langle D\rangle$, enables us to view a perturbation as small provided it is sufficiently oscillatory, but not pointwise small. In particular,

PROPOSITION 4.1. *Let δV be an $L^2(\mathbb{R}^2)$ function. Let (r,θ) denote standard polar coordinates in $\mathbb{R}^2$ and consider the $N-$ fold symmetric function $\delta V_N = \delta V(r, N\theta)$. Then,*

$$|||\;\delta V_N\;|||\;\equiv\;\|T_{\delta V_N}\|_{\mathcal{B}(L^2)}\;\le\;\frac{C}{N}, \tag{4.8}$$

where $\|A\|_{\mathcal{B}(L^2)}$ denotes the L^2 operator norm of A.

As N increases, the structure described by the potential δV_N becomes progressively more "microstructured," yet remains of fixed contrast. We have the following

THEOREM 4.1. *(a) Let V_0 denote a potential for which $|||V_0|||$ is finite. Let (E_0, ψ_0) denote a solution of the corresponding scattering resonance problem. Assume*

$$\text{dimension}\;\{\;f\in L^2:\;(\;I\;+\;T_+(E_0)T_{V_0}\;)\;\Psi\;=\;0\;\}\;=\;1 \tag{4.9}$$

and this subspace is spanned by the function

$$\Psi_0 = \langle D\rangle\chi\psi_0.$$

Then, there exists $\varepsilon_0 > 0$ such that for any potential V satisfying $|||V-V_0||| \;< \varepsilon_0$ ($\delta V \;=\; V - V_0$), there corresponds a unique solution $(E(V), \psi(V))$ of the scattering resonance problem which lies near (E_0, ψ_0).
(b) The mapping

$$V\;\mapsto\;(E,\psi)\;\in\;\mathbb{C}\times H^1(\chi(x)dx),$$

which associates to a potential V a solution $(E(V), \psi(V))$ of the scattering resonance problem, is analytic in a neighborhood $|||V-V_0||| < \varepsilon_0$ of V_0.

Theorem 4.1 enables us to expand $E(V_0+\delta V) - E(V_0)$ in δV:

$$E(V_0+\delta V) - E(V_0)\;=\;\delta E^{(1)}\;+\;\delta E^{(2)}\;+\;\cdots, \tag{4.10}$$

where $\delta E^{(j)}$ is of order j in δV. For the case, $\delta V \;=\; \delta V(r, N\theta)$, where δV has angular average zero, we have $\delta E^{(1)} = 0$ and

PROPOSITION 4.2. *Then,*

$$\delta E_N^{(2)} = \frac{\langle\ \delta V_N \overline{\psi_0}\ ,\ (-\Delta - E_0 + V_0)_+^{-1}\ \delta V_N \psi_0\ \rangle}{\langle\ V_0 \overline{\psi_0}\ ,\ (-\Delta - E_0)_+^{-2}\ V_0 \psi_0\ \rangle} \tag{4.11}$$

In the following section, we derive by multiple scale perturbation theory an asymptotic expansion of solutions of the scattering resonance problem in powers of N^{-1}. For the above $N-$ fold symmetric structure, this gives the *homogenization expansion*:

$$E\ -\ E_0\ =\ \frac{1}{N^2} E_2^{\text{homog}}\ +\ \mathcal{O}(\frac{1}{N^4});$$

see section 5. For large N, we have shown that the resonance perturbation theory of this section agrees with higher order homogenization through second order:

PROPOSITION 4.3. *Let $V \in C^2$. Then,*

$$\delta E_N^{(2)}\ =\ \frac{1}{N^2}\left(\ E_2^{\text{homog}}\ + \mathcal{O}\left(\frac{1}{N^2}\right)\ \right),\quad N \to \infty. \tag{4.12}$$

The case where V is less regular is under study. We remark that the the condition (4.9) may not be satisfied due to degeneracies arising from symmetries of the potential. Theorem 4.1 can still be seen to apply, however, as we describe in [**6**].

5. Multiple scales and homogenization expansion

In this section we derive a multiple scale expansion of solutions to the resonance problem for microstructures, giving a homogenized (averaged) theory at leading order plus systematically computable corrections. We carry out the analysis for a class of potentials V with the dependence:

$$V\ =\ V(r,\theta,N\theta)\ =\ V(r,\theta,\Theta), \tag{5.1}$$

where V is $2\pi-$periodic in θ and Θ. Thus (5.1) allows both a slow and a fast angular modulation of the index profile. In the special case where V does not depend on θ and we have $V = V(r, N\theta)$, the potential corresponds to an index profile with an N- fold symmetry. Assume $V \geq 0$ and that for $r \geq r_L$, $V \equiv 0$. We will show the following:

Homogenization Expansion: *The resonance mode problem (3.1-3.5) has, for large N, solutions with the formal expansion:*

$$\varphi(r,\theta;N)\ =\ \Phi^{(N)}(r,\theta,\Theta),\ \Theta = N\theta$$

given by

$$\begin{aligned} \Phi^{(N)} &= \Phi_0\ +\ \frac{1}{N^2}\Phi_2\ +\ \mathcal{O}(\frac{1}{N^3}) \\ E^{(N)} &= E_0\ +\ \frac{1}{N^2}E_2\ + \mathcal{O}(\frac{1}{N^3}). \end{aligned} \tag{5.2}$$

(The coefficient E_2 is referred to as E_2^{homog} in section 4.)

- *$(\Phi_0(r,\theta), E_0)$ is a non-trivial solution of the resonance problem:*

$$H_{\text{av}}\Phi_0\ =\ E_0\Phi_0 \tag{5.3}$$

where Φ_0 is subject to the outward going radiation condition (3.5), and the averaged operator is given by

$$H_{\rm av} \equiv \frac{1}{2\pi}\int_0^{2\pi} H\, d\Theta \ \equiv -\Delta_\perp \ + \ V_{\rm av}(r,\theta). \tag{5.4}$$

• $\Phi_2 = \Phi_2^{(p)} + \Phi_2^{(h)}$, *where $\Phi_2^{(p)}$ is the mean zero solution (easily derived from the Fourier series of $V(\cdot,\cdot,\Theta)$) of:*

$$\frac{1}{r^2}\partial_\Theta^2\Phi_2^{(p)} \ = \ [\ V(r,\theta,\Theta) - V_{\rm av}(r,\theta)\ ]\,\Phi_0(r,\theta). \tag{5.5}$$

and $(\Phi_2^{(h)}, E_2)$ solves:

$$(\ H_{\rm av} \ - \ E_0)\,\Phi_2^{(h)} = \left(E_2 \ + \ \frac{r^2}{2\pi}\int_0^{2\pi} \left|\partial_\Theta^{-1}[V(r,\theta,\Theta) - V_{\rm av}(r,\theta)]\right|^2 d\Theta \right)\Phi_0(r,\theta) \tag{5.6}$$

Finally, E_2 is uniquely determined by the condition that (5.6) has a solution which satisfies an outgoing radiation condition at $r=\infty$; see (3.5).

• *This yields an approximate solution of the scalar wave equation and an approximation (neglecting vector effects) of the transverse electric field, $\mathcal{E}_\perp = (\mathcal{E}_{\perp 1}, \mathcal{E}_{\perp 2})$, of Maxwell's equations:*

$$\mathcal{E}_{\perp,q}(\mathbf{x};\beta) \ \sim e^{i(\beta x_3 - \omega t)} \left[\ \Phi_0(|\mathbf{x}_\perp|,\theta;\omega) \ + \ \frac{1}{N^2}\Phi_2(|\mathbf{x}_\perp|,\theta,N\theta;\omega) \ + \ \mathcal{O}\left(\frac{1}{N^3}\right)\right]$$

Here, $\omega = ck$ and $\beta = \sqrt{k^2 n_g^2 - E} = \Re\beta + i\Im\beta$, with $\Im\beta > 0$. $\mathcal{E}_{\perp,q}(\mathbf{x}_\perp, x_3)$ decays with increasing x_3 and is therefore a "leaky mode." These modes are not square integrable.

In section 6, we present the results of numerical simulations performed using a two-term truncation of our homogenization expansion, compare them with published computations of solutions to the full problem, and discuss the question of convergence of our expansion.

5.1. Multiple scale expansion. Due to the rapidly varying coefficient in (2.12) we expect rapid variations in its solutions. We therefore explicitly introduce the fast angular variable:

$$\Theta \ = \ N\theta, \tag{5.7}$$

and view φ as a function of the three independent variables: r, θ and Θ:

$$\varphi(r,\theta) \ = \ \Phi(r,\theta,\Theta). \tag{5.8}$$

Here, we use polar coordinates $x = (r,\theta)$ and note that

$$\Delta_\perp \ = \ \Delta_r \ + \ \frac{1}{r^2}\partial_\theta^2, \tag{5.9}$$

where Δ_r denotes the radial part of the transverse Laplacian:

$$\Delta_r \ \equiv \partial_r^2 \ + \frac{1}{r}\partial_r.$$

Thus, the operator ∂_θ is replaced by $\partial_\theta + N\partial_\Theta$ and equation (2.12) becomes:

$$(-\Delta_r - \frac{1}{r^2}(\partial_\theta + N\partial_\Theta)^2 + V(r,\theta,\Theta)\ )\Phi \ = \ E\ \Phi, \tag{5.10}$$

Equivalently

$$\left(H - \frac{2N}{r^2}\partial_\theta\partial_\Theta - \frac{N^2}{r^2}\partial_\Theta^2 \right)\Phi = E\Phi \tag{5.11}$$

We seek an expansion of Φ and E in the small parameter $1/N$

$$\begin{aligned} \Phi^{(N)} &= \Phi_0 + \frac{1}{N}\Phi_1 + \frac{1}{N^2}\Phi_2 + \frac{1}{N^3}\Phi_3 + \frac{1}{N^4}\Phi_4 + \Phi_5^{(N)} \\ E^{(N)} &= E_0 + \frac{1}{N}E_1 + \frac{1}{N^2}E_2 + E_3^{(N)} \end{aligned} \tag{5.12}$$

Here, $\Phi_j = \Phi(r, 6, \Theta)$.

Substitution of (5.12) into (5.11) and equating like powers of $1/N$ we obtain a hierarchy of equations arising at each order in $1/N$. We display the first five equations of this hierarchy.

$$\mathcal{O}(N^2): \quad \frac{1}{r^2}\partial_\Theta^2\Phi_0 = 0 \tag{5.13}$$

$$\mathcal{O}(N): \quad \frac{1}{r^2}\partial_\Theta^2\Phi_1 = -\frac{2}{r^2}\partial_\theta\partial_\Theta\Phi_0 \tag{5.14}$$

$$\mathcal{O}(1): \quad \frac{1}{r^2}\partial_\Theta^2\Phi_2 = -\frac{2}{r^2}\partial_\theta\partial_\Theta\Phi_1 + (H - E_0)\,\Phi_0 \tag{5.15}$$

$$\mathcal{O}(N^{-1}): \quad \frac{1}{r^2}\partial_\Theta^2\Phi_3 = -\frac{2}{r^2}\partial_\theta\partial_\Theta\Phi_2 + (H - E_0)\,\Phi_1 - E_1\Phi_0 \tag{5.16}$$

$$\mathcal{O}(N^{-2}): \quad \frac{1}{r^2}\partial_\Theta^2\Phi_4 = -\frac{2}{r^2}\partial_\theta\partial_\Theta\Phi_3 + (H - E_0)\,\Phi_2 - E_1\Phi_1 - E_2\Phi_0 \tag{5.17}$$

5.2. Construction of the multiple scale expansion. As is typical in perturbation expansions, we must solve a hierarchy of inhomogeneous linear equations, which have a *fixed* linear operator to invert and varying inhomogeneous term. Each member of the hierarchy is of the form:

$$\frac{1}{r^2}\partial_\Theta^2 F = G(r, \theta, \Theta). \tag{5.18}$$

A necessary and sufficient condition for solvability in the space of 2π periodic in Θ functions is:

$$\int_0^{2\pi} G(r, \theta, p)\, dp = 0. \tag{5.19}$$

We now proceed with a term by term construction of the perturbation expansion.

$\mathcal{O}(N^2)$ *Terms:*

$$\frac{1}{r^2}\partial_\Theta^2\Phi_0 = 0 \tag{5.20}$$

From this it follows that Φ_0 is independent of the fast phase, Θ:

$$\Phi_0(r, \theta, \Theta) = \Phi_0(r, \theta) \tag{5.21}$$

$\mathcal{O}(N)$ *Terms:*

$$-\frac{2}{r^2}\partial_\theta\partial_\Theta\Phi_0 - \frac{1}{r^2}\partial_\Theta^2\Phi_1 = 0 \tag{5.22}$$

Using (5.21) in (5.22) we have that

$$\frac{1}{r^2}\partial_\Theta^2 \Phi_1 = 0 \tag{5.23}$$

and therefore:

$$\Phi_1(r,\theta,\Theta) = \Phi_1(r,\theta) \tag{5.24}$$

Using (5.24) we find:

$\mathcal{O}(1)$ *Terms:*

$$\frac{1}{r^2}\partial_\Theta^2 \Phi_2 = (H - E_0)\,\Phi_0 \tag{5.25}$$

Equation (5.25) has a solution in the space of 2π-periodic in Θ functions if and only if the following solvability condition holds:

$$(H_{\text{av}} - E_0)\,\Phi_0 = 0. \tag{5.26}$$

Here, H_{av} and V_{av} denote, respectively, the average operator and potential with respect to the fast angular dependence. They are given by:

$$\begin{aligned} H_{\text{av}} &= -\Delta_\perp + V_{\text{av}} = -\Delta_r - \frac{1}{r^2}\partial_\theta^2 + V_{\text{av}}(r,\theta) \\ V_{\text{av}}(r,\theta) &= \frac{1}{2\pi}\int_0^{2\pi} V(r,\theta,p)\,dp. \end{aligned} \tag{5.27}$$

Note that

$$H - H_{\text{av}} = V - V_{\text{av}}.$$

We fix $(\Phi_0(r,\theta), E_0)$ to be a scattering resonance pair of an outgoing solution and its associated *complex* eigenvalue.

Since Φ_0 satisfies (5.26), the inhomogeneous equation (5.25) can be put in the simpler form:

$$\frac{1}{r^2}\partial_\Theta^2 \Phi_2 = [\ V(r,\theta,\Theta) - V_{\text{av}}(r,\theta)\ ]\,\Phi_0(r,\theta). \tag{5.28}$$

We express the solution of (5.28) in the form

$$\Phi_2 = \Phi_2^{(p)}(r,\theta,\Theta) + \Phi_2^{(h)}(r,\theta), \tag{5.29}$$

where $\Phi_2^{(p)}(r,\theta,\Theta)$ and $\Phi_2^{(h)}(r,\theta)$ are, respectively, a particular solution of mean zero and a homogeneous solution of (5.28), which is to be determined at higher order in N^{-1}. That $\Phi_2^{(p)}(r,\theta,\Theta)$ can be chosen to be 2π-periodic in Θ follows from the mean zero property of $V(r,\theta,\Theta) - V_{\text{av}}(r,\theta)$. Also, $\Phi_2^{(p)}$ is mean zero in Θ because the constant term in its Fourier expansion is included in $\Phi_2^{(h)}$. To compute $\Phi_2^{(p)}$, we expand $V(r,\theta,\Theta) - V_{\text{av}}(r,\theta)$ in a Fourier series in Θ:

$$V(r,\theta,\Theta) - V_{\text{av}}(r,\theta) = \sum_{|j|\geq 1} \eta_j(r,\theta)\, e^{ij\Theta} \tag{5.30}$$

We have

$$\begin{aligned} \Phi_2^{(p)}(r,\theta,\Theta) &= \partial_\Theta^{-2}\,[V(r,\theta,\Theta) - V_{\text{av}}(r,\theta)] \cdot r^2\Phi_0(r,\theta) \\ &= \sum_{|j|\geq 1} (ij)^{-2}\eta_j(r,\theta)e^{ij\Theta} \cdot r^2\Phi_0(r,\theta) \end{aligned} \tag{5.31}$$

$\mathcal{O}(N^{-1})$ *Terms:*

$$
\begin{aligned}
\frac{1}{r^2}\partial_\Theta^2 \Phi_3 &= -\frac{2}{r^2}\partial_\theta\partial_\Theta \Phi_2 + (\ H - E_0\ )\,\Phi_1 - E_1\Phi_0 \\
&= -\frac{2}{r^2}\partial_\theta\partial_\Theta \Phi_2^{(p)} + (\ H - E_0\ )\,\Phi_1 - E_1\Phi_0 \qquad (5.32)
\end{aligned}
$$

The solvability condition becomes

$$
(\ H_{\text{av}} - E_0\ )\,\Phi_1 \ = \ E_1\Phi_0, \qquad (5.33)
$$

which we satisfy by taking $E_1 = 0$ and $\Phi_1 \equiv 0$.

Using the mean zero property of $\Phi_2^{(p)}$, we see that equation (5.32) becomes

$$
\partial_\Theta \Phi_3(r,\theta,\Theta) \ = \ -2\partial_\theta \Phi_2^{(p)}(r,\theta,\Theta) \qquad (5.34)
$$

$\mathcal{O}(N^{-2})$ *Terms:* Using that $E_1 = 0$, $\Phi_1 \equiv 0$ and (5.34) to solve for Φ_3 in terms of $\Phi_2^{(p)}$ we find at this order that:

$$
\begin{aligned}
\frac{1}{r^2}\partial_\Theta^2 \Phi_4 &= \frac{4}{r^2}\partial_\theta^2\Phi_2^{(p)} + (H - E_0)\,\Phi_2 - E_2\Phi_0 \\
&= \left(H - E_0 + \frac{4}{r^2}\partial_\theta^2\right)\Phi_2^{(p)} + (H - E_0)\,\Phi_2^{(h)} - E_2\Phi_0 \\
&= \left(H_{\text{av}} - E_0 + \frac{4}{r^2}\partial_\theta^2\right)\Phi_2^{(p)} + (H - H_{\text{av}})\,\Phi_2^{(h)} \\
&\quad + (H_{\text{av}} - E_0)\,\Phi_2^{(h)} - E_2\Phi_0 + (H - H_{\text{av}})\,\Phi_2^{(p)}. \qquad (5.35)
\end{aligned}
$$

The first two terms on the right hand side of (5.35) have zero average in Θ. Therefore, the solvability condition for Φ_4 becomes:

$$
\begin{aligned}
&(\ H_{\text{av}} - E_0)\,\Phi_2^{(h)} \qquad (5.36) \\
&\quad = E_2\Phi_0 - \frac{1}{2\pi}\int_0^{2\pi} (H - H_{\text{av}})\,\Phi_2^{(p)}\,dp \\
&\quad = E_2\Phi_0 - \frac{1}{2\pi}\int_0^{2\pi} [V(r,\theta,p) - V_{\text{av}}(r,\theta)]\Phi_2^{(p)}(r,\theta,p)\,dp \\
&\quad = \left(E_2 + \frac{r^2}{2\pi}\int_0^{2\pi}\left|\partial_p^{-1}[V(r,\theta,p) - V_{\text{av}}(r,\theta)]\right|^2 dp\right)\Phi_0(r,\theta),
\end{aligned}
$$

where we have used (5.31), and where ∂_p^{-1} denotes the mean-zero antiderivative. We seek an *outgoing* solution of (5.36). This determines E_2 and therewith the expansion for the scattering resonance (effective index) through order N^{-2}.

In the following section we use this truncated expansion to numerically obtain the complex effective indexes (scattering resonances) for various structures. The details of the implementation are set out in section 7. In particular, in section 7 we derive an explicit expression for E_2 (see equations (7.17,7.18,7.8)) for the class of N-fold symmetric waveguides.

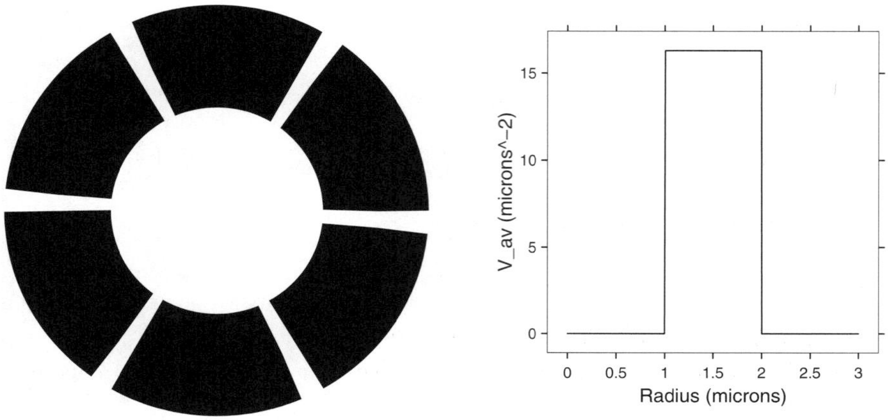

(a) (b)

FIGURE 2. (a) A ring of "air wedges" supported by $N = 6$ webs of glass. The inner and outer radii of the annulus are 1 and 2μm, respectively, while the air fill fraction within the annulus is $f = 0.9$. (b) The averaged potential $V_{\rm av}$.

6. Numerical simulations for selected structures

We now illustrate the theory of the preceding section with numerical calculations performed for a class of structures. These structures consist of an annulus of air supported with glass "webs;" see figure 2. More general structures are considered in [**7**], *e.g.* rings of circular holes, truncated hexagonal lattices of circular holes. General structures can be approximated by such "simple layered structures," of the type in figure 2. The details of the implementation are set out in section 7.

We will identify individual leaky modes, or scattering resonances, of these structures using the LP_{lm} (linearly polarized) notation [**11**] appropriate for solutions of the scalar wave equation. Usually this notation applies to guided modes, but no ambiguity will result as none of the structures we consider supports guided modes. The subscript $l \in \{0, 1, 2, ...\}$ refers to the angular dependence $e^{il\theta}$ of solutions to the averaged equation, while $m \in \{1, 2, 3, ...\}$ indexes the collection of leaky modes with fixed l. This collection is ordered by $\Re n_{\rm eff}$, with $m = 1$ corresponding to the largest such value. In the figures, we will use distinct plotting symbols to represent different values of l; the various values of m for fixed l are not explicitly displayed, but can be inferred from the monotonic relationship between m and $\Re n_{\rm eff}$.

In all of the examples, the refractive index of glass was taken to be 1.45 and the holes were assumed empty. The calculations were all performed at $\lambda_{\rm fs} = 1.55\mu$m. All of these structures support only continuous spectrum, so we report on the complex eigenvalues associated with resonances of the structures.

We computed the following sets of resonances for the webbed structures depicted in 2. We restricted attention to leaky mode solutions of the averaged structure with effective indexes $\Re n_{\rm eff} \in [1.41, 1.45]$, attenuation coefficients less than about $2\,$dB/mm, and angular index $l \in \{0, 1, 2, 3\}$. To express these quantities in our notation, recall from (2.11) that

$$\beta = \sqrt{k^2 n_g^2 - E} \tag{6.1}$$

and that the effective index is given by

$$n_{\text{eff}} = k^{-1}\,\beta \tag{6.2}$$

Also, the attenuation γ (in dB/cm, if β has units of $1/\mu$m) of a leaky mode is given by

$$\gamma \equiv 10 \log_{10} \frac{\text{Power Input}}{\text{Power Output}} = \frac{2 \times 10^5}{\ln(10)} \Im\beta. \tag{6.3}$$

The factor of 2 in (6.3) is due to the definition of γ as the attenuation coefficient for intensity, rather than field amplitude. Once this set of leaky modes of the averaged potential was found, the leading order corrections E_2 to the leading order energies E_0 were computed.

The results for three scalings and two air-fill fractions, along with various values of the periodicity N, of the ring of "air wedges" are presented in figure 3.

We also computed resonances of a structure of the type shown in figure 2 with $R_{\text{in}} = 1\mu$m, $R_{\text{out}} = 2\mu$m for the fill fractions $f = 0.8$, 0.9, and 1, with $N = 3$ and $N = 6$ holes. The calculations were performed for a range of free-space wavelengths λ_{fs} ranging from 1μm to 2μm. The structure with $f = 1$ is an idealized ring of air with no supporting structure. In figure 4 we compare the results of our theory to parallel calculations performed in [**15**] by a Fourier decomposition algorithm. We observe that the two methods agree exactly for the $f = 1$ case, as they should. Additionally, in all cases our method agrees with the Fourier calculation as $\lambda_{\text{fs}} \to 2\mu$m. Even for smaller λ_{fs} the agreement is quite good except for the $f = 0.8$, $N = 3$ case, which is not surprising given the small value of N and the fact that the width of the ring is equal to the smallest free space wavelength considered. The plots displayed in figure 4 are consistent with the expectation that approximation by the homogenization expansion improves (a) for fixed λ_{fs} and increasing N, as well as (b) for fixed N and λ_{fs} increasing.

The dashed lines in figure 4 display the attenuations of the averaged structures, without the $\mathcal{O}(N^{-2})$ correction. The necessity of including these corrections is evident, though the effect in these structures is not as dramatic as in some of the other structures we consider in [**7**].

Another result reported in [**15**] is the effective index of the first excited state above the fundamental (the LP_{11} state) for the $f = 0.9$, $N = 3$ structure at $\lambda_{\text{fs}} = 1.55\mu$m. In table 1 we compare the results of the Fourier method with truncations of our homogenization expansion to one and two terms, for both the fundamental and first excited leaky modes. We note that the one-term (averaged theory) truncation predicts the real part of the effective index well, as reported in [**2**], but again, the $\mathcal{O}(N^{-2})$ corrections are necessary to obtain good agreement of the attenuation rates.

6.1. Convergence as $N \to \infty$. In this section we describe a consistency check on our implementation of the scheme described in section 7.2 for simple layered potentials. We compute the first non-trivial corrections $N^{-2}\Phi_2$ and $N^{-2}E_2$ to the field and energy, respectively, in (5.12) and express the field and energy as our

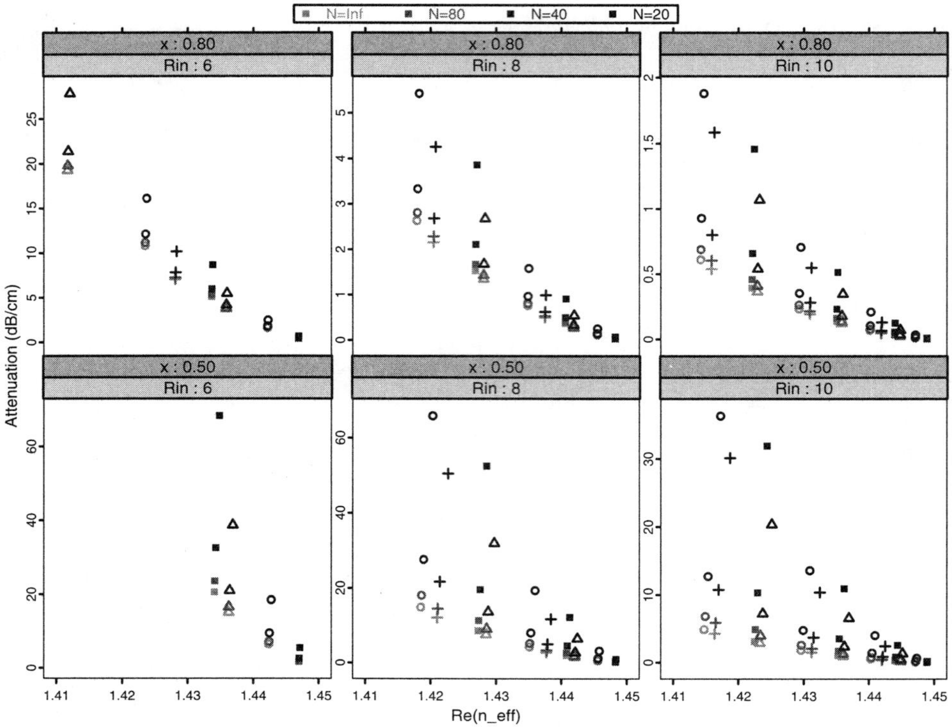

FIGURE 3. The effective indexes and attenuation coefficients of a set of the least-lossy resonances with $l \in \{0, 1, 2, 3\}$ of the structure depicted in figure 2, for two choices of fill fraction x and three different radii. The plotting symbol encodes the angular index l: $\blacksquare$, $\circ$, $\triangle$, $+$ correspond to $l = 0, 1, 2, 3$; the shade of grey encodes the periodicity N. The results of the averaged structure are labelled $N = \infty$.

approximation plus error terms:

$$\Phi = \Phi_0 + \frac{1}{N^2}\Phi_2 + \Phi_3^{(N)}$$

$$E = E_0 + \frac{1}{N^2}E_2 + E_3^{(N)}. \tag{6.4}$$

We consider by how much the mode equation (3.1) is *not* satisfied by the first two terms of the expansion (6.4). Therefore, we define the residual:

$$F_{\text{res}} \equiv \left(-\Delta + V - \left(E_0 + \frac{1}{N^2}E_2\right)\right)\left(\Phi_0 + \frac{1}{N^2}\Phi_2\right). \tag{6.5}$$

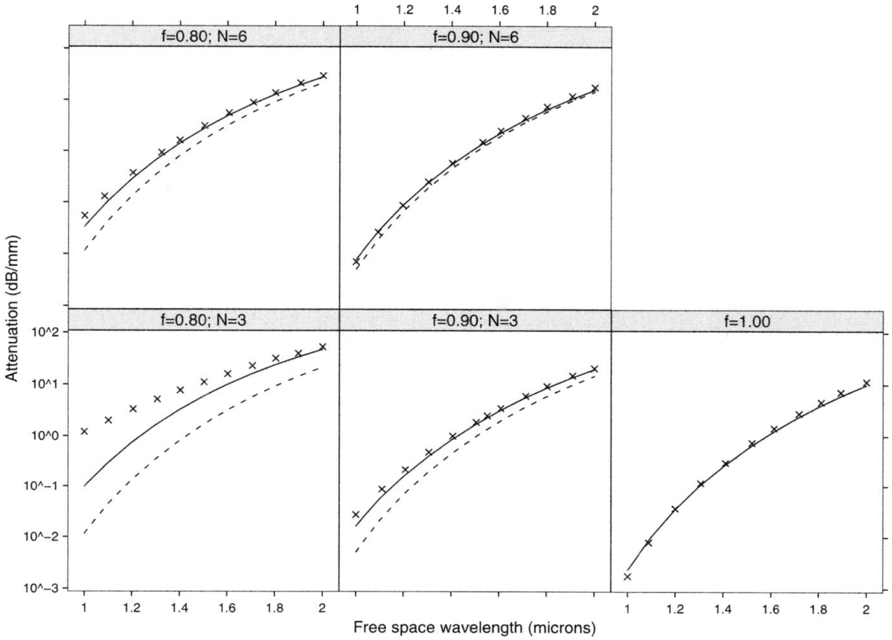

FIGURE 4. Attenuation of the lowest order (fundamental, or LP_{01}) resonance for a structure of the type shown in figure 2, with $R_{\text{in}} = 1\mu\text{m}$, $R_{\text{out}} = 2\mu\text{m}$ for the fill fractions $f = 0.8$, 0.9, and 1, with $N = 3$ and $N = 6$ holes. The calculations were performed for a range of free-space wavelengths λ_{fs} ranging from $1\mu\text{m}$ to $2\mu\text{m}$. The solid lines are the attenuations computed according to the methods of this paper, including the $\mathcal{O}(N^{-2})$ corrections; the dashed lines are the attenuations of the averaged structure; and the crosses correspond to the results presented in figure 3a of [**15**].

Substituting (6.4) into the (3.1) (3.1), we find that the residual is given by

$$(6.6) \qquad F_{\text{res}} = E_3^{(N)} \left(\Phi_0 + \frac{1}{N^2}\Phi_2 + \Phi_3^{(N)} \right) - \left(\Delta + V - \left(E_0 + \frac{1}{N^2}E_2 \right) \right) \Phi_3^{(N)}.$$

We expect F_{res} to be of order N^{-1} because $\Phi_3^{(N)}$ and $E_3^{(N)}$ are formally of order N^{-3} and the Laplacian in (6.6), when acting on the fast angular dependence, gives a factor of N^2. Expanding F_{res} in powers of $1/N$, we find that when $l > 0$ the

mode	*Averaged* n_{eff}	*atten.*	*2nd Order Homog.* n_{eff}	*atten.*	*Fourier* n_{eff}	*atten.*
LP_{01}	$1.3712+ 0.0000387i$	1.36	$1.3727+ 0.0000644i$	2.27	$1.365+ 0.000071i$	2.48
LP_{11}	$1.2468+ 0.000436i$	15.3	$1.2515+ 0.000832i$	29.3	1.255+ 0.00075i	27

TABLE 1. Comparison of results for the first two leaky modes of a ring of "air wedges" with $R_{\text{in}} = 1\mu\text{m}$, $R_{\text{out}} = 2\mu\text{m}$, fill fraction $f = 0.9$, and $N = 3$ holes. The effective indexes and attenuations (in dB/mm) of the leading term (averaged), and of the two-term truncation, of the homogenization expansion are compared with that of the Fourier expansion [**15**] (the values for the LP_{01} mode calculated by the Fourier expansion were read off of figure 3 of [**15**]).

leading term, $F_{\text{res}}^{(1)}$, is the $\mathcal{O}(1/N)$ term

$$F_{\text{res}}^{(1)} = \frac{1}{N}\frac{1}{r^2}\partial_\Theta^2 \Phi_3. \tag{6.7}$$

This may be simplified using (5.34) and (5.31) to

$$F_{\text{res}}^{(1)} = -\frac{2}{N}\left(\partial_\theta \Phi_0\right)\left(\partial_\Theta^{-1}\left(V - V_{\text{av}}\right)\right). \tag{6.8}$$

Therefore, when $l > 0$, we expect that as $N \to \infty$

$$F_{\text{res}} \to F_{\text{res}}^{(1)} \sim \mathcal{O}\left(\frac{1}{N}\right) \quad \text{and} \quad F_{\text{res}} - F_{\text{res}}^{(1)} \sim \mathcal{O}\left(\frac{1}{N^2}\right). \tag{6.9}$$

When $l = 0$ ($\Phi_0 = \Phi_0(r)$), $\partial_\theta \Phi_0$ vanishes and the leading behavior of the residual is higher order in $1/N$.

Since the potential V or its derivatives may have discontinuities, we encounter a problem when attempting to compute F_{res} using (6.6); it is possible that $\Phi_2^{(p)}$ is discontinuous as a function of r. We therefore interpret (6.6) as holding in the "weak sense" or sense of distributions, and integrate both sides against a radial test function $\phi_R(r)$ of compact support. This allows us to move the radial part of the Laplacian over to act on ϕ_R, and to define the "weak residual"

$$\begin{aligned} F_{\text{res}}^{(\text{weak})}(\theta) &= \int_0^\infty r\,dr \left(-\phi_R'' - \frac{1}{r}\phi_R' - \phi_R \frac{1}{r^2}\partial_\theta^2 + \phi_R\left(V - E_0 - \frac{1}{N^2}E_2\right)\right) \\ &\qquad \times \left(\Phi_0 + \frac{1}{N^2}\Phi_2\right) \end{aligned} \tag{6.10}$$

and its leading behavior in $1/N$

$$F_{\text{res}}^{(\text{weak},1)}(\theta) = -\frac{2}{N}\int_0^\infty r\,dr\,\phi_R\left(\partial_\theta \Phi_0\right)\left(\partial_\Theta^{-1}\left(V - V_{\text{av}}\right)\right). \tag{6.11}$$

We computed (6.10) and (6.11) for the structure of example 2 (figure 2), with $R_{\text{in}} = 10\mu\text{m}$, $R_{\text{out}} = 11\mu\text{m}$, for the LP_{11} leaky mode. We allowed N to vary from 20 to 22,000. The results are presented in figure 5. The solid line in (a) represents $\|F_{\text{res}}^{(\text{weak})}(\theta) - F_{\text{res}}^{(\text{weak},1)}(\theta)\|_\infty$ as a function of N. The scales are logarithmic on both

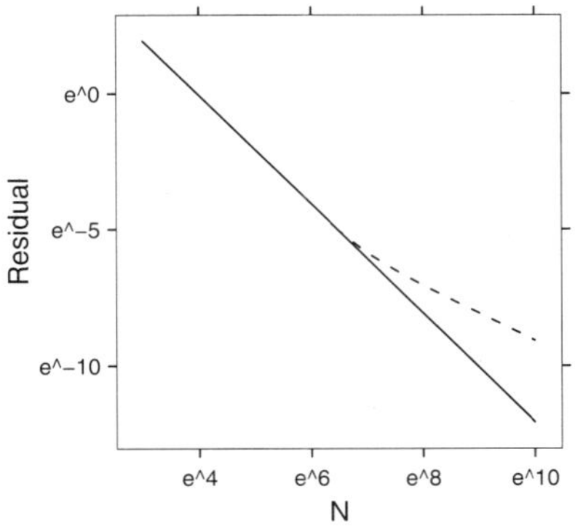

(a)

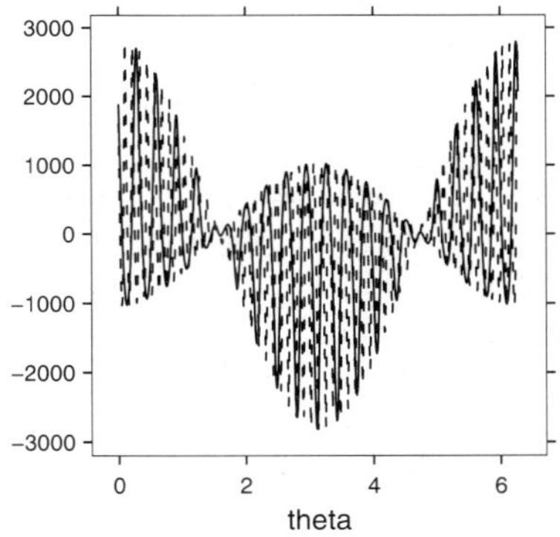

(b)

FIGURE 5. Results of the weak residual calculation. (a) $\|F_{\text{res}}^{(\text{weak})}(\theta) - F_{\text{res}}^{(\text{weak},1)}(\theta)\|_\infty$ (solid); $\|F_{\text{res}}^{(\text{weak})}(\theta)\|_\infty$ (dashed) (b) $N^2\,\Re(F_{\text{res}}^{(\text{weak})}(\theta) - F_{\text{res}}^{(\text{weak},1)}(\theta))$ for $N = 20$ (solid) and $N = 54$ (dashed).

axes, so the slope of -2 verifies that (6.9) holds over the entire range considered. The dashed line in (a) represents $\|F_{\text{res}}^{(\text{weak})}(\theta)\|_\infty$; we see that the slope changes from -2 to -1 at around $N = 1000$, which again verifies (6.9).

Figure 5 (b) shows the pointwise behavior of the weak residuals by plotting $N^2\,\Re(F_{\text{res}}^{(\text{weak})}(\theta) - F_{\text{res}}^{(\text{weak},1)}(\theta))$ vs. θ for $N = 20$ and $N = 54$. We observe that the envelope of the residuals is invariant after scaling by N^2, suggesting that the N^{-2} term dominates the residual after the leading term is subtracted, even for these moderate values of N. This behavior persists for all N we considered, up to $N = 22,000$.

7. Implementation of homogenization expansion

In this section, we give a detailed description of the procedure used to obtain the simulation results of the previous section.

We specialize to the case of a potential with no slow angular dependence, so $V(r,\theta,\Theta) = V(r,\Theta)$. In this case, the potential has an N-fold symmetry. We first, in section 7.1, sketch the expansion of solutions to the resonance problem summarized in the homogenization expansion of section 5 with no additional assumptions. This discussion pertains to N-fold symmetric microstructure, whose individual microfeatures have arbitrary geometry. Our strategy is to then approximate a general microstucture potential V by a *simple layered potential* in section 7.2. For the case of simple layered potentials, analytic expressions can be obtained which facilitate the numerical computations.

7.1. General Structures. The angularly averaged potential is a radial function, $V_{\text{av}} = V_{\text{av}}(r)$. We take

$$\Phi_0(r,\theta) = f(r)e^{il\theta}, \ l \text{ an integer}$$

Then, $f(r)$ satisfies

$$\left(-\Delta_r + \frac{l^2}{r^2} + V_{\rm av}(r) - E_0\right) f(r) \ = \ 0 \tag{7.1}$$

A solution of the resonance problem is a pair $(f(r), E_0)$, such that $f(r)$ is a nonsingular solutions of (7.1) satisfying an *outgoing radiation condition at* $r = \infty$, (3.5). As noted in section 3, E_0 is complex with $\Im E_0 < 0$. In general, the existence of solutions to the resonance problem is a subtle technical problem. In the next subsection, we analyze it in detail for the particular class of *simple* layered structures, which we define there. In this section, we posit the existence of a solution, $(f(r), E_0)$. Note that since $V_{\rm av}(r) = 0$ for $r \geq r_*$,

$$f(r) \ = \ k_* \ H_l^{(1)}(\sqrt{E_0}r), \ r \geq r_*, \tag{7.2}$$

for some constant k_*.

We next calculate the leading order correction, $N^{-2}\Phi_2$, by our perturbation scheme. In the separable case, the equation for $\Phi_2^{(h)}$ is of the form:

$$(\ H_{\rm av} - E_0 \ )\Phi_2^{(h)} = (\ E_2 + Q(r) \ ) \ \Phi_0, \tag{7.3}$$

where $Q(r)$ has compact support. Since $\Phi_0(r,\theta) = f(r)e^{il\theta}$, all of the Fourier modes of $\Phi_2^{(h)}(r,\theta)$ except the $l^{\rm th}$ vanish and so we define $\Phi_2^{(h)}(r,\theta) = U(r)e^{il\theta}$ and consider the equation

$$\left(-\Delta_r + \frac{l^2}{r^2} + V_{\rm av}(r) - E_0 \right) U(r) \ = \ (\ E_2 + Q(r) \ ) \, f(r) \tag{7.4}$$

Recall that by (2.13), both Q and $V_{\rm av}$ are zero for $r \geq r_*$ and we have

$$\left(-\Delta_r + \frac{l^2}{r^2} - E_0 \right) U(r) \ = \ E_2 \ f(r), \quad r \geq r_*. \tag{7.5}$$

We proceed with the solution of the inhomogeneous problem (7.4) for $r \geq 0$ and then obtain an expression for E_2 by imposing the outward going radiation condition.

We have assumed that V is regular at the origin. We let $\tilde{y}_1(r)$ denote the solution to the homogeneous equation associated with (7.4) that is regular at $r = 0$, and $\tilde{y}_2(r)$ denote an independent solution, which together with $\tilde{y}_1$ spans the solution set. Then $\tilde{y}_1(r) \propto r^l$, and $\tilde{y}_2(r) \propto r^{-l}$ if $l > 0$, or $\tilde{y}_1(r) \propto \ln(r)$ if $l = 0$. For $r > r_*$, an appropriate set of homogeneous solutions of (7.5) is in terms Hankel functions, $H_l^{(j)}(\sqrt{E_0}\ r)$, $j = 1, 2$. $H_l^{(1)}$ satisfies the outward going radiation condition and infinity and $H_l^{(2)}$ the inward going radiation condition at infinity.

A particular solution of (7.4) can be constructed by variation of parameters using these homogeneous solutions in the intervals:

$$R_{\rm in} = [0, r_*], \quad \text{and} \quad R_{\rm out} = (r_*, \infty).$$

Specifically, we have

$$\text{in } R_{\rm in}: \quad U(r) \ = \ E_2 \ \mathcal{G}_{\rm in}[f](r) \ + \ \mathcal{G}_{\rm in}[Qf](r) \tag{7.6}$$

$$\text{in } R_{\rm out}: \quad U(r) \ = \ \xi_* H_l^{(1)}(\sqrt{E_0}\ r) \ + \ \eta_* H_l^{(2)}(\sqrt{E_0}\ r) \ + \ E_2 \ \mathcal{G}_{\rm out}[f](r). \tag{7.7}$$

Imposing continuity of U and $\partial_r U$ at $r = r_*$ yields the system:

$$\begin{pmatrix} \xi_*(E_2) \\ \eta_*(E_2) \end{pmatrix} = E_2\, \mathcal{Y}^{-1}(r_*) \begin{pmatrix} \mathcal{G}_{\text{in}}[f](r_*) - \mathcal{G}_{\text{out}}[f](r_*) \\ E_0^{-\frac{1}{2}} \left(\partial\mathcal{G}_{\text{in}}[f](r_*) - \partial\mathcal{G}_{\text{out}}[f](r_*)\right) \end{pmatrix}$$

$$(7.8) \qquad + \ \mathcal{Y}^{-1}(r_*) \begin{pmatrix} \mathcal{G}_{\text{in}}[Qf](r_*) \\ E_0^{-\frac{1}{2}}\ \partial\mathcal{G}_{\text{in}}[Qf](r_*) \end{pmatrix}.$$

Here, $\mathcal{G}_{\text{in}}[F]$ denotes a choice of particular solution of

$$(7.9) \qquad \left(-\Delta_r + \frac{l^2}{r^2} + V_{\text{av}}(r) - E_0 \right) \mathcal{G}_{\text{in}}[F] = F(r),\ 0 \le r < r_* \ (r \in R_{\text{in}}),$$

$\mathcal{G}_{\text{out}}[F]$ denotes a choice of particular solution of

$$(7.10) \qquad \left(-\Delta_r + \frac{l^2}{r^2} - E_0 \right) \mathcal{G}_{\text{out}}[F] = F(r),\ r_* \ge r \ (r \in R_{\text{out}}),$$

and

$$\mathcal{Y}(r) = \begin{pmatrix} H_l^{(1)}(\sqrt{E_0}r) & H_l^{(2)}(\sqrt{E_0}r) \\ \partial_s H_l^{(1)}(\sqrt{E_0}r) & \partial_s H_l^{(2)}(\sqrt{E_0}r) \end{pmatrix}.$$

Since

$$W\{H_l^{(1)}(s), H_l^{(2)}(s)\} = -\frac{4i}{\pi s},$$

it follows that

$$\mathcal{Y}^{-1}(r) = \frac{i\pi r}{4} \begin{pmatrix} \partial_s H_l^{(2)}(\sqrt{E_0}r) & -H_l^{(2)}(\sqrt{E_0}r) \\ -\partial_s H_l^{(1)}(\sqrt{E_0}r) & H_l^{(1)}(\sqrt{E_0}r) \end{pmatrix}.$$

With ξ_* and η_* given by (7.8), (7.6-7.7) defines a nonsingular solution of (7.4). It remains to impose radiation condition at $r = \infty$. Consider $U(r)$ for $r \ge r_*$. Let

$$z = \sqrt{E_0}r,\ z_* = \sqrt{E_0}r_*,\ G(z) = U(r(z)).$$

Then, (7.5) becomes:

$$(7.11) \qquad z^2 G'' + zG' + (z^2 - l^2)G = \frac{E_2}{E_0} k_*\ z^2 H_l^{(1)}(z),\ r(z) \ge r_*.$$

By (7.7) and (7.2)

$$(7.12)\ G(z) = U(r(z)) = \xi_*(E_2)\, H_l^{(1)}(z) + \eta_*(E_2)\, H_l^{(2)}(z) + \frac{E_2}{E_0} k_*\, \mathcal{G}_{\text{out}}[H_l^{(1)}](z),$$

$\mathcal{G}_{\text{out}}$ can be constructed by variation of parameters. A particular solution is

$$\xi_*^{(p)}(z) H_l^{(1)}(z) + \eta_*^{(p)}(z) H_l^{(2)}(z),$$

where [1]

$$(7.13) \qquad \xi_*^{(p)}[h](z) = -\int_{z_*}^{z} \frac{H_l^{(2)}(\zeta) h(\zeta)}{\zeta^2 W\{H_l^{(1)}, H_l^{(2)}\}}\, d\zeta$$

$$(7.14) \qquad \eta_*^{(p)}[h](z) = +\int_{z_*}^{z} \frac{H_l^{(1)}(\zeta) h(\zeta)}{\zeta^2 W\{H_l^{(1)}, H_l^{(2)}\}}\cdot d\zeta$$

---

[1] Note that z can be complex-valued. The complex integral $\int_{z_*}^{z}$ is computed by parameterizing the line from z_* to z as: $t \mapsto z_* + t(z - z_*)$. Then, $\int_{z_*}^{z} F(\zeta)d\zeta = (z - z_*)\int_0^1 F(z_* + t(z - z_*))dt$.

Therefore,

$$\mathcal{G}_{\mathrm{out}}[F](z) \;=\; \left(\xi_*^{(p)}[\zeta^2 F](z) H_l^{(1)}(z) + \eta_*^{(p)}[\zeta^2 F](z) H_l^{(2)}(z) \right), \tag{7.15}$$

Using (7.15) in (7.12) we have:

$$\begin{aligned} G(z) = U(r(z)) \;&=\; \xi_*(E_2)\, H_l^{(1)}(z) \;+\; \eta_*(E_2)\, H_l^{(2)}(z) \\ &-\; \frac{i\pi}{4}\frac{E_2}{E_0} k_* \cdot \int_{z_*}^{z} H_l^{(2)}(\zeta) H_l^{(1)}(\zeta)\zeta \; d\zeta \; H_l^{(1)}(z) \\ &+\; \frac{i\pi}{4}\frac{E_2}{E_0} k_* \cdot \int_{z_*}^{z} [H_l^{(1)}(\zeta)]^2 \zeta \; d\zeta \; H_l^{(2)}(z). \end{aligned} \tag{7.16}$$

Recall that $\xi_*(E_2)$ and $\eta_*(E_2)$ are determined by the continuity conditions at $r = r_*$ and are given by (7.8). The condition determining E_2 is that the *incoming* part of $G(z)$ be identically zero. Derivation of this condition on E_2 requires the use of certain integral identities involving Hankel functions.

Note that both terms in (7.16) that are proportional to $H_l^{(1)}$ are outgoing, as can be seen by referring to the asymptotic forms (3.4). From [**8**] (page 634, equation 5.54) we have

$$\int \zeta [H_l^{(j)}(\zeta)]^2 d\zeta \;=\; \frac{\zeta^2}{2} \left([H_l^{(j)}(\zeta)]^2 - H_{l-1}^{(j)}(\zeta) H_{l+1}^{(j)}(\zeta) \right).$$

Therefore, the terms in (7.16) that are proportional to $H_l^{(2)}$ can be written as:

$$\begin{aligned} &\eta_*(E_2) H_l^{(2)}(z) \;+\; \frac{ik_*\pi}{4}\frac{E_2}{E_0} \cdot \frac{\zeta^2}{2} \left([H_l^{(1)}(\zeta)]^2 - H_{l-1}^{(1)}(\zeta) H_{l+1}^{(1)}(\zeta) \right)\Big|_{z_*}^{z} \; H_l^{(2)}(z) \\ &= \left[\eta_*(E_2) - \frac{ik_*\pi}{4}\frac{E_2}{E_0} \cdot \frac{z_*^2}{2} \left([H_l^{(1)}(z_*)]^2 - H_{l-1}^{(1)}(z_*) H_{l+1}^{(1)}(z_*) \right) \right] \; H_l^{(2)}(z) \\ &+ \;\; \frac{ik_*\pi}{4}\frac{E_2}{E_0} \cdot \frac{z^2}{2} \left([H_l^{(1)}(z)]^2 - H_{l-1}^{(1)}(z) H_{l+1}^{(1)}(z) \right) \; H_l^{(2)}(z). \end{aligned}$$

Again from (3.4) the latter term is seen to be outgoing at infinity so the condition on E_2 is:

$$\eta_*(E_2) - \frac{ik_*\pi}{4}\frac{E_2}{E_0} \cdot \frac{z_*^2}{2} \left([H_l^{(1)}(z_*)]^2 - H_{l-1}^{(1)}(z_*) H_{l+1}^{(1)}(z_*) \right) = 0. \tag{7.17}$$

We can write $\eta_*(E_2)$ as

$$\eta_*(E_2) \;= \eta_{*0} + E_2 \eta_{*1}, \tag{7.18}$$

where η_{*j} can be read off from the second component of (7.8). Equations (7.17) and (7.18) give a linear relation for E_2 which can be solved.

7.2. Simple layered structures: Leading order resonances. We now give the details of the above scheme for case of a *simple layered* structure, which we define to be one with potential $V(r,\theta,\Theta)$ that is both independent of the slow angular variable θ and is a simple function of the form

$$V(r,\Theta) = \sum_{i=1}^{L}\sum_{j=1}^{M_i} \mathbf{1}_{[r_i,r_{i+1}]}(r)\, \mathbf{1}_{[\Theta_j,\Theta_{j+1}]}(\Theta)\, V_{i,j}, \tag{7.19}$$

where L is the number of radial layers and M_i is the number of angular sectors in one period of the ith layer. $\mathbf{1}_A(\zeta)$ denotes the indicator function of the set A,

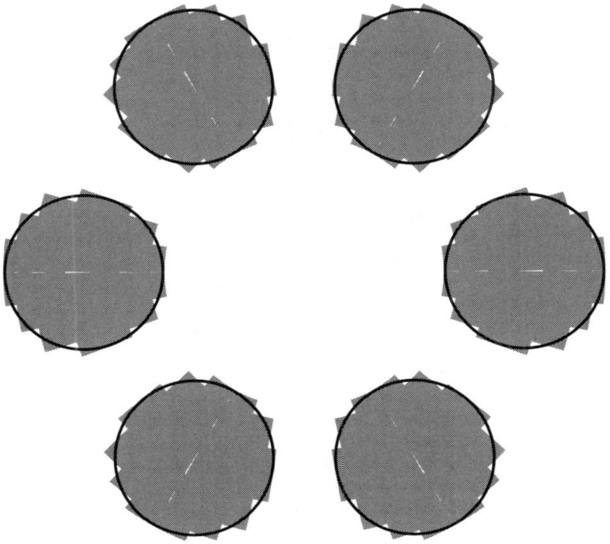

FIGURE 6. An example of an approximation by "simple layered structures" (see (7.19)) of a structure consisting of six circular air holes.

taking the value one for $\zeta \in A$ and zero for $\zeta \notin A$. By definition, $r_1 = 0$, $\Theta_1 = 0$ and $\Theta_{M_i+1} = 2\pi$. The potential in (7.19) has, as a function of Θ, M_i jumps in layer R_i and, as a function of θ NM_i jumps in layer R_i. An example of a simple layered approximation to a structure containing six circular air holes is presented in figure 6.

In the remainder of this section, we describe the calculation of the leading order resonances, those associated with the averaged potential V_{av}. In the next section, we will compute the leading corrections due to microstructure.

With the definition (7.19) of $V(r, \Theta)$, $V_{\text{av}}(r)$ is constant on each region $R_i = [r_i, r_{i+1}]$; equivalently,

$$V_{\text{av}}(r) = V_{\text{av},i} \qquad r \in R_i. \tag{7.20}$$

In the special case where the index profile, $n(r, \theta)$, takes on two values n_g and n_h and $2\pi f_i$ denotes the total angle of the annulus R_i occupied by "material h," we have by averaging V in (2.11):

$$V_{\text{av}}(r) \;=\; k^2 f_i [n_g^2 - n_h^2], \quad r \in R_i. \tag{7.21}$$

In each interval, R_i, the solution will be oscillatory or nonoscillatory depending on whether (i) $\Re E_0 > V_{\text{av},i}$ or (ii) $\Re E_0 < V_{\text{av},i}$. Accordingly, we transform equation (7.1) to Bessel's equation in case (i) and the modified Bessel equation in case (ii).

(i) $\Re E_0 > V_{\text{av},i}$*:* In this case, we set

$$z = \sqrt{E_0 - V_{\text{av},i}}\; r \quad \text{and } f(r) = g(z). \tag{7.22}$$

Then, $g(z)$ satisfies Bessel's equation [**1**]:

$$z^2 g'' + zg' + (z^2 - l^2)g = 0. \tag{7.23}$$

For $i < L$, we write

$$g(z) = \alpha_i J_l(z) + \beta_i Y_l(z), \tag{7.24}$$

where J_l and Y_l are standard Bessel functions of order l that satisfy the Wronksian relation

$$W\{J_l, Y_l\} = \frac{2}{\pi z}. \tag{7.25}$$

In the outermost region R_L, we choose to express g as a linear combination of $H_l^{(1)}$, $H_l^{(2)}$ instead of J_l, Y_l in (7.24).
(ii) $\Re E_0 < V_{\text{av},i}$*:* In this case, we set

$$z = \sqrt{V_{\text{av},i} - E_0}\; r \quad \text{and } f(r) = g(z). \tag{7.26}$$

Then, $g(z)$ satisfies the modified Bessel equation [**1**]:

$$z^2 g'' + zg' - (z^2 + l^2)g = 0. \tag{7.27}$$

In this case we write

$$g(z) = \alpha_i I_l(z) + \beta_i K_l(z), \tag{7.28}$$

where I_l and K_l are standard modified Bessel functions of order l that satisfy the Wronksian relation

$$W\{I_l, K_l\} = \frac{1}{z}. \tag{7.29}$$

Solutions of (7.1) are pieced together subject to the conditions of regularity at $r = 0$, continuity of f and f' at $r = r_i$ and the radiation (boundary) condition at $r = \infty$. We now discuss these in detail.

Boundary condition at $r = 0$*:* Since Y_l and K_l have singularities at $r = 0$, for the solution to be non-singular at $r = 0$ we require $\beta = 0$ in both cases (i) and (ii).

Radiation condition at $r = \infty$*:* This concerns the region $\{r > r_L\}$. Here, $\Re E_0 > V_{\text{av},L}$ and therefore, the solution is oscillatory. Note that the outgoing radiation condition is satisfied by the first Hankel function, $H_l^{(1)}(z)$, and therefore we require

$$\beta_L = 0. \tag{7.30}$$

In particular, for $r > r_L$,

$$g(z) = \alpha_L H_l^{(1)}(z). \tag{7.31}$$

Continuity of f *and* f' *across interfaces:* The interfaces are defined by the values of the radius $r = r_i$, which correspond to $z = z_i$ given by (7.22) in case (i) and (7.26) in case (ii). Continuity of f at interfaces means $f(r_i^-) = f(r_i^+)$, which corresponds to:

$$g(z_i^-) = g(z_i^+). \tag{7.32}$$

Continuity of f' at the interfaces means $f'(r_i^-) = f'(r_i^+)$. Since

$$\begin{aligned} \frac{dz}{dr} &= \sqrt{V_{\text{av}} - E_0}, \text{ where } \Re E_0 < V_{\text{av}} \\ \frac{dz}{dr} &= \sqrt{E_0 - V_{\text{av}}}, \text{ where } \Re E_0 > V_{\text{av}}, \end{aligned}$$

in terms of $g(z)$, continuity of derivatives is given by:

$$\left.\frac{dg}{dz}\frac{dz}{dr}\right|_{z=z_i^+} = \left.\frac{dg}{dz}\frac{dz}{dr}\right|_{z=z_i^-}. \tag{7.33}$$

We now implement the matching across interfaces. Let (y_1, y_2) denote either (J_l, Y_l) or (I_l, K_l) our fundamental sets of solutions corresponding to cases (i) and (ii), respectively, and let $\mathcal{Y}$ denote the fundamental matrix:

$$\mathcal{Y} = \begin{pmatrix} y_1 & y_2 \\ y_1' & y_2' \end{pmatrix}. \tag{7.34}$$

Note that $W\{y_1, y_2\} = \det \mathcal{Y}$.

For $r \in R_i$,

$$\begin{pmatrix} g \\ g' \end{pmatrix} = \mathcal{Y}(z) \begin{pmatrix} \alpha \\ \beta \end{pmatrix}.$$

The pair of interface matching conditions at $z = z_i$ is then

$$\left.\begin{pmatrix} 1 & 0 \\ 0 & \frac{dz}{dr} \end{pmatrix}\right|_{r=r_i^+} \left.\begin{pmatrix} g \\ g' \end{pmatrix}\right|_{z=z(r_i^+)} = \left.\begin{pmatrix} 1 & 0 \\ 0 & \frac{dz}{dr} \end{pmatrix}\right|_{r=r_i^-} \left.\begin{pmatrix} g \\ g' \end{pmatrix}\right|_{z=z(r_i^-)}.$$

Equivalently,

$$\begin{aligned} &\left.\begin{pmatrix} 1 & 0 \\ 0 & \frac{dz}{dr} \end{pmatrix}\right|_{r=r_i^+} \mathcal{Y}(z(r_i^+)) \begin{pmatrix} \alpha_i \\ \beta_i \end{pmatrix} \\ &\quad = \left.\begin{pmatrix} 1 & 0 \\ 0 & \frac{dz}{dr} \end{pmatrix}\right|_{r=r_i^-} \mathcal{Y}(z(r_i^-)) \begin{pmatrix} \alpha_{q-1} \\ \beta_{q-1} \end{pmatrix} \end{aligned} \tag{7.35}$$

or

$$\begin{pmatrix} \alpha_i \\ \beta_i \end{pmatrix} = \mathbf{T}_l(E_0) \begin{pmatrix} \alpha_{q-1} \\ \beta_{q-1} \end{pmatrix}, \tag{7.36}$$

where $\mathbf{T}_i$ denotes the transfer matrix from R_{i-1} to R_i:

$$\mathbf{T}_i = \mathcal{Y}(z(r_i^+))^{-1}\, \mathbf{J}_i\, \mathcal{Y}(z(r_i^-)), \tag{7.37}$$

where $\mathbf{J}_i$ is the "jump matrix":

$$\mathbf{J}_i = \begin{pmatrix} 1 & 0 \\ 0 & \left(\frac{dz}{dr}(r_i^+)\right)^{-1} \frac{dz}{dr}(r_i^-) \end{pmatrix}. \tag{7.38}$$

Therefore, the connection formula relating the coefficients α_1, β_1 in region R_1 with α_L, β_L in region R_L is:

$$\begin{pmatrix} \alpha_L \\ \beta_L \end{pmatrix} = \mathbf{T}(E_0) \begin{pmatrix} \alpha_1 \\ \beta_1 \end{pmatrix}, \tag{7.39}$$

where

$$\mathbf{T}(E_0) = \mathbf{T}_L(E_0) \cdots \mathbf{T}_3(E_0)\mathbf{T}_2(E_0). \tag{7.40}$$

The condition that the solution be nonsingular at $r = 0$ implies that

$$\begin{pmatrix} \alpha_1 \\ \beta_1 \end{pmatrix} = \begin{pmatrix} 1 \\ 0 \end{pmatrix}$$

or a multiple thereof. The radiation condition (7.30) is equivalent to the orthogonality of $(0, 1)$ and (α_L, β_L). Therefore, the resonance or scattering energies are determined by the following scalar transcendental equation:

$$(7.41)\qquad \begin{pmatrix} 0 & 1 \end{pmatrix} \mathbf{T}(E_0) \begin{pmatrix} 1 \\ 0 \end{pmatrix} = 0$$

or equivalently

$$(7.42)\qquad T_{21}(E_0) = 0,$$

where T_{kj} denotes the (k, j) entry of the matrix $\mathbf{T}(E_0)$.

7.3. Simple layered structures: Microstructure corrections to Φ_0 and E_0. In this subsection we construct the order N^{-2} correction to Φ_0 ($N^{-2}\Phi_2$) and to E_0 ($N^{-2}E_2$) for the special case of a simple layered structure (7.19). Recall that $\Phi_2 = \Phi_2^{(p)} + \Phi_2^{(h)}$. Here $\Phi_2^{(p)}$ is, in general, given by the Fourier series (5.31). When the potential is of the form (7.19), $\Phi_2^{(p)}$ can be explicitly calculated, which is done in appendix A. $\Phi_2^{(h)}$ solves equation (5.36) and E_2 is chosen so that (5.31) has a solution satisfying the outgoing radiation condition at $r = \infty$. We now focus on the determination of E_2 and $\Phi_2^{(h)}$ in the separable case.

Recall that the equation (5.36) for $u = \Phi_2^{(h)}$ is of the form:

$$(7.43)\qquad (\ H_{\rm av} - E_0\ )u = (\ E_2 + Q(r)\ )\ \Phi_0,$$

where $Q(r)$ has compact support; its explicit form for simple structures is presented in appendix A. Since $\Phi_0(r, \theta) = f(r)e^{il\theta}$, we define $u(r) = U(r)e^{il\theta}$ and consider the equation

$$(7.44)\qquad \left(-\Delta_r + \frac{l^2}{r^2} + V_{\rm av}(r) - E_0 \right) U(r) = (\ E_2 + Q(r)\ )\, f(r)$$

As in the previous subsection, we define

$$\begin{aligned} z &= \sqrt{E_0 - V_{\rm av}}\ r\ ,\ \Re E_0 > V_{\rm av} \\ z &= \sqrt{V_{\rm av} - E_0}\ r\ ,\ \Re E_0 < V_{\rm av} \end{aligned}$$

and

$$Q_1(z) = Q(r(z)),\quad U_1(z) = U(r(z)),\quad g(z) = f(r(z)).$$

Then,

$$\begin{aligned} z^2U_1'' + zU_1' + (z^2 - l^2)U_1 &= b_+(z) \equiv \frac{-z^2}{E_0 - V_{\rm av}}(Q_1 + E_2)g(z), \quad \Re E_0 > V_{\rm av} \\ z^2U_1'' + zU_1' - (z^2 + l^2)U_1 &= b_-(z) \equiv \frac{-z^2}{V_{\rm av} - E_0}(Q_1 + E_2)g(z), \quad \Re E_0 < V_{\rm av}. \end{aligned}$$

(7.45)

We now solve (7.45). Fix any interval R_q where $V_{\rm av}(r(z))$ is constant. The solution in this interval can be written as:

$$(7.46)\qquad U_{1,q}(z) = U_{1,q}^{(p)}(z) + \xi_q y_1(z) + \eta_q y_2(z),$$

where $U_1^{(p)}(z)$ denotes a particular solution obtained by variation of parameters:

$$U_{1,q}^{(p)}(z) = U_{1,q}^{(p)}(z; E_2) = \xi_q^{(p)}(z)\ y_1(z) + \eta_q^{(p)}(z)\ y_2(z). \tag{7.47}$$

In regions $R_1, ..., R_{L-1}$, y_1 and y_2 are taken to be Bessel or modified Bessel while

$$\text{in region } R_L: \qquad y_1 = H_l^{(1)} \text{ and } y_2 = H_l^{(2)},$$

where $H_l^{(1)}$ and $H_l^{(2)}$ are Hankel functions satisfying, respectively, outgoing and incoming radiation conditions at $r = \infty$. We have

$$\xi_q^{(p)}(z) = -\int^z \frac{y_2(\zeta)b_\pm(\zeta)}{\zeta^2 W\{y_1, y_2\}}\, d\zeta \tag{7.48}$$

$$\eta_q^{(p)}(z) = +\int^z \frac{y_1(\zeta)b_\pm(\zeta)}{\zeta^2 W\{y_1, y_2\}}\, d\zeta. \tag{7.49}$$

The Wronskian, $W\{y_1, y_2\} = \det \mathcal{Y}$ is given in (7.25) in the case of Bessel functions, (7.29) in the case of modified Bessel functions and (7.11) in the case of Hankel functions. The lower limits of integration in (7.48) and (7.49) are chosen to be z_q, the left endpoint of the interval R_q.

Boundary condition at $r = 0$: The lower limit of integration of $U_1^{(p)}$ in R_1 is $z = 0$, so we require, due to the singularity at $r = 0$ of Y_l and K_l, that $\eta_1 = 0$. In fact, for simplicity we take the homogeneous part of the solution in this region to be identically zero. Thus,

$$\begin{pmatrix} \xi_1 \\ \eta_1 \end{pmatrix} = \begin{pmatrix} 0 \\ 0 \end{pmatrix}. \tag{7.50}$$

Continuity across interfaces: We require continuity of $U_1(r)$ and $\partial_r U_1(r)$ at the interfaces $r = r_q$ ($z = z_q = z(r_q)$), for $q = 2, \ldots, L$. This gives (*cf.* (7.35))

$$\begin{pmatrix} U_{1,q}(z) \\ \partial_z U_{1,q}(z) \end{pmatrix}\Bigg|_{z=z_q^+} = \mathbf{J}_q \begin{pmatrix} U_{1,q-1}(z) \\ \partial_z U_{1,q-1}(z) \end{pmatrix}\Bigg|_{z=z_q^-}. \tag{7.51}$$

Equation (7.51) is equivalent to the following set of algebraic relations on the coefficients ξ and η:

$$\begin{aligned} \begin{pmatrix} \xi_q \\ \eta_q \end{pmatrix} &= \mathcal{Y}(z_q^+)^{-1}\left[-\mathbf{U_1^{(p)}}(z_q^+) + \mathbf{J}_q \mathbf{U_1^{(p)}}(z_q^-)\right] \\ &\qquad + \mathcal{Y}(z_q^+)^{-1}\mathbf{J}_q\mathcal{Y}(z_q^-)\begin{pmatrix} \xi_{q-1} \\ \eta_{q-1} \end{pmatrix} \end{aligned} \tag{7.52}$$

Here, $\mathbf{U_1^{(p)}} = \mathbf{U_1^{(p)}}(E_2)$ denotes the column 2-vector with components $U_1^{(p)}$ and $\partial U_1^{(p)}$. Equation (7.52) has the form

$$\begin{pmatrix} \xi_q \\ \eta_q \end{pmatrix} = \mathbf{F}_q(E_2) + \mathbf{T}_q \begin{pmatrix} \xi_{q-1} \\ \eta_{q-1} \end{pmatrix}, \tag{7.53}$$

where $\mathbf{F}_q(E_2)$ and $\mathbf{T}_q$ is the transfer matrix from (7.37), as can be read off from (7.52).

Starting with $(\xi_1, \eta_1) = (0, 0)$, given in (7.50), repeated application of (7.53) for $q = 2, \ldots, L$ yields explicit expressions for the coefficients, ξ_q and η_q in each region, ensuring that (7.46) is a non-singular solution of (7.44) for all $r < \infty$.

Radiation boundary condition at $r = \infty$: It remains to impose the condition of outward going radiation, which is where E_2 is determined. To see this, note that iteration of (7.53) and use of (7.50) yields:

$$\begin{pmatrix} \xi_L(E_2) \\ \eta_L(E_2) \end{pmatrix} = \mathbf{F}_L + \mathbf{T}_L\mathbf{F}_{L-1} + \mathbf{T}_L\mathbf{T}_{L-1}\mathbf{F}_{L-2} + \ldots + \mathbf{T}_L\mathbf{T}_{L-1}\cdots\mathbf{T}_3\mathbf{F}_2$$

(7.54)

For $r \in R_L$, $(r > r_L)$ we have:

$$U_1(z) = U_{1,L}(z) = U_{1,L}^{(p)}(z; E_2) + \xi_L(E_2)H_l^{(1)}(z) + \eta_L(E_2)H_L^{(2)}(z).$$

We now construct $U_{1,L}^{(p)}(z; E_2)$ for $r(z) \in R_L$. Noting that in this region $\Re E_0 > V_{\rm av} = 0$ and $Q_1(z) = 0$, we find from the first equation in (7.45) and (7.31) that $U_{1,L}^{(p)}$ satisfies

$$z^2U_1'' + zU_1' + (z^2 - l^2)U_1 = z^2\left(\frac{E_2}{E_0}\right)\alpha_L H_l^{(1)}(z). \tag{7.55}$$

This is precisely equation (7.11) with $k_* = \alpha_L$. We can therefore use the calculations of this previous subsection to obtain a particular solution. This gives the equation for E_2:

$$\eta_L(E_2) - \frac{i\alpha_L\pi}{4}\frac{E_2}{E_0}\frac{z_L^2}{2}\left([H_l^{(1)}(z_L)]^2 - H_{l-1}^{(1)}(z_L)H_{l+1}^{(1)}(z_L) \right) = 0, \tag{7.56}$$

where

$$\eta_L(E_2) \equiv \eta_{L0} + E_2\,\eta_{L1} \tag{7.57}$$

is read off from (7.54). E_2 is then found by solving the linear relation (7.56–7.57).

8. Summary and discussion

Motivated by the problem of light propagation in a microstructured waveguide, we have considered the scattering resonance problem for the Schrödinger equation with high contrast and "microstructured" potentials. We outline the results of the authors [**6**] in the special case of 2-dimensional N-fold symmetric potentials potentials, which correspond to waveguides whose cross-section has an N-fold symmetry.

We have further derived a systematic homogenization expansion for such microstructured cylindrical waveguides with transverse N-fold symmetry. The method of derivation, multiple scale analysis, facilitates the removal of fast scales due the rapid variations in the microstructure. Therefore, numerical implementation of the expansion does not encounter the intrinsic "stiffness" associated with problems having large separations of scales, and gives an efficient algorithm for calculation of scattering resonances in certain regimes.

We have implemented the expansion for a number of sample structures and have computed the effective indexes (scattering resonance energies) of their leaky modes. Of great importance are the imaginary parts of the effective indexes, which correspond to the radiative leakage rates, resulting from a combination of propagation and tunneling losses. In contrast to the real parts of the effective indexes, these leakage rates are very sensitive to the geometry of the microstructure in the case of low index (air) holes.

Appendix A. Calculations for simple layered structures

In this appendix we present the expressions needed to compute the leading corrections to Φ_0 and E_0 due to microstructure. There are two quantities we require:

$$\Phi_2^{(p)}(r,\theta,\Theta) = r^2\,\Phi_0(r,\theta)\,\partial_\Theta^{-2}\Big[V(r,\Theta) - V_{\text{av}}(r)\Big] \tag{A.1}$$

and

$$Q(r) = \frac{r^2}{2\pi}\int_0^{2\pi}\Big|\partial_p^{-1}\Big[V(r,p) - V_{\text{av}}(r)\Big]\Big|^2\,dp. \tag{A.2}$$

The mean-zero antiderivatives in (A.1) and (A.2) are explicitly calculable due to the form of the potential (7.19). For $r \in R_i$,

$$\partial_\Theta^{-1}\Big[V(r,\Theta) - V_{\text{av}}(r)\Big] = C_1 - V_{\text{av,i}}\,\Theta + \sum_{j=1}^{M_i}\mathbf{1}_{[\Theta_j,\Theta_{j+1}]}(\Theta)\Big(S_j + V_{i,j}\,(\Theta-\Theta_j)\Big), \tag{A.3}$$

where

$$S_j = \sum_{k=1}^{j-1} V_{i,k}\,(\Theta_{k+1} - \Theta_k) \tag{A.4}$$

and the constant C_1 is given below. Integrating again, once more for $r \in R_i$,

$$\begin{aligned}\partial_\Theta^{-2}\Big[V(r,\Theta) - V_{\text{av}}(r)\Big] &= C_2 + C_1\,\Theta - \frac{1}{2}V_{\text{av},i}\,\Theta^2 + \\ &\sum_{j=1}^{M_i}\mathbf{1}_{[\Theta_j,\Theta_{j+1}]}(\Theta)\Big(T_j + S_j\,(\Theta-\Theta_j) + \frac{1}{2}V_{i,j}\,(\Theta-\Theta_j)^2\Big)\end{aligned} \tag{A.5}$$

where

$$T_j = \sum_{k=1}^{j-1}\Big(S_k\,(\Theta_{k+1}-\Theta_k) + \frac{1}{2}V_{i,k}\,(\Theta_{k+1}-\Theta_k)^2\Big) \tag{A.6}$$

and the constants C_1, C_2 are given by

$$\begin{aligned} C_1 &= \pi V_{\text{av},i} - T_{M_i+1} \\ C_2 &= -\pi C_1 + \frac{(2\pi)^2}{6}V_{\text{av},i} - \\ &\frac{1}{2\pi}\sum_{j=1}^{M_i}\Big(T_j\,(\Theta_{j+1}-\Theta_j) + \frac{1}{2}S_j\,(\Theta_{j+1}-\Theta_j)^2 + \frac{1}{6}V_{i,j}\,(\Theta_{j+1}-\Theta j)^3\Big).\end{aligned} \tag{A.7}$$

Finally, for $r \in R_i$, the integral in (A.2) evaluates to

$$\begin{aligned}\frac{1}{2\pi}\int_0^{2\pi}\Big|\partial_p^{-1}\Big[V(r,p) - V_{\text{av}}(r)\Big]\Big|^2\,dp &= C_1^2 + \frac{1}{\pi}C_1 T_{M_i+1} \\ &+\frac{1}{2\pi}\sum_{j=1}^{M_i}\Big(S_j^2\,(\Theta_{j+1}-\Theta_j) + S_j(V_{i,j}-V_{\text{av},i})(\Theta_{j+1}-\Theta_j)^2 \\ &+\frac{1}{3}(V_{i,j}-V_{\text{av},i})^2(\Theta_{j+1}-\Theta_j)^3\Big).\end{aligned} \tag{A.8}$$

References

1. M. Abramowitz and I.E. Stegun, *Handbook of mathematical functions*, National Bureau of Standards, 1972.
2. A. Argyros, I.M. Bassett, M.A. van Eijkelenborg, M.C.J. Large, J. Zagari, N.A.P. Nicorovici, R.C. McPhedran, and C.M. de Sterke, *Ring structures in microstructured polymer optical fibers*, Optics Express **9** (2001), 813–820.
3. E. Balslev and E. Skibsted, *Resonance theory of two-body schrödinger operators*, Ann. Inst. Henri Poincaré **51** (1989), no. 2, 129–154.
4. A. Bensoussan, J.L. Lions, and G.C. Papanicolaou, *Asymptotic analysis for periodic structures*, Studies in Mathematics and its Applications, vol. 5, North-Holland, 1978.
5. C.L. Dolph, J.B. McLeod, and D. Thoe, *The analytic continuation of the resolvent kernel and scattering operator associated with the schrödinger operator*, J. Math. Anal. Appl. **16** (1966), 311–332.
6. S.E. Golowich and M.I. Weinstein, *Scattering resonances and homogenization theory*, Bell Labs preprint, 2003.
7. S.E. Golowich and M.I. Weinstein, *Homogenization expansion for resonances of microstructured photonic waveguides*, J. Opt. Soc. Am. B (2003, to appear).
8. I.S. Gradshteyn and I.M. Ryzhik, *Table of integrals, series and products*, Academic Press, 1980.
9. P.D. Hislop and I.M. Sigal, *Introduction to spectral theory*, Springer Verlag, 1996.
10. C. Kerbage and B.J. Eggleton, *Microstructured optical fibers*, Optics and Photonics News **13** (2002), 38–42.
11. D. Marcuse, *Theory of dielectric optical waveguides*, 2nd ed., Academic Press, Boston, 1991.
12. G.W. Milton, *The theory of composites*, Cambridge Monographs in Applied and Computational Mathematics, Cambridge, 2002.
13. P.M. Morse and H. Fesbach, *Methods of theoretical physics i & ii*, McGraw Hill, 1953.
14. S. Moskow and M. Vogelius, *First-order corrections to the homogenized eigenvalues of a periodic composite medium. a convergence proof*, Proc. Roy. Soc. Edinburgh **127A** (1997), 1263–1299.
15. L. Poladian, N.A. Issa, , and T.M. Monro, *Fourier decomposition algorithm for leaky modes of fibres with arbitrary geometry*, Optics Express **10** (2002), 449–454.
16. F. Santosa and M. Vogelius, *First-order corrections to the homogenized eigenvalues of a periodic composite medium*, SIAM J. Appl. Math. **53** (1993), 1636–1668.
17. N. Shenk and D. Thoe, *Resonance states and poles of the scattering matrix for perturbations of* $-\Delta$, J. Math. Anal. Appl. **37** (1972), 467–491.
18. A. W. Snyder and John D. Love, *Optical waveguide theory*, Chapman and Hall, London, 1983.
19. J. Vuckovic, M. Loncar, H. Mabuchi, and A. Scherer, *Design of photonic crystal microcavities for cavity QED*, Phys. Rev. E **65** (2001), 016608.
20. T.P. White, R.C. McPhedran, C.M. de Sterke, L.C. Botten, and M.J. Steel, *Confinement losses in microstructured optical fibers*, Opt. lett. **26** (2001), 1660–1662.

Mathematical Sciences Research, Bell Laboratories, Lucent Technologies, 700 Mountain Avenue, Murray Hill, NJ 07974

Contemporary Mathematics
Volume **339**, 2003

Asymptotic Formulas for Eigenvalues and Eigenfunctions of Periodic Magnetic Schrödinger Operators.

Yu. Karpeshina

ABSTRACT. The paper considers periodic magnetic Schrödinger operators in dimensions two and three. It discusses asymptotic formulas for Bloch eigenvalues and eigenfunctions in the high energy region and outlines a proof of finiteness of the number of gaps in the spectrum.

1. Introduction.

We study an operator

$$Hu = \left(i\nabla + \vec{a}\right)^2 u + a_0(x)u \tag{1.1}$$

in $L_2(R^d)$, where $\vec{a}(x) = (a_1(x), a_2(x), ..., a_d(x))$, $a_s(x)$, $s = 0, 1, ...d$, are real periodic functions. We assume that the functions $a_s(x)$, $s = 0, 1, ..., d$, have the same periods $\vec{b}_1, \vec{b}_2, ..., \vec{b}_d$. For the sake of simplicity, we consider the periods being orthogonal and directed along x coordinates. Without the loss of generality we assume that

$$\text{div } \vec{a}(x) = 0, \tag{1.2}$$

$$\int_Q a_s(x)dx = 0, \quad s = 0, 1, ..., d, \tag{1.3}$$

Q being an elementary cell of the periods, $Q = [0, b_1] ... \times [0, b_d]$, $b_j = |\vec{b}_j|$, $j = 1, ...$ The operator (1.1) is of great physical importance, since it describes behavior of an electron in a periodic magnetic field. Periodic magnetic potential $\vec{a}$ corresponds to a periodic magnetic field B, $B = curl\ \vec{a}$ for d=3. We consider here only periodic magnetic potentials. Unfortunately, a constant magnetic field is not of this type, its magnetic potential $\vec{a}$ being a linear function. Note that the case of a periodic B with non-periodic $\vec{a}$ is significantly more difficult than ours. Most of the results in this situation are obtained for the case of a constant magnetic field (e.g [**LL, DN, Sj**] and literature there).

1991 *Mathematics Subject Classification.* Primary 35J10, 35Q40; Secondary 35P15.

Key words and phrases. Schroedinger operator, perturbation theory, spectral properties.
The author was supported in part by NSF Grant #0201383.

We assume that functions $a_0(x)$, $a_1(x)$, ..., $a_d(x)$ are trigonometric polynomials:

$$(1.4)\qquad a_s(x) = \sum_{m\in Z^d,\ 0<|m|<R_0} a_{sm}\exp i(\vec{p}_m(0), x), \quad s = 0, 1, ..., d, \quad R_0 < \infty,$$

where $(\cdot,\cdot)$ is the inner product in R^d and $\vec{p}_m(0)$ is a vector of the dual lattice:

$$(1.5)\qquad \vec{p}_m(0) = 2\pi(m_1 b_1^{-1}, ..., m_d b_d^{-1}).$$

We use the notation

$$(1.6)\qquad \vec{A}(x) = (a_0(x), a_1(x), ..., a_d(x))\,.$$

2. Floquet decomposition.

It is well known (see e.g. [**G, RS**]) that spectral analysis of H can be reduced to analysis of a family of operators $H(t)$, $t \in K$, where K is the elementary cell of the dual lattice,

$$K = [0, 2\pi b_1^{-1}) \times ... \times [0, 2\pi b_d^{-1}).$$

The vector t is called the *quasimomentum*. The operator $H(t)$, $t \in K$, acts in $L_2(Q)$. It is described by formula (1.1) and the quasiperiodic conditions:

$$u(x_1, ...x_{j-1}, b_j, x_{j+1}, ...x_d) = \exp(it_j b_j)u(x_1, ...x_{j-1}, 0, x_{j+1}, ...x_d),$$
$$u_{x_j}(x_1, ...x_{j-1}, b_j, x_{j+1}, ...x_d) = \exp(it_j b_j)u_{x_j}(x_1, ...x_{j-1}, 0, x_{j+1}, ...x_d),$$
$$(2.1)\qquad j = 1, ..., d.$$

Each operator $H(t)$, $t \in K$, has a discrete bounded below spectrum $\Lambda(t)$:

$$\Lambda(t) = \cup_{i=1}^{\infty}\lambda_i(t), \ \lim_{i\to\infty}\lambda_i(t) = \infty,$$

eigenvalues $\lambda_i(t)$ being labeled in the increasing order here. The spectrum Λ of the operator H is the union of the spectra $\Lambda(t)$ of the operators $H(t)$: $\Lambda = \cup_{t\in K}\Lambda(t) = \cup_{i\in N, t\in K}\lambda_i(t)$. The functions $\lambda_i(t)$ are continuous, so Λ has a band structure:

$$\Lambda = \cup_{i=1}^{\infty}[q_i, Q_i], \ q_i = \min_{t\in K}\lambda_i(t), \ Q_i = \max_{t\in K}\lambda_i(t).$$

The eigenfunctions of $H(t)$ and H are simply related. Extending all the eigenfunctions of the operators $H(t)$ quasiperiodically (see (2.1)) to R^d, we obtain the complete system of generalized eigenfunctions of H.

In the case of $\vec{A} = 0$, $\vec{A}$ being defined by (1.6), the eigenvalues and eigenfunctions of the corresponding operators $H_0(t)$, $t \in K$, are naturally indexed by the points j of Z^d:

$$(2.2)\qquad \Psi_j^0(t, x) = |Q|^{-1}\exp i(\vec{p}_j(t), x), \quad \lambda_j^0(t) = p_j^2(t),$$

here and below $\vec{p}_j(t) = \vec{p}_j(0) + t$, $\vec{p}_j(0)$ being given by (1.5), $p_j^2(t) = |\vec{p}_j(t)|^2$, and $|Q|$ is the volume of Q. Note that such an alternative indexing of eigenvalues of $H_0(t)$ is not used for enumerating zones, but only for indexing eigenvalues for a fixed t.

The spectral study of (1.1) turns out to be essentially more difficult than study of the operator with just an electric potential ($\vec{a}(x) = 0$). It is related to the fact that the function $\vec{a}(x)$ is a multiplier in front of $\nabla u(x)$, while $a_0(x)$ is a multiplier in front of $u(x)$. It is well-known that perturbations of coefficients of higher order derivatives produce stronger effect on spectral properties of a differential operator.

The series of new publications concerning (1.1) has appeared in recent years (e.g. [BiSu1-BrS,DN,HH1-I,K2-Mo,Sj,So]).

3. Absolute continuity of the spectrum.

A very important property of (1.1) is absolute continuity of its spectrum. For the operator (1.1), absolute continuity is equivalent to the fact that (1.1) does not have a band degenerated into a point [**Ku**]. For small $\vec{a}$ and any dimension d absolute continuity of the spectrum was proved by R.Hempel and I.Herbst [**HH1**]. They used L. Thomas' approach [**Th**] originally developed for the Schrödinger operator with just an electric potential ($\vec{a} = 0$). It is based on the fact that the resolvent of the operator is compact and admits an analytic extension to complex values of each component t_j, $j = 1, ...d$, of the quasimomentum provided that others are real and fixed. L. Thomas [**Th**] proved that the resolvent goes to zero in the class of bounded operators when $\operatorname{Im} t_j \to \infty$ for some j, while $\operatorname{Im} t_r = 0$ for $r \neq j$. It is easy to see that this can not happen if H has a band degenerated into a point, i.e. it has an eigenvalue constant with respect to t. This contradiction proves absolute continuity of the spectrum. M.Birman and T.Suslina [**BiSu1, BiSu2**] generalized L.Thomas' approach for dimension two and a wide class of magnetic potentials $\vec{a}$. Later A.Sobolev [**So**] extended L.Thomas' approach to any dimension d and smooth magnetic potentials of arbitrary magnitude. The main point in [**So**] is the analytic extension of the resolvent to the complex plane with respect to the component of t along a distinguished direction $\vec{\alpha} \in R^d$ which depends on the vector-potential $\vec{a}$. P.Kuchment and S.Levendorskii [**KuLe**] simplified Sobolev's proof and obtained stronger results (weaker assumptions on $\vec{a}$, a validity for non-selfadjoint case). The latest paper on the subject belongs to L.Friedlander [**F**]. He obtained a new proof of absolute continuity of the spectrum for a second order self-adjoint elliptic differential operator with periodic coefficients satisfying additional symmetry conditions. In the case of operator (1.1) these conditions mean that functions $a_0(x)$ and $\vec{a}(x)$ have to be even with respect to x_1 and x_2.

4. Finiteness of the number of gaps in the spectrum.

Finiteness of the number of gaps in the spectrum of a Schrödinger operator with a periodic electromagnetic potential is a long-standing mathematical problem. It is well-known [**RS**] that the operator (1.1) in R^1, generally speaking, has an infinite number of gaps. It was noted long time ago that the multidimensional situation is different. The conjecture that the Schrödinger operator with an electric periodic potential in R^2, R^3 has only a finite number of gaps was formulated by physicists Bethe and Sommerfeld [**BeSo**]. Rigorous mathematical proofs of finiteness of the number of gaps for an electric potential in dimensions 2,3 and 4 appeared later (e.g. [**Sk1, Sk2, Sk3, V1, K1, HM, PaSo1, PaSo2**]). A recent preprint [**V2**] based on ideas of [**V3**] claims the proof for an arbitrary dimension. Finiteness of the number of gaps in the multidimensional situation is closely related to the fact that the spectrum in the multidimensional case is much "thicker", in the sense that it has infinite multiplicity [1], unlike the spectrum in the one-dimensional case, which has

[1]In this sense the thickness of the spectrum is used for proving the Bethe-Sommerfeld conjecture in [**V1, K1**]. The papers [**Sk1, Sk2, Sk3, HM, PaSo1, PaSo2**] use somewhat different approach: the validity of the Bethe-Sommerfeld conjecture is explained there by the large number of overlapping zones.

multiplicity two. This thickness of the spectrum can not be destroyed by an electric potential, however it can be destroyed by other bounded perturbations [**Sk2**]. Thus, the validity of the Bethe-Sommerfeld conjecture for a magnetic operator (1.1) is far from being obvious, even though it is proven for the operators with just electric potentials. The perturbation by a magnetic potential is much stronger than that by just an electric potential. Can it destroy even a thick spectrum, producing gaps? So far the Bethe-Sommerfeld conjecture is proven for magnetic potentials only in the case $d = 2$. The proof belongs to A.Morame (A.Mohamed). The restriction $d = 2$ is somewhat surprising: considering that the spectrum is getting thicker for higher dimensions, one can expect that the treatment of higher dimensions should be easier than that of lower ones. However, it is not the case even for electric potentials. The explanation of this difficulty lies in more complicated structure of the spectrum of the Laplacian in higher dimensions. Bloch eigenvalues $\{\lambda_i^0(t)\}_{i=1}^{\infty}$ (enumerated in the increasing order) are situated much denser in higher dimensional situations, they oscillate stronger as functions of t, and the ranges of $\lambda_i^0(t)$ are very short intervals when i is large. This makes the spectrum of the operator much more vulnerable for a perturbation, and it is more difficult to describe changes of eigenvalues $\lambda_i(t)$ even under weak perturbations.

As it has already been mentioned, A.Morame had proven the conjecture for magnetic operators in dimension two. He also obtained an asymptotic formula for the integrated density of states at high energies. The notion of the integrated density of states is closely related in his approach to the proof of Bethe-Sommerfeld conjecture. First, we describe A.Morame's results on the integrated density of states and then we will sketch the proof of the conjecture. The integrated density of states is defined by the formula:

$$D(H, \lambda) = \int_K N(H, \lambda, t)dt, \tag{4.1}$$

here $N(H, \lambda, t)$ is a counting function for Bloch eigenvalues:

$$N(H, \lambda, t) = \sum_{i:\lambda_i(t)<\lambda} 1.$$

It is easy to see that $N(H_0, \lambda, t)$, corresponding to $H_0 = -\Delta$, is just the number of the lattice points $\{\vec{p}_m(t)\}_{m\in Z^d}$ in the ball of the radius $\sqrt{\lambda}$ centered at the origin, and $D(H_0, \lambda)$ is the volume of the ball:

$$D(H_0, \lambda) = \omega_d \lambda^{d/2}, \tag{4.2}$$

ω_d being the volume of a unit ball in R^d. The notion of integrated density of states is important in physics: $D(H, \lambda + \delta\lambda) - D(H, \lambda)$ characterizes the number of possible electron states in the energy interval $[\lambda, \lambda + \delta\lambda]$. Using methods of microlocal analysis A.Morame proved that

$$D(H, \lambda)_{\lambda\to\infty} = D(H_0, \lambda) + O(\lambda^{(d-2+\varepsilon)/2}) \tag{4.3}$$

for any positive ε. Moreover, he generalized formulae (4.1) and (4.3) considering

$$D_q(H, \lambda) = \int_K e^{i(q,t)} N(H, \lambda, t)dt, \quad q = (q_1b_1, ...q_nb_n), \ q_1, ...q_n \in Z, \tag{4.4}$$

and proving that

$$D_q(H, \lambda)_{\lambda\to\infty} = D_q(H_0, \lambda) + O(\lambda^{(d-2+\varepsilon)/2}). \tag{4.5}$$

To deduce the proof of the Bethe-Sommerfeld conjecture from (4.5), A.Morame followed the scheme of B.E.J.Dahlberg and E.Trubowitz [**DaTr**], which was developed to prove the conjecture for the case of an electric periodic potential in dimension two. A short version of the proof is the following.

We define $N_+(H,\lambda)$ and $N_-(H,\lambda)$ by the formulae:

$$N_+(H,\lambda) = \sup_t N(H,\lambda,t), \quad N_-(H,\lambda) = \inf_t N(H,\lambda,t).$$

Suppose we have proved that

$$N_+(H,\lambda) > N_-(H,\lambda). \tag{4.6}$$

The inequality shows that the number of eigenvalues $\lambda_i(t)$ which are less than λ changes with t. Considering continuity of $\lambda_i(t)$ with respect to t, one can easily conclude that there is a pair (i,t) such that $\lambda_i(t) = \lambda$, i.e., λ is a point of the spectrum. Thus, for proving the Bethe-Sommerfeld conjecture it suffices to check (4.6). It is easy to see that (4.6) follows from the pair of estimates:

$$\int_K \left(N(H,\lambda,t) - \pi|K|^{-1}\lambda\right) dt = O(\lambda^{\varepsilon/2}). \tag{4.7}$$

$$\int_K \left|N(H,\lambda,t) - \pi|K|^{-1}\lambda\right| dt \geq \tilde{c}_q\lambda^{1/4}. \tag{4.8}$$

Thus, it remains to prove (4.7) and (4.8). To check (4.7), we note that

$$\int_K \left(N(H_0,\lambda,t) - \pi|K|^{-1}\lambda\right) dt = \int_K N(H_0,\lambda,t)dt - \pi\lambda = 0.$$

Using (4.3) for $d = 2$, we obtain (4.7). Let us prove (4.8). Following [**DaTr**] we make two elementary observations. First,

$$D_q(H,\lambda) = \int_K e^{i(q,t)} N(H,\lambda,t)dt = \int_K e^{i(q,t)} \left(N(H,\lambda,t) - \pi|K|^{-1}\lambda\right) dt, \quad \text{if } q \neq 0.$$

Second,

$$|D_q(H,\lambda)| \leq \int_K \left|N(H,\lambda,t) - \pi|K|^{-1}\lambda\right| dt, \quad \text{if } q \neq 0.$$

From the last estimate it follows immediately:

$$\int_K \left|N(H,\lambda,t) - \pi|K|^{-1}\lambda\right| dt \geq \frac{1}{2}|D_q(H,\lambda)| + \frac{1}{2}|D_{2q}(H,\lambda)| \tag{4.9}$$

for any $q \neq 0$. Now we estimate the right-hand side of (4.9). First, we consider the case $H = H_0$. A simple calculation gives:

$$D_q(H_0,\lambda) = \int_K e^{i(q,t)} N(H_0,\lambda,t)dt = \int_K e^{i(q,t)} \left(\sum_{r\in Z^2 : p_r^2(t)<\lambda} 1 \right) dt = \int_{B_{\sqrt{\lambda}}} e^{i(q,x)}dx,$$

$B_{\sqrt{\lambda}}$ being the disk of radius $\sqrt{\lambda}$ centered at zero. Rewriting the last integral in polar coordinates and using the stationary phase method, we obtain:

$$D_q(H_0,\lambda) = (2\pi|q|)^{-3/2}\lambda^{1/4} \sin\left(\pi/4 - \sqrt{\lambda}|q|\right) + O(\lambda^{-1/4}). \tag{4.10}$$

It is easy to see now that for some q

$$|D_q(H_0,\lambda)| + |D_{2q}(H_0,\lambda)| > c_q\lambda^{1/4}, \quad c_q > 0. \tag{4.11}$$

Considering (4.5) for $d = 2$, we get

$$|D_q(H,\lambda)| + |D_{2q}(H,\lambda)| > \tilde{c}_q \lambda^{1/4},$$

when $\varepsilon < 1/2$. Combining (4.9) and the last estimate, we obtain (4.8). Thus, the Bethe-Sommerfeld conjecture is proved for the case $d = 2$.

Note that this method of proving does not work for dimensions higher that two. Indeed, the analog of formula (4.10) for an arbitrary dimension d is:

$$D_q(H_0,\lambda) = (2\pi|q|)^{-(d+1)/2}\lambda^{(d-1)/4}\sin\left(\pi(d-1)/4 - \sqrt{\lambda}|q|\right) + O(\lambda^{(d-3)/4}).$$

Hence, the analog of (4.11) is:

$$|D_q(H_0,\lambda)| + |D_{2q}(H_0,\lambda)| > c_q \lambda^{(d-1)/4}, \quad c_q > 0.$$

Using (4.5), we get

$$|D_q(H,\lambda)| + |D_{2q}(H,\lambda)| > c_q \lambda^{(d-1)/4} + O(\lambda^{(d-2+\varepsilon)/2}). \tag{4.12}$$

Note that $(d-1)/4 < (d-2+\varepsilon)/2$ when $d > 2$. Thus, we can obtain only a trivial estimate from below for $|D_q(H,\lambda)| + |D_{2q}(H,\lambda)|$, and, therefore, the estimate of the type (4.8) cannot be reached.

Finally in this section we describe results on the band structure of the spectrum in the particular case of a strong magnetic field. The paper of R. Hempel and I. Herbst [**HH1**] is devoted to the operator $H(\mu\vec{a}) = (-i\nabla - \mu\vec{a})^2$ in $L_2(R^d)$, $d \geq 2$, where $\vec{a} \in C^1(R^d, R^d)$ and $\mu \to \infty$. Letting $M = \{x : B(x) = 0\}$, where B is the magnetic field associated with $\vec{a}$, and $M_{\vec{a}} = \{x : \vec{a}(x) = 0\}$, they prove that $H(\mu\vec{a})$ converges to the Dirichlet Laplacian on the closed set M in the strong resolvent sense, as $\mu \to \infty$, provided the set $M \setminus M_{\vec{a}}$ has measure zero. The authors use the result on resolvent to show that there exist periodic vector potentials $\vec{a}$ with the property that $H(\mu\vec{a})$ has spectral gaps for μ large enough.

5. Perturbation formulae.

The next natural question is how Bloch eigenvalues and eigenfunctions of (1.1) look like. Bloch eigenvalues and eigenfunctions of the unperturbed operator $H_0(t)$ are given by formulae (2.2). What happens with $\Psi_j^0(t)$, $\lambda_j^0(t)$ under the perturbation? It is easier to answer this question in the high energy region, since one can use energy as a large parameter and try to construct an asymptotic expansion. First, let us discuss the case of just an electric potential ($\vec{a} = 0$). It was proven in [**K1**] that in this case most of perturbed eigenvalues and eigenfunctions are close to unperturbed ones for $d = 2, 3$. Physically, this closeness of perturbed eigenfunctions to unperturbed ones means weak diffraction inside the crystal. It is well known that there are cases of strong diffraction inside the crystal (see e.g. [**Ma**]) and they are extremely important for applications. However, strong diffraction happens rarely: most plane waves are only weakly diffracted inside the crystal. To state the result rigorously we first introduce an unperturbed isoenergetic surface, i.e., the set of all such t that $H_0(t)$ has an eigenvalue equal to a given λ:

$$S_0(\lambda) = \{t, t \in K : \exists j \in Z^d : \lambda_j^0(t) = \lambda\}.$$

Considering that $\lambda_j^0(t) = p_j^2(t)$, we can easily figure out that $S_0(\lambda)$ is a sphere of the radius $\sqrt{\lambda}$ "packed" into the box K: to obtain $S_0(\lambda)$ we take the intersection of the cell K with the union of all the spheres of the radius $\sqrt{\lambda}$ centered at the points $\{\vec{p}_m(0)\}_{m\in Z^d}$. It is proven [**K1**] that for any λ large enough there is a rich subset

$\tilde{S}_0(\lambda)$ of $S_0(\lambda)$ such that if $t \in \tilde{S}_0(\lambda)$, then the perturbations of $\lambda_j^0(t)$ and $\Psi_j^0(t)$ are small. The rigorous statement of the result is the following:

THEOREM 1. *Suppose $d = 2, 3$ and $\vec{a}(x) = 0$. Then, for any positive δ and any sufficiently large λ, $\lambda > \lambda_0(a_0(x), \delta)$, there is a subset $\tilde{S}(\lambda, \delta)$ of $S_0(\lambda)$ with the following properties:*

1) The measure of $\tilde{S}(\lambda, \delta)$ is asymptotically close to that of $S_0(\lambda)$ in the following sense: [2]

$$1 - \frac{s\left(\tilde{S}(\lambda, \delta)\right)}{s(S_0(\lambda))} < c\lambda^{-\delta},$$

$s(\cdot)$ being the curve length for $d = 2$ and the surface area for $d = 3$. For every $t \in \tilde{S}(\lambda, \delta)$ there is a single $j \in Z^d$ such that $p_j^2(t) = \lambda$ and

$$\min_{m \in Z^d, m \neq j} |p_j^2(t) - p_m^2(t)| > 2k^{-d+2-4\delta}. \tag{5.1}$$

2) If $t \in \tilde{S}(\lambda, \delta)$, then the operator $H(t)$ has a simple eigenvalue $\lambda_j(t)$ close to λ,

$$\lambda_j(t) = \lambda + O(\lambda^{-1+8\delta}). \tag{5.2}$$

This eigenvalue is unique: there are no other eigenvalues in the interval $(\lambda - \lambda^{-4\delta}, \lambda + \lambda^{-4\delta})$ if $d = 2$ and in the interval $(\lambda - \lambda^{-1/2-4\delta}, \lambda + \lambda^{-1/2-4\delta})$ if $d = 3$.

3) An eigenfunction $\Psi_j(t, x)$ corresponding to $\lambda_j(t)$ is close to the plane wave $\Psi_j^0(t, x)$:

$$\left\|\Psi_j(t, x) - \Psi_j^0(t, x)\right\|_{L_\infty} = O(\lambda^{-1/2+8\delta}). \tag{5.3}$$

Note that the inequality (5.1) means that $\lambda = p_j^2(t)$ is a simple eigenvalue of $H_0(t)$ and the distance from $p_j^2(t)$ to the rest of eigenvalues is greater than $2k^{-d+2-4\delta}$. The set $\tilde{S}(\lambda, \delta)$ is defined by a system of inequalities, (5.1) being the first of them.

It is proven in [**K1**] that the asymptotic formulae (5.2), (5.3) are differentiable with respect to t in the $(\lambda^{-(d-1)/2-4\delta})$-neighborhood of $\tilde{S}(\lambda, \delta)$. The isoenergetic surface of H is defined as the set of t for which $H(t)$ has an eigenvalue equal to a given λ. Using (5.2) and its differentiability with respect to t one can show [**K1**] that there is a single piece of the perturbed isoenergetic surface in the $(\lambda^{-3/2+9\delta})$-neighborhood of each component of $\tilde{S}(\lambda, \delta)$. In other words, the perturbation of H_0 by an electric potential changes $\tilde{S}(\lambda, \delta)$ only a little bit. The validity of the Bethe-Sommerfeld conjecture obviously follows from the last observation, since existence of an isoenergetic surface for a given λ implies that λ belongs to the spectrum.

A natural question is whether Theorem 1 can be generalized to the case of a nonzero magnetic potential. The answer is no. It turns out that perturbed eigenvalues and eigenfunctions for $\vec{a} \neq 0$ are not close to unperturbed ones – the perturbation is too big. To understand what is really happening when $\vec{a} \neq 0$ we introduce certain *model functions* . They have three basic properties: first, they are constructed explicitly; second they satisfy equations for eigenfunctions with a high accuracy; and third, they are almost orthogonal to each other.

---

[2] The measure of $S_0(\lambda)$ is obviously equal to that of a sphere of radius $\sqrt{\lambda}$.

We construct now a set of model functions corresponding to the case of a relatively weak diffraction. The case of a strong diffraction is described in [**K3**] for $d=2$. In the weak diffraction case we are looking for a model function in the form

$$\hat{\Psi}_j(t,x) = e^{i(p_j(t),x)+S_j(t,x)}. \tag{5.4}$$

We calculate the function $S_j(t,x)$ below. The function $S_j(t,x)$ is generally speaking not small, thus $\hat{\Psi}_j(t,x)$ is not close to a plane wave. However, the real part of $S_j(t,x)$ is small. This means that the introduction of $S_j(t,x)$ changes the phase of the wave without a significant change of its amplitude. This situation we classify as the case of a weak diffraction. The construction of model functions (5.4) includes finding an approximate solution of an eikonal equation by perturbative methods. Let us obtain formulae for $S_j(t,x)$.

Let $0<\beta<1/2$, $0<2\varepsilon<1/2-\beta$, k be a large parameter and $Z_0(k,\beta,\varepsilon)$ be the set of $j\in Z^d$ satisfying the following conditions:

(1)

$$k/4 < p_j(t) < 7k/4. \tag{5.5}$$

(2) For any $q\in Z^d : 0<|q|\le k^{\varepsilon}$:

$$|(\vec{p}_j(t),\vec{p}_q(0))| > k^{1/2+\beta}. \tag{5.6}$$

If $j\in Z_0(k,\beta,\varepsilon)$, we define the model function $\hat{\Psi}_j(x)$ by the formula (5.4). We want $\hat{\Psi}_j(x)$ to satisfy the equation for eigenfunctions $H(t)\hat{\Psi}_j(x)=\hat{\lambda}_j(t)\hat{\Psi}_j(x)$ with some constant $\hat{\lambda}_j(t)$ as precisely as possible. This gives us an asymptotic equation to define a periodic function $S_j(t,x)$:

$$2i\,(\vec{p}_j(t),\nabla S_j(t,x)) + \Delta S_j(t,x) + (\nabla S_j(t,x),\nabla S_j(t,x)) + 2i\,(\nabla S_j(t,x),\vec{a}(x)) =$$

$$2\,(\vec{p}_j(t),\vec{a}(x)) + (\vec{a}(x),\vec{a}(x)) + a_0(x) + \hat{\lambda}_j(t) - p_j^2(t) + \epsilon_N(x), \tag{5.7}$$

where ∇ and Δ are taken with respect to x, the function $\epsilon_N(x)$ admits the estimate:

$$\|\epsilon_N(x)\|_0 < c_{\vec{A}} k^{1-2(N+2)\beta}, \tag{5.8}$$

N is a fixed integer greater than -1, larger N obviously corresponding to a better approximation.

To satisfy this equation we look for $S_j(t,x)$ in the form of a sum:

$$S_j(t,x) = \sum_{n=0}^{N+1} S_j^{(n)}(t,x), \tag{5.9}$$

where

$$S_j^{(0)}(t,x) = -\sum_{r\in Z^d\setminus\{0\}} \frac{2(\vec{p}_j(t),\vec{a}_r)+\alpha a_{0r}}{p_{j+r}^2(t)-p_j^2(t)} e^{i(\vec{p}_r(0),x)}, \tag{5.10}$$

$\vec{a}_r$, a_{0r} being the Fourier coefficients of $\vec{a}(x)$ and $a_0(x)$ correspondingly. Next,

$$S_j^{(1)}(t,x) = \sum_{r\in Z^d\setminus\{0\}} \frac{\tilde{a}_{jr}^{(1)}(t)}{p_{j+r}^2(t)-p_j^2(t)} e^{i(\vec{p}_r(0),x)}, \tag{5.11}$$

where $\tilde{a}_{jr}^{(1)}(t)$ are the Fourier coefficients of the function

$$\tilde{a}_j^{(1)}(t,x) = 2i\left(\nabla S_j^{(0)}(t,x),\vec{a}(x)\right) + \left(\nabla S_j^{(0)}(t,x),\nabla S_j^{(0)}(t,x)\right) - (\vec{a}(x),\vec{a}(x)). \tag{5.12}$$

The functions $S_j^{(n)}(t,x)$, $n \geq 2$ are defined by the recurrent formula:

$$S_j^{(n)}(t,x) = \sum_{r \in Z^d \setminus \{0\}} \frac{\tilde{a}_{jr}^{(n)}(t)}{p_{j+r}^2(t) - p_j^2(t)} e^{i(\vec{p}_r(0),x)}, \tag{5.13}$$

where $a_{jr}^{(n)}(t)$ are the Fourier coefficients of the function

$$\tilde{a}_j^{(n)}(t,x) = 2i\left(\nabla S_j^{(n-1)}(t,x), \vec{a}(x)\right) + 2\left(\nabla S_j^{(n-1)}(t,x), \sum_{s=0}^{n-2} \nabla S_j^{(s)}(t,x)\right)$$

$$+ \left(\nabla S_j^{(n-1)}(t,x), \nabla S_j^{(n-1)}(t,x)\right). \tag{5.14}$$

The following lemma shows that expansion (5.9) has an asymptotic character for large k. We denote by $\|f\|_\nu$ the following norm of a function $f(x)$:

$$\|f\|_\nu = |f_0| + \sum_{r \in Z^d \setminus \{0\}} |f_r||r|^\nu,$$

f_r being the Fourier coefficients of $f(x)$.

LEMMA 1. *Suppose $0 < \beta < 1/2$, $0 < 2\varepsilon < 1/2 - \beta$, $j \in Z_0(k,\beta,\varepsilon)$, $\vec{A}$ is a trigonometric polynomial of the length R_0 and $N \in \mathbf{N}$. Then, the functions $S_j^{(n)}(x)$, n=0,1,...,N+1, and $\tilde{a}_j^{(n)}(x)$, $n = 1,...,N+1$, satisfy the following estimates:*

$$\left\|S_j^{(n)}\right\|_\nu \leq c_{n,\nu}\left\|\vec{A}\right\|_{\nu+n}^{n+1} k^{1/2-(1+2n)\beta}, \tag{5.15}$$

$$\left\|Re\ S_j^{(n)}\right\|_\nu \leq \tilde{c}_{n,\nu}\left\|\vec{A}\right\|_{\nu+n+2}^{n+1} k^{-2\beta(n+1)}, \tag{5.16}$$

$$\left\|\tilde{a}_j^{(n)}\right\|_\nu \leq c_{n,\nu}\left\|\vec{A}\right\|_{\nu+n}^{n+1} k^{1-2n\beta}, \tag{5.17}$$

$$\left\|Im\ \tilde{a}_j^{(n)}\right\|_\nu \leq \tilde{c}_{n,\nu}\left\|\vec{A}\right\|_{\nu+n+2}^{n+1} k^{1/2-(1+2n)\beta}, \tag{5.18}$$

where $c_{n,\nu} = (5b_0)^{n(n+1+\nu)+1}$, $b_0 = \max\{b_1,...,b_d\}$, $\tilde{c}_{n,\nu} = 2c_{n,\nu}$, $\tilde{c}_{n,\nu} = 4b_0c_{n,\nu}$ The estimates (5.15) – (5.18) hold for any $\nu = 0,1,..m$ and k large enough, namely $k : k^\varepsilon > 2^{N+2}R_0b_0^2$, $k^{2\beta} > (5b_0)^{N+2+\nu}\|\vec{A}\|_{\nu+N}R_0 + b_0^2$.

COROLLARY 1.

$$\|S_j\|_\nu \leq c_\nu \left\|\vec{A}\right\|_\nu k^{1/2-\beta}, \tag{5.19}$$

$$\|Re\ S_j\|_\nu \leq c_\nu \left\|\vec{A}\right\|_{\nu+2} k^{-2\beta}, \tag{5.20}$$

$$\left\|e^{S_j} - 1\right\|_0 < c\left\|\vec{A}\right\|_2 k^{-2\beta}, \tag{5.21}$$

Next, we show that $S_j(t,x)$ satisfies an eikonal equation with a high accuracy.

LEMMA 2. *Under the conditions of Lemma 1 the function $S_j(t,x)$ defined by (5.9) – (5.14) satisfies the relations (5.7), (5.8), where*

$$\hat{\lambda}_j(t) = p_j^2(t) + \sum_{n=1}^{N+1} \tilde{a}_{j0}^{(n)}. \tag{5.22}$$

COROLLARY 2.

$$\left|\hat{\lambda}_j(t) - p_j^2(t)\right| < c_b \|\vec{A}\|_1^2 k^{1-2\beta}. \tag{5.23}$$

Estimate (5.23) follows from (5.22) and (5.17).

From the previous lemma we obtain that $\hat{\Psi}_j(t,x)$ satisfies an equation for eigenfunctions with a high accuracy.

LEMMA 3. *Under the conditions of Lemma 1 the function $\hat{\Psi}_j(t,x)$ satisfies the equation:*

$$H\hat{\Psi}_j = \hat{\lambda}_j(t)\hat{\Psi}_j + \delta\hat{\Psi}_j, \tag{5.24}$$

where $\hat{\lambda}_j(t)$ is defined by (5.22), $\delta\hat{\Psi}_j$ obeys the estimate

$$\|\delta\hat{\Psi}_j\|_0 \leq c_{\vec{A}} k^{1-2(N+2)\beta}. \tag{5.25}$$

Let us consider the surface $\hat{S}_0(\lambda)$ of such t that there is a $\hat{\lambda}_j(t)$ equal to λ, $k^2/2 < \lambda < 3k^2/2$:

$$\hat{S}_0(\lambda) = \{t \in K : \exists j \in Z_0(k,\beta,\varepsilon) : \hat{\lambda}_j(t) = \lambda\}.$$

One can think about $\hat{S}_0(\lambda)$ as the "isoenergetic surface" for model functions $\hat{\Psi}_j(t,x)$. It is easy to see that the surface consists of a number of pieces, numbered by the index j. Each piece has the form of a slightly distorted fragment of a sphere of radius k.

The deviation of $\hat{S}_0(\lambda)$ from the exact spherical surface is no larger than $k^{-2\beta}$. The holes come from the restriction $j \in Z_0(k,\beta,\varepsilon)$. However, the holes are relatively small, the measure of $\hat{S}_0(\lambda)$ is close to that of the whole sphere if $\beta + d\varepsilon < 1/2$:

$$1 - \frac{s\left(\hat{S}_0(\lambda)\right)}{s(S_0(\lambda))} < c\lambda^{-(1-2\beta-2d\varepsilon)/4}.$$

It turns out that at least for $d=2$ and $d=3$ most of eigenfunctions of H are close to model functions $\hat{\Psi}_j(t)$.

THEOREM 2. *Suppose $d=2$ or $d=3$. Then, for $0<\beta<1/6$, and $d\varepsilon < 1/2-3\beta$, $0 < 8\delta < 1/2 - 3\beta - d\varepsilon$ and any sufficiently large λ, $\lambda > \lambda_0(\vec{A}(x),\beta,\varepsilon,\delta)$, there is a subset $\hat{S}(\lambda,\delta)$ of $\hat{S}_0(\lambda)$ with the following properties:*

1) The set $\hat{S}(\lambda,\delta)$ has a measure asymptotically close to that of $S_0(\lambda)$:

$$1 - \frac{s\left(\hat{S}(\lambda,\delta)\right)}{s(S_0(\lambda))} < \lambda^{-\delta}.$$

For every $t \in \hat{S}(\lambda,\delta)$ there is a single $j \in Z^d$ such that $\hat{\lambda}_j(t) = \lambda$ and

$$\min_{m \in Z_0(k,\beta,\varepsilon), m \neq j} |\hat{\lambda}_j(t) - \hat{\lambda}_m(t)| > 2k^{-d+2-4\delta}.$$

2) If $t \in \tilde{S}(\lambda, \delta)$, then the operator $H(t)$ has a simple eigenvalue $\lambda_j(t)$ close to $\hat{\lambda}_j(t)$,

$$\lambda_j(t) = \hat{\lambda}_j(t) + O(\lambda^{1/2-\beta(N+2)}). \tag{5.26}$$

This eigenvalue is unique: there are no other eigenvalues in the interval ϵ, here $\epsilon = (\hat{\lambda}_j - \lambda^{-4\delta}, \hat{\lambda}_j + \lambda^{-4\delta})$ if $d = 2$ and $\epsilon = (\hat{\lambda}_j - \lambda^{-1/2-4\delta}, \hat{\lambda}_j + \lambda^{-1/2-4\delta})$ if $d = 3$.

3) An eigenfunction $\Psi_j(t,x)$ corresponding to $\lambda_j(t)$ is close to the model function $\hat{\Psi}_j(t,x)$:

$$\left\| \Psi_j(t,x) - \hat{\Psi}_j(t,x) \right\|_{L_\infty} = O(\lambda^{1-\beta(N+2)}). \tag{5.27}$$

The proof of the theorem will be published elsewhere.

It can be proven also that the asymptotic formulae (5.26), (5.27) are differentiable with respect to t in the $(\lambda^{-(d-1)/2-4\delta})$-neighborhood of $\hat{S}(\lambda, \delta)$. The isoenergetic surface of H is defined as a set of t for which $H(t)$ has an eigenvalue equal to a given λ. Considering that $\nabla\lambda_j(t) \approx \nabla\hat{\lambda}_j(t) \approx \vec{p}_j(t)$ in the $(\lambda^{-(d-1)/2-4\delta})$-neighborhood of $\hat{S}(\lambda, \delta)$, using (5.26) and applying Implicit Function Theorem we conclude that there is a single piece of the isoenergetic surface of $H(t)$ in the $(\lambda^{-\beta(N+2)})$-neighborhood of each component of $\hat{S}(\lambda, \delta)$. In other words, the surface $\hat{S}(\lambda, \delta)$ describes a part of the isoenergetic surface of $H(t)$ with a high accuracy. The validity of the Bethe-Sommerfeld conjecture obviously follows from the last observation, since existence of an isoenergetic surface for a given λ implies that λ belongs to the spectrum. Thus, the following corollary of Theorem 2 holds.

COROLLARY 3. *When $d = 2, 3$ the operator H has only a finite number of gaps in the spectrum.*

To prove Theorem 2, we introduce a system of model functions $\hat{\Psi}_j(t)$, $j \in Z^d$ $(d = 2, 3)$, $\hat{\Psi}_j(t)$ being defined by formula (5.4) for $j \in Z_0(k, \beta, \varepsilon)$. Functions $\hat{\Psi}_j(t)$, such that $\vec{p}_j(t)$ is from the annulus $J = \{\vec{p}_j(t) : k/4 < p_j(t) < 7k/4\}$, but $j \notin Z_0(k, \beta, \varepsilon)$, are described in **[K3]** for $d = 2$. They correspond to the case of strong diffraction. The model functions corresponding to strong diffraction look rather different from (5.4). They can be described in terms of eigenfunctions of a Hill operator in dimension one. However, they also are approximate solutions of equations for eigenfunctions of operator H. The whole system of functions $\hat{\Psi}_j(t)$, $\vec{p}_j(t) \in J$ (both weak and strong diffraction cases) play the main role in the proof of Theorem 2. They form almost orthogonal system which is complete in the high energy region in the following sense: any eigenfunction of $H(t)$ with an eigenvalue λ close enough to k^2, say $\lambda : |\lambda - k^2| < k^2/2$ can be represented as a linear combination of $\hat{\Psi}_j(t)$, $\vec{p}_j(t) \in J$ with a very high accuracy. We need model functions corresponding to strong diffraction to ensure completeness. The weak diffraction functions (5.4) alone are not sufficient for this due to the restriction $j \in Z_0(k, \beta, \varepsilon)$; the easiest way to see that they are not sufficient is just to assume $\vec{A} = 0$: then, $\hat{\Psi}_j(t)$ are just plane waves $|Q|^{-1/2}e^{i(\vec{p}_j(t),x)}$. Obviously, plane waves with $j \in Z_0(k, \beta, \varepsilon)$ do not exhaust all plane waves with eigenvalues $\lambda : |\lambda - k^2| < k^2/2$. The strong diffraction functions play important role in the proof of Theorem 2. Namely, to prove that $\lambda_j(t)$ is a unique eigenvalue in the interval ϵ, we need to keep track not only of the weak diffraction type eigenvalues, but also of the strong diffraction type eigenvalues, since we have to make sure that there are no eigenvalues other than $\lambda_j(t)$ in the interval ϵ. Uniqueness of the eigenvalue is crucial for proving (5.27). If $\vec{p}_j(t) \notin J$, then functions $\hat{\Psi}_j(t)$ are chosen to be a basis in the orthogonal complement of the functions $\hat{\Psi}_j(t), \vec{p}_j(t) \in J$ in $L_2(K)$. They do not play important role in considerations since they correspond to energies which are far away from $\lambda = k^2$. For technical reasons it is convenient to choose them as eigenfunctions of the operator $(I - P)H(I - P)$, P being the projection on the linear envelope of $\hat{\Psi}_j(t)$, $\vec{p}_j(t) \in J$. After constructing the system of model functions $\left\{\hat{\Psi}_j(t)\right\}_{j \in Z^d}$ we define a model operator $\hat{H}(t)$, $t \in K$, by the following formula:

$$\hat{H}(t)\hat{\Psi}_j(t) = \hat{\lambda}_j(t)\hat{\Psi}_j(t),$$

here $\hat{\lambda}_j(t)$, $j \in J$, are "eigenvalues" corresponding to model functions $\hat{\Psi}_j(t)$, and $\hat{\lambda}_j(t)$, $j \notin J$ are just eigenvalues of $(I - P)H(I - P)$. Next, we consider $H(t)$ as a perturbation of $\hat{H}(t)$. We prove that, if $t \in \hat{S}(\lambda, \delta)$, then the perturbation series for the resolvent $(H(t) - z)^{-1}$ converges for all z on a small circle surrounding the point $z = \lambda$. The series for the resolvent has an asymptotic character in the high energy region. The perturbation series for the resolvent immediately gives us asymptotic formulae (5.26), (5.27) for $\lambda_j(t)$ and $\Psi_j(t)$. It also gives us differentiability of these formula with respect to t, which leads to the conclusion that there is a piece of the isoenergetic surface of the operator $H(t)$ in the $(k^{-2\beta(N+2)})$-neighborhood of $\hat{S}(\lambda, \delta)$, the neighborhood being obviously of a very small size.

The method of model functions will have additional applications. For example, it allows to obtain the following asymptotic formula for the integrated density of states in dimension two, which is stronger that (4.3), obtained in [**Mo**]:

$$D(H, \lambda)_{\lambda \to \infty} = D(H_0, \lambda) + O(\lambda^{-1/4+\varepsilon})$$

for any positive ε. The proof will be published elsewhere.

References

[BeSo] Bethe, G.; Sommerfeld, A. *Elektronentheorie der Metalle*, Springer Verlag, Berlin-New York, 1967.

[BiSu1] Birman, M.Sh.; Suslina, T.A. *The two-dimensional periodic magnetic Hamiltonian is absolutely continuous.* Algebra i Analiz **9** (1997), 1, 32 –48; translation in St.Petersburg Math. J, **9**, (1998), 1, 21 –32.

[BiSu2] Birman, M.Sh.; Suslina, T.A. *Absolute continuity of the two-dimensional periodic magnetic Hamiltonian with discontinuous vector-valued potential.* Algebra i Analiz **10** (1998), 4, translation in St.Petersburg Math. J, **10**, (1999), 4, 1 – 26.

[BiSu3] Birman, M.Sh.; Suslina, T.A. *Periodic magnetic Hamiltonian with a variable metric. The problem of absolute continuity.* Algebra i Analiz **11** (1999), 2; translation in St.Petersburg Math. J, **11** (2000), 2.

[BrS] Brüning, J.; Sunada, T. *On the spectrum of periodic elliptic operators.* Nagoya Math. J. **126** (1992), pp 159 –171.

[DaTr] Dahlberg, B.E.J., Trubowitz, E. *A Remark on Two Dimensional Periodic Potentials.* Comment. Math. Helvetici, **57** (1982), pp. 130 –134.

[DN] Dubrovin, D.A.; Novikov, S.P. *Ground states in a periodc field. Magnetic Bloch functionsand vector bundles.* Soviet Math. Dokl. **22** (1980), 240 – 244.

[F] Friedlander, L. *On the spectrum of a class of second order periodic elliptic differential operators.* Comm. Math. Phys. **229** (2002), 1, 49–55.

[G] Gel'fand, I.M. *Expansion in Eigenfunctions of an Equation with Periodic Coefficients.* Dokl. Akad. Nauk SSSR, **73** (1950), pp. 1117-1120. (in Russian)

[HM] Helffer, B., Mohamed A. *Asymptotic of the density of states for the Schrödinger Operator with Periodic Electric Potential.* Duke Math. J. **92** (1998), 1, pp 1–60.

[HH1] Hempel, R.; Herbst, I.*Strong magnetic fields, Dirichlet boundaries and spectral gaps.* Comm. Math. Phys. **164** (1995), 237 – 259.

[HH2] Hempel, R.; Herbst, I.*Bands and gaps for periodic magnetic Hamiltonians.* Partial differential operators and mathematical physics. (Holzhau, 1994), 175–184, Oper. Theory Adv. Appl., 78, Birkhuser, Basel, 1995.

[I] Iwatsuka, A. *On Schrödinger operators with magnetic fields.* In Lecture Notes in Mathematics, **1450**, pp 157-172, Springer-Verlag, Berlin 1990.

[K1] Karpeshina Yu.E. *Perturbation Theory for the Schrödinger operator with a periodic potential.* Springer-Verlag, Lecture Notes in Mathematics, # 1663, 1997, 352 pp.

[K2] Karpeshina Yu.E. *On the Periodic Magnetic Schrödinger Operator in R^d. Eigenvalues and Model Functions.* Birkhuser, Operator Theory: Advances and Applications. **132** (2002), pp 219 – 231

[K3] Karpeshina Yu.E. *System of Basic Functions for the Two-Dimensional Periodic Magnetic Schrödinger Operator* preprint mp-arc 01-340, 2001, 51 pp.

[Ku] Kuchment, P. *Floquet theory for partial differential equations.* Birkhäuser, Basel, 1993.

[KuLe] Kuchment, P., Levendorskii, S. *On absolute continuity of spectra of periodic elliptic operators.* in Mathematical Results in Quantum Mechanics, J. Dittrich, P.Exner et al (Editors), Operator Theory. Advances and Applications, **108**, 1999, 291 –297, Bürkhäuser.

[LL] Landau, L.D., Livshitz, E.M. *Quantum mechanics,* Pergamon Press, London, 1958.

[Ma] Madelung, O., *Introduction to Solid State Theory.* Berlin, New-York, Springer-Verlag, 1978.

[Mo] Mohamed, A. *Asymptotic of the density of states for the Schrdinger operator with periodic electromagnetic potential.* J. Math. Phys. 38 (1997), no. 8, 4023–4051.

[PaSo1] Parnovski, L., Sobolev, A. V. *Lattice points, perturbation theory and the periodic polyharmonic operator.* Ann. Henri Poincare **2** (2001), 3, 573–581.

[PaSo2] Parnovski, L., Sobolev, A. V. *On the Bethe-Sommerfeld conjecture for the polyharmonic operator.* Duke Math. J., **107**, 2, 209-238.

[RS] Reed, M., Simon, B. *Methods of Modern Mathematical Physics.*, Vol IV, Academic Press, 3rd ed., New York – San Francisco – London (1987).

[Sj] Sjöstrand, J. *Microlocal analysis for periodic magnetic Schrödinger equations and related questions.* in Microlocal Analysis and Applications, Lecture Notes in Physics, **1495**, pp 237 –332, Springer – Verlag, Berlin, 1991.

[Sk1] Skriganov, M.M. *Proof of the Bethe-Sommerfeld Conjecture in Dimension Two.* DAN SSSR, **248** (1979), 1, pp. 49-52; English transl. in Soviet Math. Dokl., **20** (1979), 5, 956 – 959.

[Sk2] Skriganov, M.M. *Geometric and Arithmetic Methods in the Spectral Theory of Multidimensional Periodic Operators.* Trudy Mat. Inst. Steklov., **171**, pp. 1 – 121. (1985); Engl. transl.: Proc. Steklov Inst. Math., **171**, 2 (1987).

[Sk3] Skriganov, M.M. *The Spectrum Band Structure of the Three-Dimensional Schrödinger Operator with a Periodic Potential.* Invent. Math., **80** (1985) pp. 107 –121.

[So] Sobolev, A. *Absolute continuity of the periodic magnetic Schrödinger operator.* Inventiones Mathematicae **137** (1999), 1, 85 – 112.

[Th] Thomas, L.E. *Time-dependent approach to scattering from impuriries in a crystal.* Commun. Math. Phys. **33** (1973), pp 335 – 343.

[V1] Veliev, O.A. *Asymptotic Formulae for Eigenvalues of a Periodic Schrödinger Operator and Bethe-Sommerfeld Conjecture.* Functional. Anal. i Prilozhen., **21** (1987), no. 2, 1–15; Engl. transl.: Functional Anal. Appl., **21** (1987), pp. 87 –99.

[V2] Veliev O.A.*The Periodic Multidimensional Schrodinger Operator, Part 2, Asymptotic Formulae for Bloch Functions and Fermi Surfaces.* preprint, mp-arc 01-463, 32 pp.

[V3] Veliev, O.A. *Asymptotic Formulae for Bloch Functions of Multidimensional Periodic Schrödinger operator and Some of Their Applications.* Spectral Theory of Operators and its Applications, **9**, Baku, 1989, pp. 59 –76 (in Russian).

DEPARTMENT OF MATHEMATICS, UNIVERSITY OF ALABAMA AT BIRMINGHAM, BIRMINGHAM, AL, 35294

E-mail address: `karpeshi@math.uab.edu`

Contemporary Mathematics
Volume **339**, 2003

On guided waves in photonic crystal waveguides

Peter Kuchment and BengSeong Ong

ABSTRACT. The paper addresses the issue of existence of modes guided by linear defects in photonic crystals. Such modes can be created in spectral gaps of the bulk materials and are evanescent in the bulk.

1. Introduction

A photonic crystal (or a PBG material, where PBG stands for "photonic band-gap") is a periodic dielectric medium in which the frequency spectrum of electromagnetic waves has a gap. In other words, monochromatic electromagnetic waves of certain frequencies do not exist in this medium. Due to their high promise for applications, such materials have been intensively studied since the idea was suggested in 1987 [**23, 12**] (see, e.g., the recent books [**11, 13, 19**] and the survey of mathematical problems of this area [**16**]). Engineering photonic crystalline waveguides is one of the important suggested applications of PBG materials. The idea is that introducing a linear defect (a "waveguide") into an otherwise periodic PBG material, one can create waves through it at frequencies prohibited in the bulk. Numerical and experimental studies have shown that such superior quality guides can be efficiently created (e.g., [**11, 13, 17, 19**]). There also exist natural acoustic analogs of PBG materials and corresponding waveguides.

Analytic studies of such waveguides, however, are essentially absent, even in the case of a straight waveguide. The first questions one can ask are existence of modes at the frequencies in the band gap of the bulk material, their essential confinement to the guide (i.e., their exponential decay into the bulk), and structure of the arising spectrum in the gap (e.g., absence of

---

1991 *Mathematics Subject Classification.* Primary 35P99, 35Q60; Secondary 35Q72, 78A48.

Key words and phrases. photonic crystal, waveguide, spectrum.

Work of both authors was partially supported by the NSF Grants DMS 0296150 and 9971674. Work of the first author was also supported by the NSF Grant DMS 0072248.

bound states, absolute continuity of the spectrum). In this article we address some of these basic issues and derive simple initial results for the case of a straight linear defect. These results in fact do not use the periodicity of the bulk PBG material. Although one can obtain analogous statements for the case of full Maxwell equation, in this paper we only address the scalar model. This covers in particular the cases of TE and TM polarized electromagnetic waves in $2D$ photonic waveguides and waves in $2D$ and $3D$ acoustic waveguides.

We now describe the mathematical model studied in this text. Let $\varepsilon_0(x)$ and $\rho_0(x)$ be bounded positive measurable functions in $\mathbb{R}^d$ separated from zero, i.e. $0 < c_0 \leq \varepsilon_0(x), \rho_0(x) \leq c_1 < \infty$. It is usually assumed in the photonic crystal theory that both functions are periodic with respect to a lattice $\Gamma \subset \mathbb{R}^d$, but this is not required for the basic results we obtain in this text.

One can think that the whole space $\mathbb{R}^d$ is filled with a dielectric or acoustic material with properties described by the functions ε_0 and ρ_0 (the physical interpretation of these functions depends on whether one deals with electromagnetic or acoustic case, but this will be of no importance for our study). In the case of periodic functions this models a photonic or acoustic crystal.

The operator A_0 is the self-adjoint realization of

$$-\frac{1}{\rho_0(x)}\nabla\cdot\frac{1}{\varepsilon_0(x)}\nabla$$

in $L_2(\mathbb{R}^d, \rho_0(x)\,dx)$ defined by means of its quadratic form

$$\int \varepsilon_0^{-1}|\nabla u|^2 dx \tag{1}$$

with the domain $H^1(\mathbb{R}^d)$.

We will also consider a "defect" strip

$$S_l = \{x = (x_1, x') \in \mathbb{R}^d \,|\, x \in \mathbb{R},\, x' \in l\Omega\},$$

where Ω is the unit ball centered at the origin in $\mathbb{R}^{d-1}$ (the ball can be easily replaced by other bounded domains) and $l\Omega$ is the ball of radius $l > 0$. We can now introduce the perturbed medium, for which

$$\varepsilon(x) = \left\{ \begin{array}{ll} \varepsilon > 0 & \text{for } x \in S_l \\ \varepsilon_0(x) & \text{for } x \notin S_l \end{array} \right., \quad \rho(x) = \left\{ \begin{array}{ll} \rho > 0 & \text{for } x \in S_l \\ \rho_0(x) & \text{for } x \notin S_l \end{array} \right..$$

Physically, a linear homogeneous defect S_l is introduced into the original medium.

We define the perturbed operator A that corresponds to the modified medium analogously to the background operator A_0. This operator is self-adjoint in the weighted L_2-space $L_2(\mathbb{R}^d, \rho(x)\,dx)$, which will from now on be denoted $L_{2,\rho}$.

Our goal is to show that for any gap (α, β) in the spectrum $\sigma(A_0)$ of the unperturbed medium and under appropriate conditions on the parameters l, ρ, and ε of the line defect, spectrum of the perturbed medium arises in the gap. One naturally wants to interpret this as existence of guided waves in the "photonic waveguide" S_l, while this would require proving some additional properties. These are first of all confinement of the wave to the guide (i.e., evanescence into the bulk), which is also proven below, and non-existence of bound states (i.e., the fact that the wave is actually propagating along the waveguide), which we have not been able to establish yet.

Our results complement the discussions of the preprint [**1**] by H. Ammari and F. Santosa, where spectral properties of the TM mode in a 2D PBG waveguides were studied in a situation of a linear defect aligned with a periodicity axis of an otherwise periodic medium (and hence Floquet theory [**15, 18**] was applicable). In particular, exponential confinement of the guided modes for this particular case was proven there. Some of the constructions presented here are similar to the ones used in [**6**]–[**8**] for localized defects.

2. Formulation of the results

Our main results are the following theorems.

THEOREM 1. *Let $G = (\alpha, \beta)$ be a non-empty finite gap in the spectrum of the "background medium" operator A_0 (in particular, $\alpha > 0$). Assume that for some $\delta \in (0, \frac{\beta-\alpha}{2})$ the following inequality is satisfied:*

$$l^4 \delta^2 \rho^2 \varepsilon^2 > \nu_{d-1}, \tag{2}$$

where $\nu_{d-1} > 0$ is the lowest eigenvalue of the bi-harmonic operator Δ^2 in Ω with Dirichlet boundary conditions.

Then any interval of length 2δ in the gap G contains at least one point of the spectrum $\sigma(A)$ of the perturbed operator.

This theorem guarantees that when (2) is satisfied, defect modes in the spectral gaps of the background medium exist, and the corresponding spectrum forms a δ-net in the gap. One is tempted to associate these modes with the guided waves. In order to do so one needs first to establish their confinement to the waveguide (i.e., their evanescent nature in the bulk of the material) and secondly, that they do not correspond to discrete spectrum (bound states). The first of these tasks is achieved in the next simple result, while the second is much harder and still awaits its solution. Before formulating the theorem, we remind the reader about existence of the so called generalized eigenfunction expansions. In the case of the operators we consider, for almost all points $z \in \mathbb{R}$ with respect to the spectral measure, such a generalized eigenfunction $u \in H^1_{loc}$ of the operator A has the properties $(1+|x|)^{-N}u(x) \in L_2(\mathbb{R}^d)$ and $(1+|x|)^{-N}\nabla u(x) \in L_2(\mathbb{R}^d)$ for some $N > 0$. The eigenvalue problem $Au = zu$ is satisfied in the distributional sense. This is a well known fact for elliptic operators with smooth coefficients [**3**], while for operators of the type we study one can find the corresponding results in [**14**]. We will use the polynomial boundedness condition in the following form:

$$\|u\|_{L_2(K+x)} + \|\nabla u\|_{L_2(K+x)} \leq C_K(1+|x|)^N \tag{3}$$

for any compact K and $x \in \mathbb{R}^d$. We summarize (3) as polynomial growth of order N.

THEOREM 2. *Let u be a polynomially bounded generalized eigenfunction of A that corresponds to a value z in a gap G of $\sigma(A_0)$, then it decays exponentially away from the defect strip S_l.*

The exact meaning of the exponential decay will be provided in the reformulation of this theorem given in Theorem 3 of the next section.

3. Proofs of the results

We adopt the following notations: the norm and inner product in $L_2(\mathbb{R}^d)$ will be denoted by $\|\cdot\|$ and $\langle\cdot,\cdot\rangle$ respectively, while those in the weighted space $L_2(\mathbb{R}^d;\rho(x)dx)$ will be denoted with the subscript ρ: $\|\cdot\|_\rho$, $\langle\cdot,\cdot\rangle_\rho$. Notice that the norms $\|\cdot\|$, $\|\cdot\|_\rho$, and $\|\cdot\|_{\rho_0}$ are equivalent.

3.1. Proof of Theorem 1. Let $(\mu-\delta,\mu+\delta)$ be a sub-interval of the gap G. The idea of the proof is that under the conditions of the theorem one is able to provide an approximate eigenfunction $u(x)$ for the operator A, such that $\|u\|_\rho = 1$ and

$$\|Au-\mu u\|^2_\rho < \delta^2. \tag{4}$$

Finding such an approximate eigenfunction would imply immediately the statement of the theorem. For functions u supported in the defect strip, where $\rho(x) = \rho$, this is equivalent to finding a function u such that $\|u\| = 1$ and

$$\|Au-\mu u\|^2 < \delta^2. \tag{5}$$

So, let us construct such a function.

Let a real valued function $\phi(x') \in C_0^\infty(\Omega)$ have unit L_2-norm and $\phi_l(x') = l^{(1-d)/2}\phi(x'/l)$. We will also need real valued functions $\psi(x_1) \in C_0^\infty(\mathbb{R})$ with a unit L_2-norm and $\psi_n(x_1) = n^{-1/2}\psi(x_1/n)$ for $n > 0$. It is clear that the functions ϕ_l, ψ_n still have unit norms in the corresponding spaces. Introducing $k = \sqrt{\mu\rho\varepsilon}$, we consider the function

$$u_{l,n}(x) = \phi_l(x')\psi_n(x_1)e^{ikx_1}, \tag{6}$$

which will be our try for an approximate eigenfunction.

Instead of estimating the left hand side of (5), we will estimate $\|\varepsilon\rho\,(Au - \mu u)\|^2$. Taking into account that the function u is supported inside the defect, the needed inequality (5) can be also rewritten as

$$\|\Delta u + \mu\varepsilon\rho u\|^2 < \delta^2\rho^2\varepsilon^2. \tag{7}$$

Let us calculate directly the left hand side of (7) for the function u introduced above, understanding all norms as the norms in L_2:

$$\begin{aligned} \|\Delta u + \mu\varepsilon\rho u\|^2 &= \|(\Delta_{x'}\phi_l)\psi_n + \phi_l\psi_n'' + 2ik\phi_l\psi_n'\|^2 \\ &= \|(\Delta_{x'}\phi_l)\psi_n + \phi_l\psi_n''\|^2 + 4\mu\rho\varepsilon\,\|\phi_l\psi_n'\|^2 . \end{aligned} \tag{8}$$

We used the definition of k and the condition that functions ϕ and ψ are real valued.

Let $\gamma > 0$. Using the inequality

$$(a+b)^2 \le (1+\gamma)\,a^2 + \left(1 + \frac{1}{\gamma}\right) b^2$$

and normalization of the functions, we can estimate the last expression from above to get the following upper bound for the expression in question:

$$\begin{aligned} \|\Delta u + \mu\varepsilon\rho u\|^2 \le \tfrac{1+\gamma}{l^{d+3}}\,\|\Delta\phi\left(\tfrac{x'}{l}\right)\|^2_{L_2(l\Omega)} \\ + \tfrac{1+\frac{1}{\gamma}}{n^5}\,\|\psi''\left(\tfrac{x_1}{n}\right)\|^2_{L_2(\mathbb{R})} + \tfrac{4\mu\rho\varepsilon}{n^3}\,\|\psi'\left(\tfrac{x_1}{n}\right)\|^2_{L_2(\mathbb{R})} . \end{aligned} \tag{9}$$

By changing variables to $(x_1/n, x'/l)$, this expression reduces to

$$\frac{1+\gamma}{l^4}\,\|\Delta\phi(x')\|^2_{L_2(\Omega)} + \frac{1+\frac{1}{\gamma}}{n^4}\,\|\psi''(x_1)\|^2_{L_2(\mathbb{R})} + \frac{4\mu\rho\varepsilon}{n^2}\,\|\psi'(x_1)\|^2_{L_2(\mathbb{R})} . \tag{10}$$

Since n can be chosen arbitrarily large (without changing the defect strip), the last two terms can be made arbitrarily small (uniformly with respect to k on any finite interval). Hence, one needs to control only the first term by an appropriate choice of a test function ϕ. In other words, one is interested in

$$\nu_{d-1} = \inf \|\Delta\phi(x')\|^2_{L_2(\Omega)},$$

where the *infimum* is taken over real functions in $C_0^\infty(\Omega)$ of unit L_2-norm. This is then the lowest eigenvalue of the bi-harmonic operator Δ^2 with Dirichlet boundary conditions in Ω. In particular, $\nu_{d-1} > 0$. Now, due to arbitrariness of $\gamma > 0$ our condition boils down to

$$\frac{\nu_{d-1}}{l^4} < \delta^2\rho^2\varepsilon^2 \tag{11}$$

or

$$l^4\delta^2\rho^2\varepsilon^2 > \nu_{d-1}, \tag{12}$$

which proves the statement of the theorem.

3.2. Proof of Theorem 2. Let G be a spectral gap of A_0 and $z \in \sigma(A) \cap G$. Let also $u(y)$ be a polynomially growing generalized eigenfunction for the operator A that corresponds to z (in the meaning explained in the introduction). Let $x = (x_1, x') \in \mathbb{R}^d$ and $\chi_x(y)$ be the characteristic function of the cube $\{y \,|\, |y_j - x_j| \leq 1\}$ centered at x.

We now give a more precise formulation of Theorem 2:

THEOREM 3. *There exist positive constants C_1 and $C(z)$ such that*

$$\|\chi_x u\| \leq C_1 \left(1 + |x_1|\right)^N e^{-C(z)dist(x,S_l)}, \tag{13}$$

where N is the order of polynomial growth of u.

REMARK 4. *One might be concerned with the fact that albeit the eigenfunction decays exponentially away from the defect strip, the factor in front of the expression grows polynomially along the strip. However, for a generalized eigenfunction that grows polynomially one cannot expect anything better. In the periodic situation, using Floquet-Bloch theory, one can guarantee absence of this growth (see the comments at the end of the paper).*

Proof. Define the sesqui-linear form

$$Q[\varphi, w] := < \nabla\varphi, \frac{1}{\varepsilon_0}\nabla w > -z < \varphi, w >_{\rho_0}$$

with the domain $H^1(\mathbb{R}^d)$.

Let $R(z) = (A_0 - z)^{-1}$ and $\varphi := R(z)\chi_x u$. We use here that z is not in the spectrum of A_0. Note that $\varphi \in D(A_0)$.

Let $p = \max(2\text{dist}(x, S_l), 1)$ and $\xi_x(y)$ be a nonnegative smooth cutoff function that depends on y_1 only, is supported in $(x_1 - (p+1), x_1 + (p+1))$ and such that it is equal to 1 on $[x_1 - p, x_1 + p]$. We assume further that

$\xi_x(y) \leq 1$ and $|\nabla\xi_x(y)| = |\xi_x'(y_1)| \leq C$ for some constant C and all $x, y \in \mathbb{R}^d$. For simplicity of notation, we drop the subscript x in $\xi = \xi_x$. Note that $\xi u \in H^1(\mathbb{R}^d)$. Using $w = \xi u$, one gets

$$Q[\varphi, \xi u] = < A_0\varphi, \xi u >_{\rho_0} - < z\varphi, \xi u >_{\rho_0} = < \chi_x u, \xi u >_{\rho_0} = \|\chi_x u\|^2_{\rho_0} \,.$$

This means that our goal should be to estimate $Q[\varphi, \xi u]$ from above. On the other hand, using the equality $Au = zu$, one gets

$$Q[\varphi, \xi u] = \langle \nabla\varphi, \tfrac{1}{\varepsilon_0}\nabla(\xi u)\rangle - \langle \varphi, \xi z u\rangle_{\rho_0}$$
$$= \langle \nabla\varphi, \tfrac{1}{\varepsilon_0}\nabla(\xi u)\rangle - \langle \varphi, \xi A u\rangle_{\rho_0} + \langle \varphi, \tfrac{\rho}{\rho_0}\xi A u\rangle_{\rho_0} - \langle \varphi, \tfrac{\rho}{\rho_0}\xi A u\rangle_{\rho_0}$$

Simple algebraic transformations and easily justifiable integration by parts allow one to rewrite the last sum as

$$\langle \nabla\varphi, \xi\tilde{\varepsilon}\nabla u\rangle + \langle \varphi, \xi\rho\tilde{\rho}Au\rangle_{\rho_0} + \langle \nabla\xi, \frac{u}{\varepsilon_0}\nabla\varphi - \frac{\varphi}{\varepsilon}\nabla u\rangle. \tag{14}$$

In these calculations we used the notations

$$\tilde{\varepsilon}(x) = \frac{1}{\varepsilon_0(x)} - \frac{1}{\varepsilon(x)},\ \tilde{\rho}(x) = \frac{1}{\rho_0(x)} - \frac{1}{\rho(x)}.$$

Notice that both these functions are supported inside the strip S_l.

Our last task in proving the theorem is to estimate from above the terms in (14). In order to do so, we need an auxiliary statement concerning the exponential decay of the resolvent, which is a result of [**2, 6**]:

LEMMA 5. [**2, 6**] *There exist a positive number m_z that depends only on the distance of the point z from the gap edges, such that for a positive constant C the following estimates hold for the local $L_2(\mathbb{R}^d)$-norm of the resolvent $R(z)$:*

$$\begin{aligned} \|\chi_u R(z)\chi_v\| &\leq Ce^{-m_z|u-v|} \\ \|\chi_u \nabla R(z)\chi_v\| &\leq Ce^{-m_z|u-v|} \end{aligned} \tag{15}$$

for any $u, v \in \mathbb{R}^d$. Here the norms in the left hand side are the operator norms in $L_2(\mathbb{R}^d)$.

We can now get the needed estimates. Let $V = [x_1 - p - 1, x_1 + p + 1] \times l\Omega$. This is a compact domain that can be covered by the union of p fixed size domains $V_j = [a_j, a_j + 2] \times l\Omega$ and which contains the supports of $(\xi\tilde{\varepsilon})$ and $(\xi\tilde{\rho})$. Also note that $\mathrm{dist}(x, V_j) \geq \mathrm{dist}(x, S_l)$. Now using the lemma above and (3) we get

(16)
$$\begin{aligned}|\langle \nabla\varphi, \xi\tilde{\varepsilon}\nabla u\rangle| &\leq \|\chi_V \nabla\varphi\| \, \|\xi\tilde{\varepsilon}\nabla u\| \leq C \, \|\sum_j \chi_{V_j} \nabla R(z)\chi_x u\| \, \|\sum_j \chi_{V_j} \nabla u\| \\ &\leq Cp^2(|x_1| + p + 1)^{2N} e^{-m_z \mathrm{dist}(x,\mathcal{S}_l)} \leq C(|x_1| + 1)^{2N} e^{-(m_z - \eta)\mathrm{dist}(x,\mathcal{S}_l)}\end{aligned}$$

We used here that $p = \max(2\mathrm{dist}(x,\mathcal{S}_l), 1)$. We also denoted by C different constants.

Analogously,

(17)
$$|\langle \varphi, \xi\rho\tilde{\rho}Au\rangle_{\rho_0}| \leq C|z| \sum_j \|\chi_{V_j}\varphi\| \, \|\xi\tilde{\rho}u\| \leq C(1 + |x_1|)^{2N} e^{-(m_z - \eta)\mathrm{dist}(x,\mathcal{S}_l)}$$

Let us move now to estimating the last term in (14). Denote by $a > 0$ a number such that shifts of $l\Omega$ by vectors aj with $j \in \mathbb{Z}^{d-1}$ cover the whole space $\mathbb{R}^{d-1}$. We denote

$$W_j := ([x_1 - p - 1, x_1 - p] \cup [x_1 + p, x_1 + p + 1]) \times (l\Omega + aj).$$

Then $W_j = W_0 + (0, aj)$. Notice that $W = \cup_j W_j$ covers $\mathrm{supp}\, \nabla\xi$ and $\mathrm{dist}(x, W_j) \geq C_1(p + |j|) - C_2$.

We are now ready to estimate the last term of (14) from above. We proceed as before, using the lemma, the polynomial growth of u, and uniform boundedness of $\nabla\xi$.

$$\begin{aligned}|\langle \nabla\xi, \tfrac{u}{\varepsilon_0}\nabla\varphi\rangle| &\leq C\sum_j \|\chi_{W_j} u\| \, \|\chi_{W_j} \nabla R(z)\chi_x u\| \\ &\leq C\sum_j (|x_1| + p + |j| + 1)^{2N} e^{-m_z \mathrm{dist}(x,W_j)} \\ &\leq C(|x_1| + p + 1)^{2N} e^{-m_z^1 \mathrm{dist}(x,\mathcal{S}_l)} \sum_j (1 + |j|)^{2N} e^{-m_z^2 |j|} \\ &\leq C(|x_1| + 1)^{2N} e^{-(m_z^1 - \eta)\mathrm{dist}(x,\mathcal{S}_l)}\end{aligned} \tag{18}$$

where m_z^1 and m_z^2 are positive constants.

The expression $|\langle \nabla\xi, \frac{\varphi}{\varepsilon}\nabla u\rangle|$ is estimated analogously. Combining the above estimates, we get

$$\|\chi_x u\|^2 \leq C \, \|\chi_x u\|^2_{\rho_0} = Q[\varphi, \xi u] \leq C_\eta (1 + |x_1|)^{2N} e^{-m_z^1 \mathrm{dist}(x,\mathcal{S}_l)}.$$

This finishes the proof of the theorem.

4. Remarks

In this section we present some comments concerning the results of this paper and possible further developments.

(1) The sufficient conditions for existence of "guided" modes provided in Theorem 1 are certainly not necessary. It would be interesting to obtain similar results under weaker conditions on the guide.

The number ν_{d-1} that enters the conditions can be estimated numerically. For instance, when $d = 2$, one can easily write a secular equation which gives an approximate value of ν_1 of about 31.29.

(2) Theorem 1 provides existence of a δ-net of the defect spectrum inside the gap. One wonders how much of the gap the spectrum occupies. When does it fill the whole gap? Can it have gaps on its own? There seem to be no rigorous results available concerning these questions.

(3) Results of [**2**] on improved Combes-Thomas resolvent estimates show that the exponential decay constant m_z, which clearly depends on the distance of the point z from the spectrum, behaves as $\sqrt{(z-\alpha)(\beta-z)}$.

(4) One is interested in eliminating the polynomially growing factor in the exponential decay estimate (13). Its appearance is due to polynomial growth of generalized eigenfunctions, which in general might not be possible to get rid of.

In the case when the background medium (i.e., the functions $\rho_0(x)$ and $\varepsilon_0(x)$) is periodic with respect to a lattice Γ in $\mathbb{R}^d$ and when the linear defect is directed along one of the lattice vectors, one can apply Floquet-Bloch theory [**15, 18**] that shows that one has generalized (Bloch) eigenfunctions with $N = 0$. Then one obtains exponential decay estimates that do not change along the waveguide.

(5) The results of the paper have their analogs for the full Maxwell system. The authors intend to present those elsewhere.

(6) As it has already being mentioned, in order to have the full right to call the discovered modes "guided", one needs to show that they do not correspond to point spectrum (i.e., to bound states). Here the most treatable case should be of a periodic medium with a linear defect aligned along one of the lattice vectors. In this situation one can apply the Floquet-Bloch theory with respect to the axial variable of the waveguide and hope to use standard techniques applied in the case of Schrödinger operators with periodic potential (e.g., [**4, 9, 15, 18**]). This happens to be not an easy task. Even in the case of "hard wall" periodic waveguides, when waves are contained in a periodic waveguide by Dirichlet, Neumann, or

more general boundary conditions, this problem is non-trivial. Although it has been considered for rather long time [**5, 15**], the first real advances are very recent [**10, 20, 21, 22**]. The case of photonic crystal waveguides is more complex, due to absence of complete confinement of the waves, which decay exponentially into the bulk, but do not vanish completely.

(7) The next issue is engineering bent waveguides in such a way that there is no significant reflection back at bends. Physics analysis and experiments (e.g., [**13, 11, 17, 19**]) show that this is possible. To the best of our knowledge, no rigorous analysis of this problem is available yet.

Acknowledgments

The authors wish to express their gratitude to H. Ammari, M. Birman, D. Dobson, P. Exner, L. Friedlander, F. Santosa, R. Shterenberg, A. Sobolev, and T. Suslina for discussions and information.

This research was partly sponsored by the NSF through the Grants DMS 0296150, 9971674, and 0072248. The authors thank the NSF for this support. The content of this paper does not necessarily reflect the position or the policy of the federal government, and no official endorsement should be inferred.

References

[1] H. Ammari and F. Santosa, Guided waves in a photonic bandgap structure with a line defect, preprint 2002.

[2] J. M. Barbaroux, J. M. Combes, and P. D. Hislop, Localization near band edges for random Schrödinger operators, Helv. Phys. Acta 70(1997), 16-43.

[3] Ju. M. Berezanskii, Expansions in Eigenfunctions of Selfadjoint Operators, AMS, Providence, RI 1968.

[4] M. Sh. Birman and T. A. Suslina, Periodic magnetic Hamiltonian with a variable metric. The problem of absolute continuity, Algebra i Analiz 11(1999), no.2. English translation in St. Petersburg Math J. 11(2000), no.2, 203-232.

[5] V. I. Derguzov, on the discreteness of the spectrum of the periodic boundary-value problem associated with the study of periodic wave guides, Siberian Math. J. 21(1980), 664-672.

[6] A. Figotin and A. Klein, Localization of classical waves I: Acoustic waves, Commun. Math. Phys. 180(1996), 439-482.

[7] A. Figotin and A. Klein, Localization of classical waves II: Electromagnetic waves, Commun. Math. Phys. 184(1997), no.2, 411-441.

[8] A. Figotin and A. Klein, Localized Classical Waves Created by Defects, J. Stat. Phys. 86(1997), no. 1-2, 165-177.

[9] L. Friedlander, On the spectrum of a class of second order periodic elliptic differential operators, Commun. Math. Phys. 229(2002), 49–55.

[10] L. Friedlander, Absolute continuity of the spectra of periodic waveguides, in this volume.

[11] J. D. Joannopoulos, R. D. Meade, and J. N. Winn, Photonic Crystals, Molding the Flow of Light, Princeton Univ. Press, Princeton, N.J. 1995.

[12] S. John, Strong localization of photons in certain disordered dielectric superlattices, Phys. Rev. Lett. 58: 2486, (1987).

[13] S. G. Johnson and J. D. Joannopoulos, *Photonic Crystals, The road from Theory to Practice*, Kluwer Acad. Publ., Boston 2002.

[14] A. Klein, A. Koines, and M. Seifert, Generalized eigenfunctions for waves in homogeneous media, to appear in J. Funct. Anal.

[15] P. Kuchment, Floquet Theory for Partial Differential Equations, Birkhäuser Verlag, Basel 1993.

[16] P. Kuchment, "The Mathematics of Photonics Crystals," Ch. 7 in Mathematical Modeling in Optical Science, Bao, G., Cowsar, L. and Masters, W.(Editors), SIAM, Philadelphia 2001.

[17] J. T. Londergan, J. P. Carini, and D. P. Murdock, Binding and Scattering in Two-Dimensional Systems, Springer Verlag, Berlin 1999.

[18] M. Reed and B. Simon, Methods of Modern Mathematical Physics, IV: Analysis of Operators, Academic Press, New York: 1978.

[19] K. Sakoda, Optical Properties of Photonic Crystals, Springer Verlag, Berlin 2001.

[20] A. V. Sobolev, J. Walthoe, Absolute continuity in periodic waveguides, Proc. London Math. Soc. (3)85(2002), 717–741.

[21] R. G. Shterenberg, Schrödinger operator in a periodic waveguide on the plane and quasi-conformal mappings, Zap. Nauchn. Sem. POMI, p. 292 (2003) (in Russian).

[22] R. G. Shterenberg and T. A. Suslina, Absolute continuity of the spectrum of a magnetic Schrödinger operator in a two-dimensional periodic waveguide, Algebra i Analiz 14(2002), 159–206; English transl. in St. Petersburg Math. J. 14(2003).

[23] E. Yablonovitch, Inhibited Spontaneous Emission in Solid-State Physics and Electronics, Phys. Rev. Lett. 58(1987), 2059–2062.

DEPARTMENT OF MATHEMATICS, TEXAS A&M UNIVERSITY, COLLEGE STATION, TX 77843-3368

E-mail address: kuchment@math.tamu.edu, bsong@math.tamu.edu

Contemporary Mathematics
Volume **339**, 2003

The Design and Photofabrication of Random Achromatic Optical Diffusers for Uniform Illumination

Tamara A. Leskova, Alexei A. Maradudin, Eugenio R. Méndez, and Javier Muñoz-Lopez

Abstract. Methods are described for designing one- and two-dimensional random surfaces that scatter electromagnetic waves with a specified angular dependence of the intensity of the scattered field. These methods are illustrated by using them to design surfaces that scatter light with a uniform intensity within a specified region of scattering angles and produce no scattering outside this region. They are validated by computer simulations, and methods of fabricating such surfaces on photoresist are described.

1. Introduction

In many practical situations it is desirable to have optical diffusers with specific light scattering properties. For example, a nonabsorbing diffuser that scatters light uniformly within a specified range of scattering angles, and produces no scattering outside this range is a useful optical element. We will call it a *band-limited uniform diffuser*. Such an element could have applications, for example, in projection systems, where one wishes to illuminate a screen with uniform intensity but not to waste light by illuminating outside the boundaries of the screen. Band-limited uniform diffusers can also be useful in microscope illumination systems, in the fabrication of displays and projection screens, and in Fourier transform holography.

The design of band-limited uniform diffusers has been considered by several authors. Kurtz [**1**] proposed a method for producing quasi-band-limited uniform diffusers based on a one-dimensional two-level random height distribution on a surface. An improvement to this approach was reported by Kurtz *et al.* [**2**], who employed a method based on the fabrication of deterministic microelement arrays with an appropriate slope distribution. A few years later Nakayama and Kato [**3**] considered the case of one-dimensional pseudo random diffusers, and Kowalczyk

1991 *Mathematics Subject Classification.* 60G50,82C4,74J20,78A46,81U40.

Key words and phrases. random surfaces, scattering, band-limited uniform diffusers, Kirchhoff approximation, inverse problems.

The research of T.A.L. and A.A.M. was supported in part by Army Research Office Grant 19-02-1-0256. The research of E.R.M. was supported in part by CONACYT (México) through grant 32254-E. The research of J.M.-L. was supported by a scholarship from CONACYT (México).

[**4**] reported some theoretical work on the more general case of a two-dimensional random surface.

More recently, diffractive optical elements that scatter light uniformly over specified angular regions have become commercially available [**5**]. These elements, however, are not random and possess the desired characteristics over only a relatively narrow range of wavelengths. In other words, they are not achromatic.

Another kind of diffuser, whose design is based on a randomized microlenslet concept, is also produced commercially [**6**]. Although these holographic light-shaping diffusers are achromatic, and possess characteristics that approximate the desired ones, the scattering distribution they produce is not uniform, and they do not have a well-defined maximum angle of scattering.

Despite the interest in this subject, until recently there were no clear procedures reported in the literature for designing and fabricating random surfaces that behave as band-limited uniform diffusers, and it was unclear what kind of surface statistics were required for the production of such an optical element. In several recent articles the present authors and their colleagues have addressed the question of how to design random surfaces that scatter [**7, 8, 9, 10, 11, 12**] or transmit [**13, 14**] light in a specified manner. The approach used in this work was based on expressing the surface profile function of the random surface as a superposition of equally spaced identical trapezoidal grooves whose amplitudes were assumed to be independent identically distributed random deviates drawn from a probability density function determined in such a way that the intensity of the light scattered from, or transmitted through, the resulting surface had the specified angular distribution. It has been shown that such surfaces can be fabricated on photoresist [**8, 12**] and produce the specified angular distribution of the intensity of the scattered light [**12, 15**].

In this paper we present methods for designing one- and two-dimensional random surfaces that scatter light in a prescribed manner, that are simpler to implement than the one employed in Refs. [**7, 8, 9, 10, 11, 12, 13, 14**], both theoretically and experimentally. Like the latter method, the new methods are based on the geometrical optics limit of the Kirchhoff approximation, a single-scattering approximation, for the scattering of scalar plane waves from impenetrable surfaces. However, the results obtained by these methods have a significantly wider range of applicability. Although the general results obtained in this paper can be used to design and fabricate one- and two-dimensional random surfaces that scatter light with more general angular distributions of the intensity of the scattered light, due to length restrictions we will illustrate them by applying them only to the design of surfaces that behave as band-limited uniform diffusers.

We have chosen to work with random surfaces in designing optical diffusrs that scatter light in a prescribed fashion because, as we will see, the use of such surfaces leads to a precise algorithm for designing them, something that we have not been able to find in dealing with deterministic surfaces. The fact that an experimentalist fabricates only a single realization of such a random surface from an ensemble of an infinite number of possible realizations of that surface, and hence produces a deterministic surface, is unimportant if the surface fabricated is large enough and the incident field illuminates a large enough area of that surface. In this case the mean scattered intensity can be regarded as a spatial average of the actual scattered intensity over an area of the surface that contains many realizations of the surface roughness. The assumption that this spatial average is equivalent to the ensemble average we use in the design of our random surfaces is difficult to prove rigorously because these surfaces are not stationary, but is lent support by the results of experiments caried out on them.

The outline of this paper is as follows. In Section 2 we present a method for designing one-dimensional random surfaces that scatter light in a specified fashion, while in Section 3 we extend this method to the design of two-dimensional random surfaces with specified scattering properties. These approaches are tested by means of computer simulations, and methods for fabricating such surfaces on photoresist are described. Finally, in Section 4 we present the main conclusions drawn from this work.

2. One-Dimensional Surfaces

We begin a discussion of designer surfaces by considering one-dimensional random surfaces, for which the theory is simpler than it is for two-dimensional surfaces, but still provides an introduction to the approaches that have been used in dealing with the latter surfaces.

The physical system we initially consider consists of vacuum in the region $x_3 > \zeta(x_1)$ and a perfect conductor in the region $x_3 < \zeta(x_1)$. The surface profile function $\zeta(x_1)$ is assumed to be a single-valued function of x_1 that is differentiable and constitutes a random process, but not necessarily a stationary one.

The surface $x_3 = \zeta(x_1)$ is illuminated from the vacuum region by an $s-$polarized plane wave of frequency ω, whose plane of incidence is the x_1x_3-plane. The single nonzero component of the electric field $E_2^>(x_1, x_3; t) = E_2^>(x_1, x_3|\omega)\exp(-i\omega t)$, in the region $x_3 > \zeta(x_1)_{max}$ is the sum of an incident plane wave and a superposition of outgoing scattered waves,

$$E_2^>(x_1, x_3|\omega) = \exp[ikx_1 - i\alpha_0(k)x_3] + \int_{-\infty}^{\infty} \frac{dq}{2\pi} R(q|k)\exp[iqx_1 + i\alpha_0(q)x_3], \tag{2.1}$$

where $\alpha_0(q) = [(\omega/c)^2 - q^2]^{\frac{1}{2}}$, with $Re\alpha_0(q) > 0$, $Im\alpha_0(q) > 0$.

The scattering amplitude $R(q|k)$ plays a central role in the present work. This is because the differential reflection coefficient, $\partial R/\partial\theta_s$, which is defined in such a way that $(\partial R/\partial\theta_s)d\theta_s$ is the fraction of the total time-averaged incident flux that is scattered into the angular interval $(\theta_s, \theta_s + d\theta_s)$, is given in terms of $R(q|k)$ by

$$\frac{\partial R}{\partial\theta_s} = \frac{1}{L_1}\frac{\omega}{2\pi c}\frac{\cos^2\theta_s}{\cos\theta_0}|R(q|k)|^2, \tag{2.2}$$

where L_1 is the length of the x_1-axis covered by the random surface, while θ_0 and θ_s are the angles of incidence and scattering measured counterclockwise and clockwise from the $+x_3$-axis, respectively, and are related to the wave-numbers k and q by

$$k = (\omega/c)\sin\theta_0, \quad q = (\omega/c)\sin\theta_s. \tag{2.3}$$

As we are concerned with the scattering of light from a randomly rough surface, it is the mean differential reflection coefficient that we need to calculate. It is defined by

$$\left\langle \frac{\partial R}{\partial \theta_s} \right\rangle = \frac{1}{L_1} \frac{\omega}{2\pi c} \frac{\cos^2\theta_s}{\cos\theta_0} \langle |R(q|k)|^2 \rangle, \tag{2.4}$$

where the angle brackets denote an average over the ensemble of realizations of the surface profile function.

An application of Green's second integral identity in the plane [**16**] to the region $x_3 > \zeta(x_1)$ yields the following expression for the scattering amplitude [**17**]

$$R(q|k) = \frac{-i}{2\alpha_0(q)} \int_{-\infty}^{\infty} dx_1 F(x_1) \exp[-iqx_1 - i\alpha_0(q)\zeta(x_1)], \tag{2.5}$$

where the source function $F(x_1)$ is the un-normalized normal derivative of the electric field in the vacuum region, evaluated on the surface $x_3 = \zeta(x_1)$,

$$F(x_1) = \left(-\zeta'(x_1) \frac{\partial}{\partial x_1} + \frac{\partial}{\partial x_3} \right) E_2^{>}(x_1, x_3|\omega) \Big|_{x_3=\zeta(x_1)}. \tag{2.6}$$

In a rigorous solution of the scattering problem $F(x_1)$ is obtained as the solution of an inhomogeneous Fredholm integral equation of the second kind [**17**]. The inhomogeneous term in this equation yields the Kirchhoff approximation to $F(x_1)$, which we adopt here for simplicity. It is given by

$$\begin{aligned} F(x_1) &= 2\left(-\zeta'(x_1) \frac{\partial}{\partial x_1} + \frac{\partial}{\partial x_3} \right) E_2^{>}(x_1, x_3|\omega)_{inc} \Big|_{x_3=\zeta(x_1)}, \\ &= -2i[k\zeta'(x) + \alpha_0(k)] \exp[ikx_1 - i\alpha_0(k)\zeta(x_1)], \end{aligned} \tag{2.7}$$

and represents the assumption that the scattering occurs from the plane tangent to the surface at each point. It is a single-scattering approximation. When Eq. (2.7) is substituted into Eq. (2.5), and an integration by parts is carried out, the scattering amplitude is obtained in the form [**8**]

$$\begin{aligned} R(q|k) &= -\frac{(\omega/c)^2 + \alpha_0(q)\alpha_0(k) - qk}{\alpha_0(q)[\alpha_0(q) + \alpha_0(k)]} \\ &\quad \times \int_{-\infty}^{\infty} dx_1 \exp[-i(q-k)x_1] \exp\{-i[\alpha_0(q) + \alpha_0(k)]\zeta(x_1)\}. \end{aligned} \tag{2.8}$$

We note that Eq. (2.8) has been obtained on the assumption that the incident light is s polarized. If we had assumed instead that the incident light is p polarized we would have obtained the same expression for the scattering amplitude, but with the opposite sign. Since we see from Eq. (2.4) that $R(q|k)$ enters the expression for the mean differential reflection coefficient in the form $|R(q|k)|^2$, the latter is indifferent to the sign of $R(q|k)$. This result means that within the Kirchhoff approximation the mean differential reflection coefficient is the same whether it is $s-$polarized light or $p-$polarized light that is scattered from the surface $x_3 = \zeta(x_1)$ of a perfect conductor.

The substitution of Eqs. (2.3) and (2.8) into Eq. (2.4) yields the latter in the form

$$\left\langle \frac{\partial R}{\partial \theta_s} \right\rangle = \frac{1}{L_1} \frac{\omega}{2\pi c} \frac{1}{\cos\theta_0} \left[\frac{1+\cos(\theta_s+\theta_0)}{\cos\theta_s + cos\theta_0} \right]^2 \times \int_{-\infty}^{\infty} dx_1 \int_{-\infty}^{\infty} dx_1' \exp[-i(q-k)(x_1 - x_1')] \times \langle \exp\{-ia[\zeta(x_1) - \zeta(x_1')]\}\rangle \quad (2.9)$$

where, to simplify the notation, we have defined $a = \alpha_0(q) + \alpha_0(k) = (\omega/c)(\cos\theta_s + \cos\theta_0)$.

Our goal in the present work is to determine the surface profile function $\zeta(x_1)$ that produces a specified form for $\langle \partial R/\partial\theta_s \rangle$ as a function of θ_s and θ_0. As it stands, the expression for $\langle \partial R/\partial\theta_s \rangle$ as a function of $\zeta(x_1)$ given by Eq. (2.9) is too complicated for us to invert it to obtain $\zeta(x_1)$ in terms of $\langle \partial R/\partial\theta_s \rangle$. To simplify Eq. (2.9) we pass to the geometrical optics limit of the Kirchhoff approximation. This limit is obtained by making the change of variable $x_1' = x_1 + u$ in Eq. (2.9), expanding the difference $\zeta(x_1) - \zeta(x_1+u)$ in powers of u, and retaining only the leading nonzero term:

$$\left\langle \frac{\partial R}{\partial \theta_s} \right\rangle = \frac{1}{L_1} \frac{\omega}{2\pi c} \frac{1}{\cos\theta_0} \left[\frac{1+\cos(\theta_s+\theta_0)}{\cos\theta_s + \cos\theta_0)} \right]^2 \times \int_{-\infty}^{\infty} dx_1 \int_{-\infty}^{\infty} du \exp[i(q-k)u] \langle \exp[iau\zeta'(x_1)]\rangle. \quad (2.10)$$

Equation (2.10) is the starting point for the determination of the surface profile function $\zeta(x_1)$ that produces a specified form of $\langle \partial R/\partial\theta_s \rangle$.

To evaluate the double integral in Eq. (2.10) we proceed as follows. We define a set of equally-spaced points along the x_1-axis by $x_n = nb$, where b is a characteristic length and $n = 0, \pm 1, \pm 2, \ldots$. Within the interval $nb \leq x_1 \leq (n+1)b$ the surface profile function $\zeta(x_1)$ is given by

$$\zeta(x_1) = a_n x_1 + b_n \qquad nb \leq x_1 \leq (n+1)b, \quad (2.11)$$

where the $\{a_n\}$ are independent identically distributed random deviates. Therefore, the probability density function (pdf) of a_n,

$$f(\gamma) = \langle \delta(\gamma - a_n) \rangle, \quad (2.12)$$

is independent of n. The derivative of the surface profile function (2.11) is

$$\zeta'(x_1) = a_n \qquad nb < x_1 < (n+1)b. \quad (2.13)$$

In order that the surface be continuous at $x_1 = (n+1)b$, the relation

$$b_{n+1} = b_n - (n+1)(a_{n+1} - a_n)b \quad (2.14)$$

must be satisfied. From this recurrence relation the $\{b_n\}$ can be determined from a knowledge of the $\{a_n\}$, provided that an initial value, e.g. that of b_0, is specified. With no loss of generality we can assume that $b_0 = 0$, and in what follows we do so. The solution of Eq. (2.14) is then

$$b_n = (a_0 + a_1 + \ldots + a_{n-1} - na_n)b \qquad n \geq 1 \quad (2.15a)$$

and

$$b_{-n} = -(a_{-1} + a_{-2} + \ldots + a_{-(n-1)} - (n-1)a_{-n})b \quad n \geq 1. \tag{2.15b}$$

The double integral in Eq. (2.10) can now be evaluated. We write it as

$$\begin{aligned}
&\int_{-\infty}^{\infty} dx_1 \int_{-\infty}^{\infty} du \exp[i(q-k)u]\langle \exp[iau\zeta'(x_1)]\rangle \\
&= \int_{-\infty}^{\infty} du \exp[i(q-k)u] \sum_{n=-N}^{N-1} \int_{nb}^{(n+1)b} dx_1 \langle \exp(iaua_n)\rangle \\
&= \frac{2\pi L_1}{a} f\left(\frac{k-q}{a}\right) \\
&= \frac{2\pi L_1}{(\omega/c)(\cos\theta_0 + \cos\theta_s)} f\left(\frac{\sin\theta_0 - \sin\theta_s}{\cos\theta_0 + \cos\theta_s}\right).
\end{aligned} \tag{2.16}$$

In obtaining this result we have assumed that the length L_1 is equal to $2Nb$.

On combining Eqs. (2.10) and (2.16) we obtain

$$\left\langle \frac{\partial R}{\partial \theta_s} \right\rangle = \frac{[1+\cos(\theta_0+\theta_s)]^2}{\cos\theta_0(\cos\theta_0+\cos\theta_s)^3} f\left(\frac{\sin\theta_0 - \sin\theta_s}{\cos\theta_0 + \cos\theta_s}\right). \tag{2.17}$$

Thus, we find that the mean differential reflection coefficient is expressed in terms of the pdf of the random deviate a_n.

We now make the change of variable

$$\frac{\sin\theta_0 - \sin\theta_s}{\cos\theta_0 + \cos\theta_s} = -\gamma, \tag{2.18}$$

so that

$$\cos\theta_s = \frac{(1-\gamma^2)\cos\theta_0 - 2\gamma\sin\theta_0}{1+\gamma^2}, \qquad \sin\theta_s = \frac{(1-\gamma^2)\sin\theta_0 + 2\gamma\cos\theta_0}{1+\gamma^2}. \tag{2.19}$$

It follows from these results and Eq. (2.17) that

$$f(-\gamma) = \frac{\cos\theta_0(\cos\theta_0+\cos\theta_s)^3}{[1+\cos\theta_0\cos\theta_s - \sin\theta_0\sin\theta_s]^2} \left\langle \frac{\partial R}{\partial \theta_s} \right\rangle (\gamma, \theta_0), \tag{2.20}$$

where $\langle \partial R/\partial\theta_s\rangle(\gamma,\theta_0)$ is the expression for $\langle \partial R/\partial\theta_s\rangle(\theta_s)$ in which its dependence on θ_s has been replaced by its dependence on γ and θ_0 through the use of Eqs. (2.19). After some algebra Eq. (2.20) becomes

$$f(\gamma) = \frac{2}{1+\gamma^2} \frac{\cos\theta_0}{\cos\theta_0 + \gamma\sin\theta_0} \left\langle \frac{\partial R}{\partial \theta_s} \right\rangle (-\gamma, \theta_0). \tag{2.21}$$

It should be noted that this result does not depend explicitly on the wavelength of the incident light.

The result expressed by Eq. (2.21) also implies that if $\langle \partial R/\partial\theta_s\rangle(\theta_s)$ is to be given by the same expression for any angle of incidence, a different ensemble of random surfaces has to be generated for each angle of incidence.

Thus, if we wish to design a perfectly conducting one-dimensional random surface which, when illuminated by $s-$ or $p-$polarized light at an angle of incidence θ_0, produces a specified form of $\langle \partial R/\partial\theta_s\rangle(\theta_s)$, we must first replace θ_s by γ and θ_0 in the latter function by the use of Eqs. (2.19), and then determine $f(\gamma)$ from Eq. (2.21). A long sequence of $\{a_n\}$ is determined from the result for $f(\gamma)$, e.g. by the rejection method [**18**], and the corresponding sequence of $\{b_n\}$ is obtained from Eqs. (2.15). The surface profile function $\zeta(x_1)$ is then constructed on the basis of Eq. (2.11).

A result for the mean differential reflection coefficient is readily obtained in the Kirchhoff approximation,

$$\left\langle \frac{\partial R}{\partial \theta_s} \right\rangle = \frac{b}{2N} \frac{\omega}{2\pi c} \frac{[1+\cos(\theta_s+\theta_0)]^2}{\cos\theta_0(\cos\theta_s+\cos\theta_0)^2} \langle |r(\theta_s|\theta_0)|^2 \rangle, \tag{2.22}$$

where

$$\begin{aligned} r(\theta_s|\theta_0) &= \sum_{n=-N}^{N-1} \exp\Big(-i\frac{\omega}{c}\{(\cos\theta_s+\cos\theta_0)b_n \\ &\quad +[\sin\theta_s - \sin\theta_0 + (\cos\theta_s+\cos\theta_0)a_n](n+\frac{1}{2})b\}\Big) \\ &\quad \times \operatorname{sinc}\frac{\omega}{c}[\sin\theta_s - \sin\theta_0 + (\cos\theta_s+\cos\theta_0)a_n]\frac{b}{2}, \end{aligned} \tag{2.23}$$

and $\operatorname{sinc}x = \sin x/x$.

The procedure now is to generate a large number N_p of realizations of the random surface, and for each realization to calculate the corresponding value of $|r(\theta_s|\theta_0)|^2$. The N_p values of $|r(\theta_s|\theta_0)|^2$ so obtained are summed, and the result is divided by N_p to yield the average indicated in Eq. (2.26). It should be noted that $\langle |r(\theta_s|\theta_0)|^2\rangle$ must increase linearly with N if the result for $\langle \partial R/\partial\theta_s\rangle$ is to be independent of N. This was found to be the case in all of our calculations.

We now seek to design a random surface that gives rise to a mean differential reflection coefficient that is a constant in the angular interval $|\theta_s| < \theta_m < \pi/2$, and vanishes for $|\theta_s| > \theta_m$. In this case $\langle \partial R/\partial\theta_s\rangle$ is given by

$$\left\langle \frac{\partial R}{\partial \theta_s} \right\rangle = A\theta(\theta_s+\theta_m)\theta(\theta_m-\theta_s), \tag{2.24}$$

where $\theta(z)$ is the Heaviside unit step function. Since $\sin\theta_s$ is a monotonically increasing function of θ_s for $-\pi/2 \le \theta_s \le \pi/2$, we can rewrite Eq. (2.24) in the form

$$\begin{aligned} \left\langle \frac{\partial R}{\partial \theta_s} \right\rangle &= A\theta(\sin\theta_s+\sin\theta_m)\theta(\sin\theta_m-\sin\theta_s) \\ &= A\theta\left(\gamma+\tan\frac{\theta_0+\theta_m}{2}\right)\theta\left(\tan\frac{\theta_m-\theta_0}{2}-\gamma\right). \end{aligned} \tag{2.25}$$

On substituting Eq. (2.25) into Eq. (2.21) we obtain for the pdf of a_n

$$f(\gamma) = \frac{2A}{1+\gamma^2}\frac{\cos\theta_0}{\cos\theta_0+\gamma\sin\theta_0}\theta\left(\gamma+\tan\frac{\theta_m-\theta_0)}{2}\right)\theta\left(\tan\frac{\theta_m+\theta_0}{2}-\gamma\right). \tag{2.26}$$

The coefficient A in this expression is obtained from the normalization condition on $f(\gamma)$,

$$\int_{-\infty}^{\infty} d\gamma f(\gamma) = 1, \tag{2.27}$$

and is given by

$$A = \frac{1}{2\cos\theta_0}\left[\theta_m \cos\theta_0 + \sin\theta_0 \ell n \frac{\cos\left(\frac{\theta_m - \theta_0}{2}\right)}{\cos\left(\frac{\theta_m + \theta_0}{2}\right)}\right]^{-1}. \tag{2.28}$$

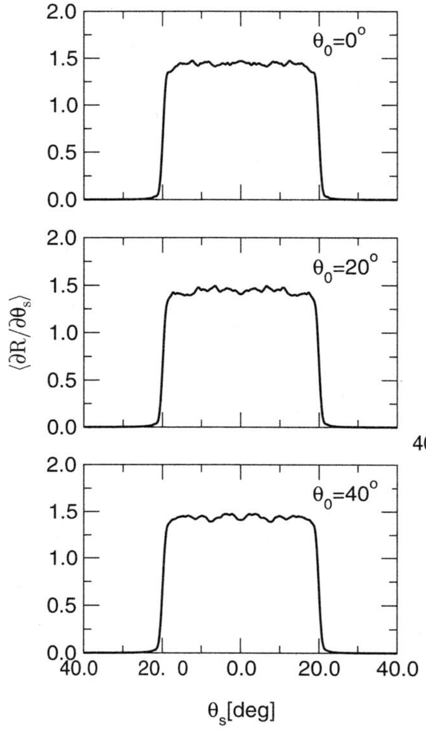

FIGURE 1. The mean differential reflection coefficient $\langle \partial R/\partial\theta_s \rangle$ estimated from $N_p = 30,000$ realizations of the surface profile function for the case of a band–limited uniform diffuser, for which $f(\gamma)$ is given by Eqs. (2.26) and (2.28). The calculations have been carried out for five angles of incidence on the basis of a rigorous computer simulation approach for the scattering of $s-$polarized light from a one–dimensional perfectly conducting surface. The parameters employed are $\theta_m = 20°$, $b = 22\mu$m, $\lambda = 0.6328\mu$m, and $N = 100$.

In Fig. 1 we present results for the mean differential reflection coefficient $\langle \partial R/\partial\theta_s \rangle$ as a function of θ_s for the band-limited uniform diffuser defined by Eqs. (2.24), (2.26), and (2.28), for five angles of incidence. The parameters used in obtaining these results, which were calculated by a rigorous computer simulation approach[**17**] that takes multiple-scattering processes of all orders into account and conserves energy, were $\theta_m = 20°$, $b = 220\mu m$, $\lambda = 632.8nm$, $N = 100$, and $b_0 = 0$. The results obtained for $N_p = 30,000$ realizations of the surface profile function were used in calculating the average in Eq. (2.4). We see that the random surface generated by our approach indeed acts as the same band-limited uniform diffuser for each angle of incidence: there is virtually no scattered intensity for scattering angles greater than 20° in magnitude, and for values of $|\theta_s|$ smaller than 20° the scattered intensity is very nearly uniform, and the height of the mean differential reflection coefficient is independent of the angle of incidence.

In the remainder of this section we will restrict ourselves to the case of normal incidence. In this case the mean differential reflection coefficient (2.24) takes the form

$$\left\langle \frac{\partial R}{\partial \theta_s} \right\rangle = \frac{\theta(\theta_m - |\theta_s|)}{2\theta_m} \tag{2.29}$$

so that the pdf $f(\gamma)$ becomes

$$f(\gamma) = \frac{1}{2\tan^{-1}\gamma_m} \frac{\theta(\gamma_m - |\gamma|)}{1+\gamma^2}, \tag{2.30}$$

where $\gamma_m = \tan(\theta_m/2)$.

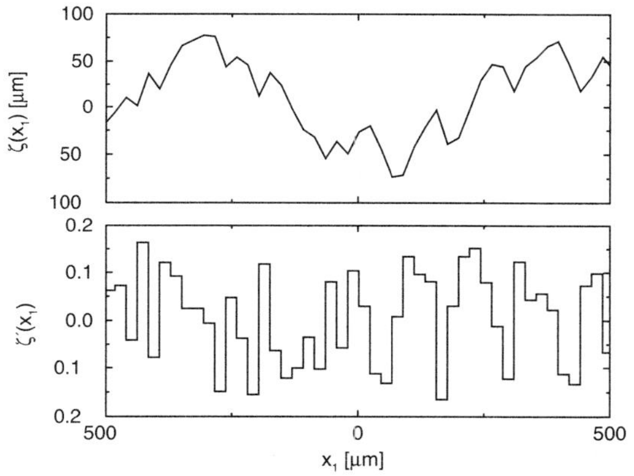

FIGURE 2. (a) A segment of one realization of the surface profile function $\zeta(x_1)$ for the case of a one–dimensional band–limited uniform diffuser, for which $f(\gamma)$ is given by Eq. (2.30). The parameters employed in obtaining this surface are $\theta_m = 20°$, $b = 22\mu$m, and $N = 1000$. (b) the derivative $\zeta'(x_1)$ of this surface profile.

A segment of the surface profile function $\zeta(x_1)$ and its derivative calculated by the approach proposed here are plotted in Figs. 2(a) and 2(b), respectively, for the case of a band-limited uniform diffuser, for which $f(\gamma)$ is given by Eq. (2.30). The parameters used in generating these functions were $\theta_m = 20°$, $b = 22\mu m$, and $N = 1000$.

In Fig. 3 we present plots of $\langle \partial R/\partial\theta_s \rangle$ as a function of θ_s calculated on the basis of the Kirchhoff approximation Eqs. (2.22) -(2.23) for the band-limited uniform diffuser defined by Eqs. (2.29) and (2.30). The parameters used in obtaining the result plotted in Fig. 3(a) were $\theta_m = 20°$, $b = 22\mu m$, $\lambda = 0.6328\mu m$, and $N = 1000$. The parameters used in obtaining the result plotted in Fig. 3(b) were $\theta_m = 20°$, $b = 220\mu m$, $\lambda = 0.6328\mu m$, and $N = 1000$. Thus, although the value of the characteristic length b was significantly larger in the calculations that led to Fig.

3(b) than it was in obtaining Fig. 3(a), the overall length of the surface, $L_1 = 2Nb$, was the same in both cases. The results obtained for $N_p = 100,000$ realizations of the surface profile function were used in calculating the average appearing in Eq. (2.26). We see that the random surface generated by our approach behaves as a band-limited uniform diffuser. We also see that the larger the characteristic length b is, the sharper are the corners of $\langle \partial R/\partial \theta_s \rangle$ and the more uniform is the scattered intensity in the region $|\theta_s| < 20°$. This appears to be a general result: The larger is b, the closer to the ideal $\langle \partial R/\partial \theta_s \rangle$ is the calculated one. Additional calculations, whose results are not shown, suggest that, for a given value of b, increasing the length of the surface L_1 also brings the calculated $\langle \partial R/\partial \theta_s \rangle$ closer to the ideal one.

Although the derivation of the pdf $f(\gamma)$ given by Eq. (2.30) was based on a single-scattering approximation and the assumption of s polarization, the random surfaces generated by its use retain the scattering properties for which they were designed when multiple scattering is taken into account and the incident light is p−polarized. To demonstrate this, we present in Fig. 4 plots of $\langle \partial R/\partial \theta_s \rangle$ as functions of θ_s for the scattering of both $p-$ and $s-$polarized light from a one-dimensional perfectly conducting random surface designed to act as a band-limited uniform diffuser with $\theta_m = 20°$, obtained by means of rigorous computer simulation calculations [**17**] that take into account multiple–scattering processes of all orders. We see that the surface displays the scattering property for which it was designed for both polarizations of the incident light.

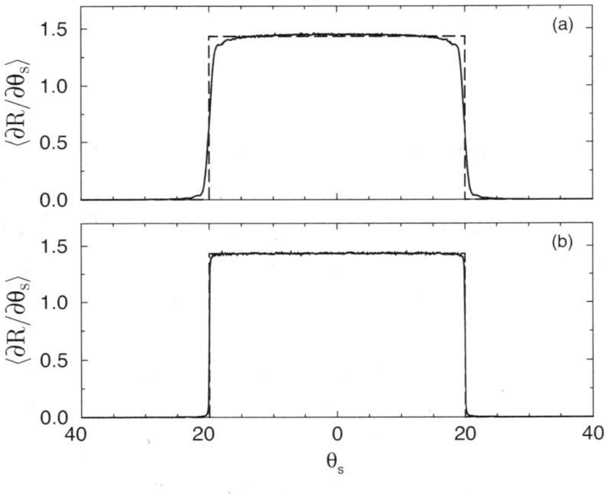

FIGURE 3. The mean differential reflection coefficient $\langle \partial R/\partial \theta_s \rangle$ estimated from $N_p = 100,000$ realizations of the surface profile function for the case of a band–limited uniform diffuser. The calculations have been carried out on the basis of the Kirchhoff approximation for scattering from a perfectly conducting surface. The parameters employed are: (a) $\theta_m = 20°$, $b = 22\mu$m, $\lambda = 0.6328\mu$m, and $N = 1000$; (b) $\theta_m = 20°$, $b = 220\mu$m, $\lambda = 0.6328\mu$m, and $N = 100$.

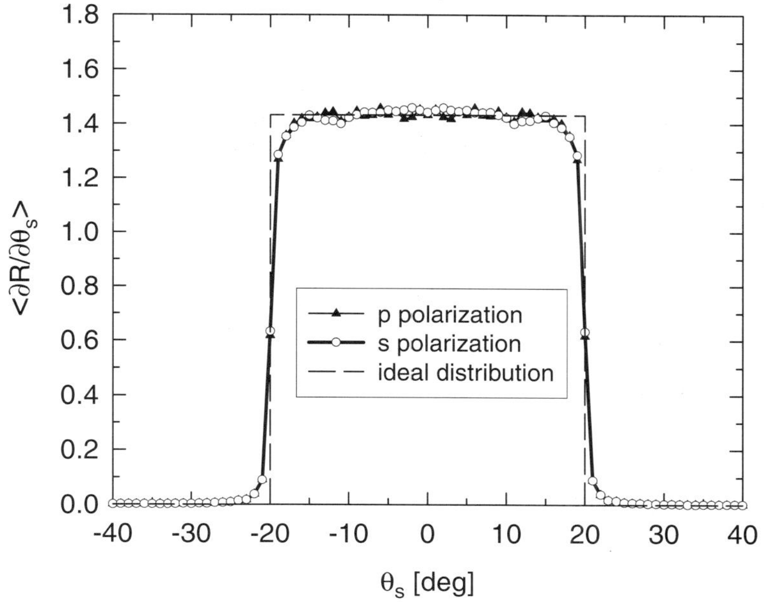

FIGURE 4. The mean differential reflection coefficient $\langle \partial R/\partial\theta_s \rangle$ estimated from $N_p = 20,000$ realizations of the surface profile function for the case of a band–limited uniform diffuser, for which $f(\gamma)$ is given by Eq. (2.30). The calculations have been carried out on the basis of a rigorous computer simulation approach for the scattering of $s-$ and $p-$polarized light from a one–dimensional perfectly conducting surface. The parameters employed are $\theta_m = 20°$, $b = 22\mu$m, $\lambda = 0.6328\mu$m.

In Fig. 5 we present the results of a rigorous computer simulation calculation of $\langle \partial R/\partial\theta_s \rangle$ as a function of θ_s for $s-$polarized light incident normally on a one-dimensional random silver surface designed to act as a band-limited uniform diffuser with $\theta_m = 20°$. Results are presented for three wavelengths of the incident light: (a) $\lambda = 632.8nm$ (He-Ne laser); (b) $\lambda = 532nm$ (the second harmonic of the YAG laser); (c) $\lambda = 442nm$ (He-Cd laser). These wavelengths cover the entire visible region of the optical spectrum. We see that even when the scattering medium is a finitely conducting metal, a surface ruled on it in accordance with Eqs. (2.11), (2.15), and (2.30) still acts as a band-limited uniform diffuser, although with a weak dependence of the scattering pattern on the wavelength of the incident light. We attribute this dependence to the strong wavelength dependence of the dielectric function of silver in the range of wavelengths considered.

To fabricate a one-dimensional random surface of the kind studied here, a single realization of a profile function $\zeta_0(x)$, generated according to Eqs. (2.11), (2.15), and (2.21), is used to fabricate a slit of mean width d shown in Fig. 6. A good quality optical system is used to form an incoherent demagnified image of the slit on a photoresist plate. The intensity image on the photoresist plate can be expressed

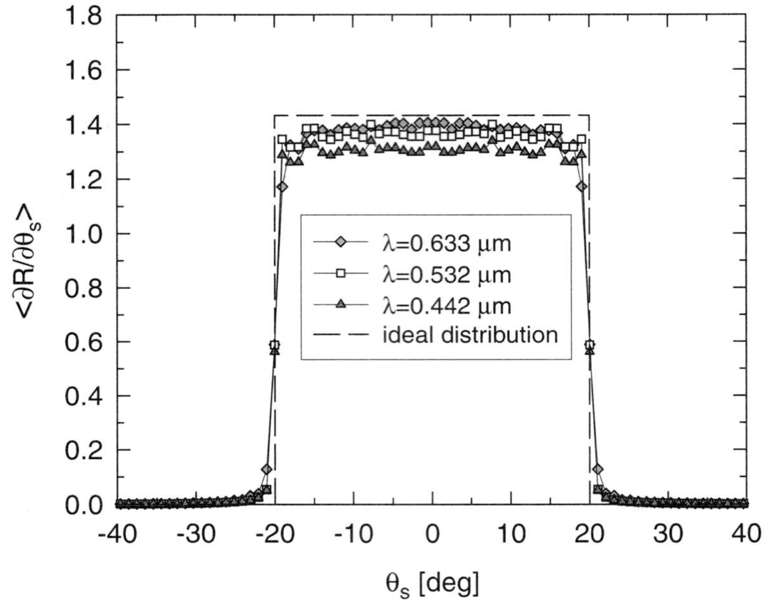

FIGURE 5. The mean differential reflection coefficient $\langle \partial R/\partial\theta_s\rangle$ estimated from $N_p = 40,000$ realizations of the surface profile function for the case of a band–limited uniform diffuser, for which $f(\gamma)$ is given by Eq. (2.30). The calculations have been carried out for three wavelengths of the incident, $s-$polarized, light on the basis of a rigorous computer simulation approach for scattering from a one–dimensional metallic surface. The parameters employed are $\theta_m = 20°$, $b = 22\mu$m.

as

$$I(x_1, x_3) = I_0\theta(x_2 + d)\theta(\zeta(x_1) - x_2), \tag{2.31}$$

where I_0 is a constant, the coordinates x_1 and x_2 are fixed on the plate, and $\zeta(x_1)$ is a scaled version of the mask profile. The plate is then scanned in the x_2 direction at a constant speed v, producing a total exposure of the form

$$E(x_1) = K \int_{-T/2}^{T/2} I(x_1, x_2 + vt)dt, \tag{2.32}$$

where K is a constant related to the sensitivity of the photoresist, $T = L_2/v$ is the time it takes to execute the scan, and the total scan length L_2 is assumed to be greater than the physical size of the plate. In this case substitution of Eq. (2.34) into Eq. (2.35) gives

$$E(x_1) = E_0 + \alpha\zeta(x_1), \tag{2.33}$$

where $E_0 = KI_0d/v$ and $\alpha = KI_0/v$. Thus, the exposure has a linear dependence on the height of the numerically generated realization of the surface profile function. Assuming that the relation between exposure and the resulting height on the surface

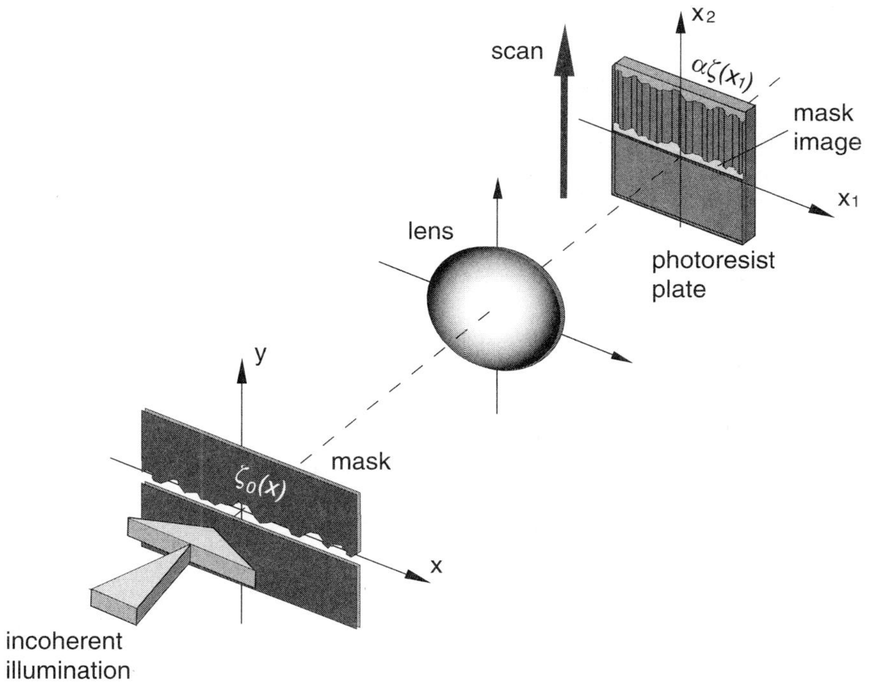

FIGURE 6. Schematic diagram of the proposed experimental arragement for the fabrication of surfaces with $f(\gamma)$ given by Eq. (2.21). The mask is imaged on the photoresist plate, which is then scanned along x_2.

is linear, the surface on the developed plate will have the desired properties. It should be noted that the vertical scale of the resulting profile can be adjusted by varying I_0 and the speed v of the scan.

3. Two-Dimensional Surfaces

A convenient starting point for the generation of two-dimensional random surfaces with specified scattering properties is the expression for the mean differential reflection coefficient in the case of the scattering of a scalar plane wave from a two-dimensional Dirichlet surface. The physical system we consider therefore consists of vacuum in the region $x_3 > \zeta(\mathbf{x}_\parallel)$, where $\mathbf{x}_\parallel = (x_1, x_2, 0)$ is a position vector in the plane $x_3 = 0$, and the scattering medium in the region $x_3 < \zeta(\mathbf{x}_\parallel)$. The surface profile function $\zeta(\mathbf{x}_\parallel)$ is assumed to be a single-valued function of $\mathbf{x}_\parallel$, that is differentiable with respect to x_1 and x_2, and constitutes a random process, but not necessarily a stationary one.

The surface $x_3 = \zeta(\mathbf{x}_\parallel)$ is illuminated from the vacuum side by a scalar plane wave, $\psi(\mathbf{x}, t)_{inc} = \hat{\psi}(\mathbf{x}|\omega)_{inc} \exp(-i\omega t)$, where

$$\hat{\psi}(\mathbf{x}|\omega)_{inc} = \exp[i\mathbf{k}_\parallel \cdot \mathbf{x}_\parallel - i\alpha_0(k_\parallel)x_3]. \tag{3.1}$$

In this expression $\mathbf{k}_\parallel = (k_1, k_2, 0)$ is a two-dimensional wave vector, and $\alpha_0(k_\parallel) = [(\omega/c)^2 - k_\parallel^2]^{\frac{1}{2}}$, with $Re\alpha_0(k_\parallel) > 0$, $Im\alpha_0(k_\parallel) > 0$. The scattered field in the far field, $\psi(\mathbf{x}, t)_{sc} = \hat{\psi}(\mathbf{x}|\omega)_{sc} \exp(-i\omega t)$, is a superposition of outgoing waves,

$$\hat{\psi}(\mathbf{x}|\omega)_{sc} = \int \frac{d^2 q_\parallel}{(2\pi)^2} R(\mathbf{q}_\parallel|\mathbf{k}_\parallel) \exp[i\mathbf{q}_\parallel \cdot \mathbf{x}_\parallel + i\alpha_0(q_\parallel)x_3], \tag{3.2}$$

in the region $x_3 > \zeta(\mathbf{x}_\parallel)_{max}$. The total field in the vacuum region is assumed to satisfy the Dirichlet boundary condition

$$[\psi(\mathbf{x}, t)_{inc} + \psi(\mathbf{x}, t)_{sc}]\Big|_{x_3 = \zeta(\mathbf{x}_\parallel))} = 0 \tag{3.3}$$

on the surface $x_3 = \zeta(\mathbf{x}_\parallel)$.

The differential reflection coefficient $\partial R/\partial\Omega_s$ is defined in such a way that $(\partial R/\partial\Omega_s)d\Omega_s$ is the fraction of the total time-averaged incident flux that is scattered into the element of solid angle $d\Omega_s$ about a given scattering direction. It is given in terms of the scattering amplitude by

$$\frac{\partial R}{\partial\Omega_s} = \frac{1}{S}\left(\frac{\omega}{2\pi c}\right)^2 \frac{\cos^2\theta_s}{\cos\theta_0} |R(\mathbf{q}_\parallel|\mathbf{k}_\parallel)|^2, \tag{3.4}$$

where S is the area of the plane $x_3 = 0$ covered by the random surface. In this expression the wave vectors $\mathbf{q}_\parallel$ and $\mathbf{k}_\parallel$ must be expressed in terms of the polar and azimuthal angles of scattering (θ_s, ϕ_s) and incidence (θ_0, ϕ_0), respectively, according to

$$\mathbf{q}_\parallel = (\omega/c)\sin\theta_s(\cos\phi_s, \sin\phi_s, 0) \tag{3.5a}$$

$$\mathbf{k}_\parallel = (\omega/c)\sin\theta_0(\cos\phi_0, \sin\phi_0, 0). \tag{3.5b}$$

As we are concerned with the scattering of a wave from a random surface, it is the mean differential reflection coefficient that we must calculate. It is given by

$$\left\langle \frac{\partial R}{\partial\Omega_s} \right\rangle = \frac{1}{S}\left(\frac{\omega}{2\pi c}\right)^2 \frac{\cos^2\theta_s}{\cos\theta_0} \langle |R(\mathbf{q}_\parallel|\mathbf{k}_\parallel)|^2 \rangle, \tag{3.6}$$

where the angle brackets denote an average over the ensemble of realizations of the surface profile function.

The application of Green's second integral identity to the region $x_3 > \zeta(\mathbf{x}_\parallel)$ yields the following expression for the scattering amplitude $R(\mathbf{q}_\parallel|\mathbf{k}_\parallel)$[**12**]

$$R(\mathbf{q}_\parallel|\mathbf{k}_\parallel) = \frac{1}{2i\alpha_0(q_\parallel)} \int d^2x_\parallel F(\mathbf{x}_\parallel) \exp[-i\mathbf{q}_\parallel \cdot \mathbf{x}_\parallel - i\alpha_0(q_\parallel)\zeta(\mathbf{x}_\parallel)], \tag{3.7}$$

where the source function $F(\mathbf{x}_\parallel)$ is defined by

$$\begin{aligned} F(\mathbf{x}_\parallel) &= \left(-\frac{\partial\zeta(\mathbf{x}_\parallel)}{\partial x_1}\frac{\partial}{\partial x_1} - \frac{\partial\zeta(\mathbf{x}_\parallel)}{\partial x_2}\frac{\partial}{\partial x_2} + \frac{\partial}{\partial x_3} \right) \\ &\quad \times \left[\hat{\psi}(\mathbf{x}|\omega)_{inc} + \hat{\psi}(\mathbf{x}|\omega)_{sc} \right] \Big|_{x_3 = \zeta(\mathbf{x}_\parallel)} . \end{aligned} \tag{3.8}$$

In the Kirchhoff approximation, which we adopt here for simplicity, the source function is [**12**]

$$F(\mathbf{x}_\parallel) = 2\left(-\frac{\partial\zeta(\mathbf{x}_\parallel)}{\partial x_1}\frac{\partial}{\partial x_1} - \frac{\partial\zeta(\mathbf{x}_\parallel)}{\partial x_2}\frac{\partial}{\partial x_2} + \frac{\partial}{\partial x_3}\right)\hat{\psi}(\mathbf{x}|\omega)_{inc}\Big|_{x_3=\zeta(\mathbf{x}_\parallel)} \tag{3.9a}$$

$$= 2\left(-ik_1\frac{\partial\zeta(\mathbf{x}_\parallel)}{\partial x_1} - ik_2\frac{\partial\zeta(\mathbf{x}_\parallel)}{\partial x_2} - i\alpha_0(k_\parallel)\right)\exp[i\mathbf{k}_\parallel\cdot\mathbf{x}_\parallel - i\alpha_0(k_\parallel)\zeta(\mathbf{x}_\parallel)]. \tag{3.9b}$$

This approximation is equivalent to the assumption that the scattering takes place from the plane tangent to the surface at each point.

Substitution of Eq. (3.9b) into Eq. (3.6) yields the result that

$$R(\mathbf{q}_\parallel|\mathbf{k}_\parallel) = -\frac{1}{\alpha_0(q_\parallel)}\int d^2x_\parallel\left(k_1\frac{\partial\zeta(\mathbf{x}_\parallel)}{\partial x_1} + k_2\frac{\partial\zeta(\mathbf{x}_\parallel)}{\partial x_2} + \alpha_0(k_\parallel)\right) \times\exp[-i(\mathbf{q}_\parallel - \mathbf{k}_\parallel)\cdot\mathbf{x}_\parallel - i(\alpha_0(q_\parallel)+\alpha_0(k_\parallel))\zeta(\mathbf{x}_\parallel)]. \tag{3.10}$$

After an integration by parts this expression becomes [**12**]

$$R(\mathbf{q}_\parallel|\mathbf{k}_\parallel) = -\frac{(\omega/c)^2 + \alpha_0(q_\parallel)\alpha_0(k_\parallel) - \mathbf{q}_\parallel\cdot\mathbf{k}_\parallel}{\alpha_0(q_\parallel)[\alpha_0(q_\parallel)+\alpha_0(k_\parallel)]} \times\int d^2x_\parallel \exp[-i(\mathbf{q}_\parallel - \mathbf{k}_\parallel)\cdot\mathbf{x}_\parallel - i(\alpha_0(q_\parallel)+\alpha_0(k_\parallel))\zeta(\mathbf{x}_\parallel)]. \tag{3.11}$$

It follows from Eqs. (3.5) and (3.11) that the mean differential reflection coefficient (3.6) can be written in the form

$$\left\langle\frac{\partial R}{\partial\Omega_s}\right\rangle = \frac{1}{S}\left(\frac{\omega}{2\pi c}\right)^2\frac{[1+\cos\theta_0\cos\theta_s - \sin\theta_0\sin\theta_s\cos(\phi_0-\phi_s)]^2}{\cos\theta_0(\cos\theta_0+\cos\theta_s)^2} \times\int d^2x_\parallel\int d^2x'_\parallel \exp[-i(\mathbf{q}_\parallel - \mathbf{k}_\parallel)\cdot(\mathbf{x}_\parallel - \mathbf{x}'_\parallel)] \times\langle\exp[-i(\omega/c)(\cos\theta_0+\cos\theta_s)(\zeta(\mathbf{x}_\parallel)-\zeta(\mathbf{x}'_\parallel))]\rangle. \tag{3.12}$$

The result (3.11) for $R(\mathbf{q}_\parallel|\mathbf{k}_\parallel)$ was obtained on the assumption that the total field in the vacuum region satisfies the Dirichlet boundary condition (3.3) on the surface $x_3 = \zeta(\mathbf{x}_\parallel)$. If, instead, it is assumed that the total field in the vacuum region satisfies the Neumann boundary condition

$$\left[-\frac{\partial\zeta(\mathbf{x}_\parallel)}{\partial x_1}\frac{\partial}{\partial x_1} - \frac{\partial\zeta(\mathbf{x}_\parallel)}{\partial x_2}\frac{\partial}{\partial x_2} + \frac{\partial}{\partial x_3}\right][\psi(\mathbf{x},t)_{inc} + \psi(\mathbf{x},t)_{sc}]\Big|_{x_3=\zeta(\mathbf{x}_\parallel)} = 0 \tag{3.13}$$

on the surface $x_3 = \zeta(\mathbf{x}_\parallel)$, the corresponding scattering amplitude $R(\mathbf{q}_\parallel|\mathbf{k}_\parallel)$ is given by Eq. (3.11), but with the opposite sign. Since $R(\mathbf{q}_\parallel|\mathbf{k}_\parallel)$ enters the mean differential reflection coefficient $\langle\partial R/\partial\Omega_s\rangle$ only in the form $|R(\mathbf{q}_\parallel|\mathbf{k}_\parallel)|^2$, it follows that the result given by Eq. (3.12) is valid for a two-dimensional Neumann surface as well as for a two-dimensional Dirichlet surface.

The result given by Eq. (3.12) simplifies greatly in the case of normal incidence, $\theta_0 = 0$, $\mathbf{k}_\parallel = 0$,

$$\left\langle\frac{\partial R}{\partial\Omega_s}\right\rangle = \frac{1}{S}\left(\frac{\omega}{2\pi c}\right)^2\int d^2x_\parallel\int d^2x'_\parallel \exp[-i\mathbf{q}_\parallel\cdot(\mathbf{x}_\parallel - \mathbf{x}'_\parallel)] \times\langle\exp[-ia(\zeta(\mathbf{x}_\parallel)-\zeta(\mathbf{x}'_\parallel))]\rangle, \tag{3.14}$$

where $a = (\omega/c)(1 + \cos\theta_s)$, and we will restrict ourselves to this case in what follows. As our aim is to find a function $\zeta(\mathbf{x}_\parallel)$ that yields a prescribed form for $\langle \partial R/\partial\Omega_s \rangle$, the expression given by Eq. (3.14) is too complicated to work with. We therefore pass to the geometrical optics limit of the Kirchhoff approximation, which is obtained by making the change of variables $\mathbf{x}'_\parallel = \mathbf{x}_\parallel - \mathbf{u}_\parallel$ in Eq. (3.14), expanding the difference $\zeta(\mathbf{x}_\parallel) - \zeta(\mathbf{x}_\parallel - \mathbf{u}_\parallel)$ in powers of $\mathbf{u}_\parallel$, and keeping only the linear term. In this way we obtain the result

$$\begin{aligned}\left\langle \frac{\partial R}{\partial \Omega_s} \right\rangle &= \frac{1}{S}\left(\frac{\omega}{2\pi c}\right)^2 \int d^2u_\parallel \exp(-i\mathbf{q}_\parallel \cdot \mathbf{u}_\parallel) \\ &\quad \times \int d^2x_\parallel \langle \exp[-ia\mathbf{u}_\parallel \cdot \nabla\zeta(\mathbf{x}_\parallel)] \rangle. \qquad (3.15)\end{aligned}$$

Equation (3.15) is our starting point for the design of two-dimensional random surfaces with specified scattering properties.

3.1. Band-Limited Uniform Diffusers in a Rectangular Domain. Perhaps the simplest two-dimensional random surface that scatters light in a prescribed fashion is one that scatters light with uniform intensity within a rectangular region and produces no scattering outside this region. In this case it is convenient to represent the surface profile function $\zeta(\mathbf{x}_\parallel)$ in the form

$$\zeta(\mathbf{x}_\parallel) = \zeta_1(x_1) + \zeta_2(x_2), \qquad (3.16)$$

where $\zeta_1(x_1)$ and $\zeta_2(x_2)$ are statistically independent random processes. In this case Eq. (3.15) becomes

$$\left\langle \frac{\partial R}{\partial \Omega_s} \right\rangle = \frac{1}{L_1 L_2}\left(\frac{\omega}{2\pi c}\right)^2 I_1(q_1) I_2(q_2), \qquad (3.17)$$

where L_1 and L_2 are the lengths of the x_1- and x_2-axes covered by the random surface, and where $(j = 1, 2)$

$$I_j(q_j) = \int_{-\infty}^{\infty} dx_j \int_{-\infty}^{\infty} du_j \exp(-iq_j u_j) \langle \exp[-iau_j \zeta'_j(x_j)] \rangle. \qquad (3.18)$$

We assume for $\zeta_j(x_j)$ the form

$$\zeta_j(x_j) = a_{n_j} x_j + b_{n_j} \qquad \begin{aligned} &n_j b \le x_j \le (n_j + 1)b \\ &n_j = 0, \pm 1, \pm 2, \ldots, \end{aligned} \qquad (3.19)$$

where the $\{a_{n_j}\}$ are independent identically distributed random deviates, b is a characteristic length, and the $\{b_{n_j}\}$ are obtained from the requirement that $\zeta_j(x_j)$ be a continuous function of x_j. This condition leads to the recurrence relation

$$b_{n_j+1} = b_{n_j} - (n_j + 1)(a_{n_j+1} - a_{n_j})b \qquad (3.20)$$

for determining the $\{b_{n_j}\}$. In applying it, it is convenient to set $b_0 = 0$, and we do so. Due to the independence of the $\{a_{n_j}\}$, and the fact that they are identically distributed, the probability density function (pdf) of a_{n_j},

$$f_j(\gamma_j) = \langle \delta(\gamma_j - a_{n_j}) \rangle, \qquad (3.21)$$

is independent of n_j. The derivative $\zeta'_j(x_j)$ of $\zeta_j(x_j)$ with respect to x_j is given by

$$\zeta'_j(x_j) = a_{n_j} \qquad n_j b < x_j < (n_j + 1)b. \qquad (3.22)$$

When the result given by Eq. (3.19) is substituted into Eq. (3.18), the latter equation becomes

$$\begin{aligned} I_j(q_j) &= \int_{-\infty}^{\infty} du_j \exp(-iq_ju_j) \sum_{n_j=-N_j}^{N_j-1} \int_{n_jb}^{(n_j+1)b} dx_j \langle \exp[-iau_ja_{n_j}]\rangle \\ &= L_j \frac{2\pi}{a} f_j\left(-\frac{q_j}{a}\right), \end{aligned} \tag{3.23}$$

where we have set $L_j = 2N_jb$. On combining Eqs. (3.17) and (3.23) we obtain the result

$$\left\langle \frac{\partial R}{\partial \Omega_s} \right\rangle = \frac{1}{(1+\cos\theta_s)^2} f_1\left(-\frac{q_1}{a}\right) f_2\left(-\frac{q_2}{a}\right). \tag{3.24}$$

Thus, if we wish $\langle \partial R/\partial \Omega_s \rangle$ to have the form

$$\left\langle \frac{\partial R}{\partial \Omega_s} \right\rangle = \frac{\theta(q_1^{(m)} - |q_1|)}{2(cq_1^{(m)}/\omega)} \frac{\theta(q_2^{(m)} - |q_2|)}{2(cq_2^{(m)}/\omega)}, \tag{3.25}$$

we must set

$$\frac{1}{1+\cos\theta_s} f_1\left(-\frac{q_1}{a}\right) = \frac{\theta(q_1^{(m)} - |q_1|)}{2(cq_1^{(m)}/\omega)} \tag{3.26a}$$

and

$$\frac{1}{1+\cos\theta_s} f_2\left(-\frac{q_2}{a}\right) = \frac{\theta(q_2^{(m)} - |q_2|)}{2(cq_2^{(m)}/\omega)}. \tag{3.26b}$$

In the limit of small scattering angles, where we can replace $1+\cos\theta_s$ by 2, these equations yield the result that $(j = 1, 2)$

$$f_j(\gamma_j) = \frac{\theta(\gamma_j^{(m)} - |\gamma_j|)}{2\gamma_j^{(m)}}, \tag{3.27}$$

where $\gamma_j^{(m)} = cq_j^{(m)}/(2\omega) \cong \theta_j^{(m)}/2$, which defines the angles $\theta_j^{(m)}$.

An expression for $\langle \partial R/\partial \Omega_s \rangle$ in the Kirchhoff approximation, but without passing to the geometrical optics limit of it, is readily obtained,

$$\left\langle \frac{\partial R}{\partial \Omega_s} \right\rangle = \frac{b^2}{4N_1N_2} \left(\frac{\omega}{2\pi c}\right)^2 \langle |r_1(q_1)|^2\rangle \langle |r_2(q_2)|^2\rangle. \tag{3.28}$$

where

$$\begin{aligned} r_j(q_j) &= \sum_{n_j=-N_j}^{N_j-1} \exp(-iab_{n_j}) \exp[-i(q_j + aa_{n_j})(n_j + \frac{1}{2})b] \\ &\quad \times \mathrm{sinc}(q_j + aa_{n_j})\frac{b}{2}, \end{aligned} \tag{3.29}$$

with $\mathrm{sinc}x = \sin x/x$.

In Fig. 7 we present a plot of $\langle \partial R/\partial \Omega_s \rangle$ calculated on the basis of Eqs. (3.28) and (3.29). The wavelength of the incident light was $\lambda = 0.6328\mu m$. The roughness parameters employed were $b = 200\mu m$, $\theta_1^{(m)} = 18°$, and $\theta_2^{(m)} = 22°$. The results for $N_p = 20,000$ realizations of the surface profile function were averaged to obtain this result. It is seen that the angular dependence of the intensity of the scattered light

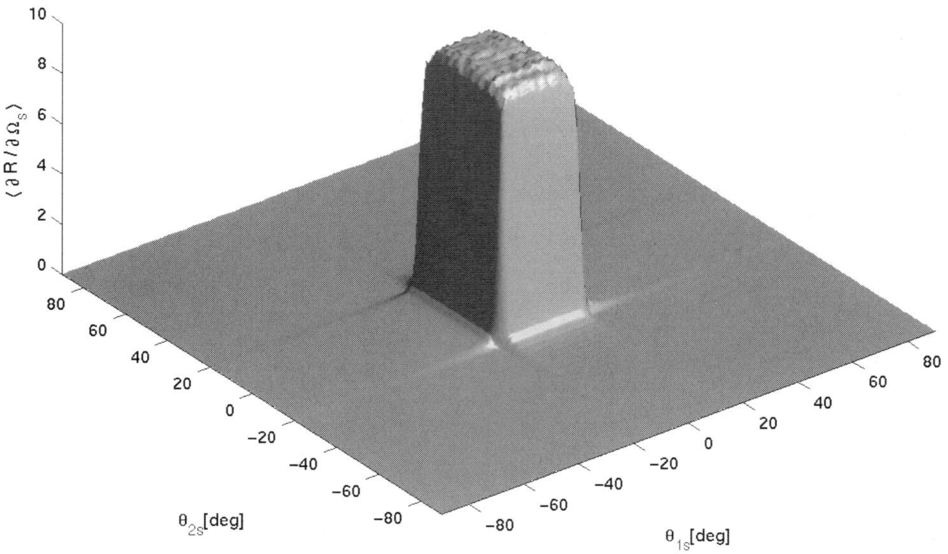

FIGURE 7. The mean differential reflection coefficient $\langle \partial R/\partial\Omega_s \rangle$ estimated from $N_p = 20,000$ realizations of the surface profile function for the case of normal incidence and a wavelength $\lambda = 0.6328\mu\text{m}$. The calculations were carried out for a two–dimensional Dirichlet surface that behaves as a band–limited uniform diffuser within a rectangular region of scattering angles on the basis of the Kirchhoff approximation given by Eqs. (3.28) and (3.29). The parameters employed are $b = 200\mu\text{m}$, $\theta_1^{(m)} = 18°$, and $\theta_2^{(m)} = 22°$.

is very close to the desired result. There is almost no scattered intensity outside the range $-18° < \theta_{1s} < 18°$, $-22° < \theta_{2s} < 22°$, where $\theta_{js} = cq_j/\omega$, and inside it the distribution of intensity is quite uniform.

Since the two-dimensional random surface considered in this subsection is a superposition of two one-dimensional random surfaces whose generators are perpendicular to each other, Eq. (3.16), they can be fabricated on photoresist by two applications of the method described at the end of Section 2 for producing one-dimensional surfaces.

3.2. Band-Limited Uniform Diffuser in a Circular Domain. The approach described in the preceding subsection is convenient to use in designing two-dimensional random surfaces that scatter electromagnetic waves in a prescribed fashion within a rectangular region of scattering angles. It is not useful, however, if the region of scattering angles is not rectangular. In this subsection we describe a method for designing a two-dimensional randomly rough Dirichlet surface which, when illuminated normally by a scalar plane wave, scatters it with a prescribed circularly symmetric distribution of intensity. We illustrate this method by applying it to the design of a two-dimensional random surface that scatters a scalar plane wave with uniform intensity within a circular region and produces no scattering outside this region.

The idea that underlies this approach is that if we wish to have a two-dimensional random surface that scatters a normally incident scalar plane wave in a circularly symmetric fashion, then the random surface itself should be circularly symmetric. Consequently we assume that the surface profile function $\zeta(\mathbf{x}_{\|})$ is a function of the radial coordinate $r = |\mathbf{x}_{\|}|$ alone, $\zeta(\mathbf{x}_{\|}) = H(r)$, and we choose $H(r)$ to have the form

$$H(r) = a_n r + b_n \qquad nb \leq r \leq (n+1)b, \qquad n = 0, 1, 2 \ldots \tag{3.30}$$

where the $\{a_n\}$ are independent identically distributed random deviates, b is a characteristic length, and the $\{b_n\}$ are determined in such a way as to make $H(r)$ a continuous function of r,

$$b_n = b_0 + (a_0 + a_1 + \cdots + a_{n-1} - na_n)b \quad n \geq 1. \tag{3.31}$$

We seek the probability density function (pdf) of a_n, $f(\gamma) = \langle \delta(\gamma - a_n) \rangle$ such that the surface profile defined by Eqs. (3.30) and (3.31) yields a mean differential reflection coefficient of a prescribed form. Due to the independence of the $\{a_n\}$ the pdf $f(\gamma)$ is independent of n.

For the form of the surface profile function we have chosen, its gradient is given by

$$\nabla\zeta(\mathbf{x}_{\|}) = \hat{\mathbf{x}}_{\|} a_n \qquad nb < r < (n+1)b. \tag{3.32}$$

The double integral in Eq. (3.15) therefore becomes

$$\begin{aligned}
&\int d^2u_{\|} \exp(-i\mathbf{q}_{\|} \cdot \mathbf{u}_{\|}) \int d^2x_{\|} \langle \exp[-ia\mathbf{u}_{\|} \cdot \nabla\zeta(\mathbf{x}_{\|})] \rangle \\
= &\int d^2u_{\|} \exp(-i\mathbf{q}_{\|} \cdot \mathbf{u}_{\|}) \sum_{n=0}^{N-1} \int_{nb}^{(n+1)b} dx_{\|} x_{\|} \int_{-\pi}^{\pi} d\phi_x \langle \exp[-iaa_n u_{\|} \cos(\phi_u - \phi_x)] \rangle \\
= &\; S \frac{2\pi}{a} \frac{1}{q_{\|}} f\left(\frac{q_{\|}}{a}\right),
\end{aligned} \tag{3.33}$$

where $S = \pi(Nb)^2$. On combining Eqs. (3.15) and (3.33) we find that the mean differential reflection coefficient is given by

$$\left\langle \frac{\partial R}{\partial \Omega_s} \right\rangle = \frac{2\pi}{a} \left(\frac{\omega}{2\pi c}\right)^2 \frac{1}{q_{\|}} f\left(\frac{q_{\|}}{a}\right). \tag{3.34}$$

Thus, the mean differential reflection coefficient is given in terms of the pdf of a_n. If we make the change of variable $\gamma = q_{\|}/a = \tan(\theta_s/2)$, we find that $f(\gamma)$ is given by

$$f(\gamma) = 8\pi \frac{\gamma}{(1+\gamma^2)^2} \left\langle \frac{\partial R}{\partial \Omega_s} \right\rangle (\gamma) \qquad \gamma \geq 0. \tag{3.35}$$

We apply Eq. (3.35) to the design of a surface that acts as a band-limited uniform diffuser within the circular region $0 \leq \theta_s \leq \theta_m$ and $0 \leq \phi_s \leq 2\pi$. For such a surface

$$\begin{aligned}\left\langle \frac{\partial R}{\partial \theta_s} \right\rangle &= \frac{\theta(\theta_m - \theta_s)\theta(\theta_s)}{4\pi \sin^2(\theta_m/2)} \\ &= \frac{\theta(\tan(\theta_m/2) - \tan(\theta_s/2))\theta(\tan(\theta_s/2))}{4\pi \sin^2(\theta_m/2)} \\ &= \frac{1}{4\pi}\frac{1+\gamma_m^2}{\gamma_m^2}\theta(\gamma_m - \gamma)\theta(\gamma), \end{aligned} \tag{3.36}$$

where $\theta(z)$ is the Heaviside unit step function, and $\gamma_m = \tan(\theta_m/2)$. It follows from Eq. (3.35) that $f(\gamma)$ in this case is given by

$$f(\gamma) = 2\frac{1+\gamma_m^2}{\gamma_m^2}\frac{\gamma}{(1+\gamma^2)^2}\theta(\gamma_m - \gamma)\theta(\gamma). \tag{3.37}$$

Because rigorous calculations of scattering from two-dimensional random surfaces are time consuming, we illustrate this approach to the design of random surfaces that scatter light in a circularly symmetric fashion by performing the calculations of the mean differential reflection coefficient in the Kirchhoff approximation, but without going to the geometrical optics limit of it. The mean differential reflection coefficient in this approximation is given by

$$\left\langle \frac{\partial R}{\partial \Omega_s} \right\rangle = \frac{1}{S}\left(\frac{\omega}{2\pi c}\right)^2 \langle |r(\mathbf{q}_\|)|^2 \rangle, \tag{3.38}$$

where the area S is $S = \pi(Nb)^2$, and

$$r(\mathbf{q}_\|) = 2\pi \sum_{n=0}^{N-1} \exp(-iab_n) \int_{nb}^{(n+1)b} dr r J_0(q_\| r)\exp(-iaa_n r), \tag{3.39}$$

where $J_0(z)$ is a Bessel function, $a = (\omega/c)(1+\cos\theta_s)$, and $q_\| = (\omega/c)\sin\theta_s$.

The procedure now is to generate a large number N_p of realizations of the random surface, and for each realization to calculate the corresponding value of $|r(\mathbf{q}_\|)|^2$. The N_p values of $|r(\mathbf{q}_\|)|^2$ so obtained are summed and the sum then divided by N_p to yield the average indicated in Eq. (3.38). It should be noted that the average $\langle r(\mathbf{q}_\|)|^2\rangle$ must increase quadratically with N in order to produce a mean differential reflection coefficient that is independent of N. This is indeed found to be the case for the example studied.

An example of the surface profile function $H(r)$ and its derivative calculated by the approach proposed here is plotted in Fig. 8 for the circular uniform diffuser defined by Eq. (3.36). The parameters used in generating these functions were $\theta_m = 20°$, $b = 200\lambda$, and $N = 20$. In addition, we set $b_0 = 0$ in Eq. (3.31). Due to the form of $f(\gamma)$ given by Eq. (3.37) each slope a_n is positive, with the result that the surface increases monotonically with increasing r.

In Fig. 9 we present a plot of $\langle \partial R/\partial \Omega_s \rangle$ as a function of θ_s for this circular uniform diffuser obtained by the approach proposed in this paper. The parameters used in obtaining the result plotted in this figure were the same as those assumed in obtaining the surface profile function plotted in Fig. 9. The results obtained for $N_p = 8000$ realizations of the surface profile function were used in calculating the average indicated in Eq. (3.38). We see from this figure that the random surface

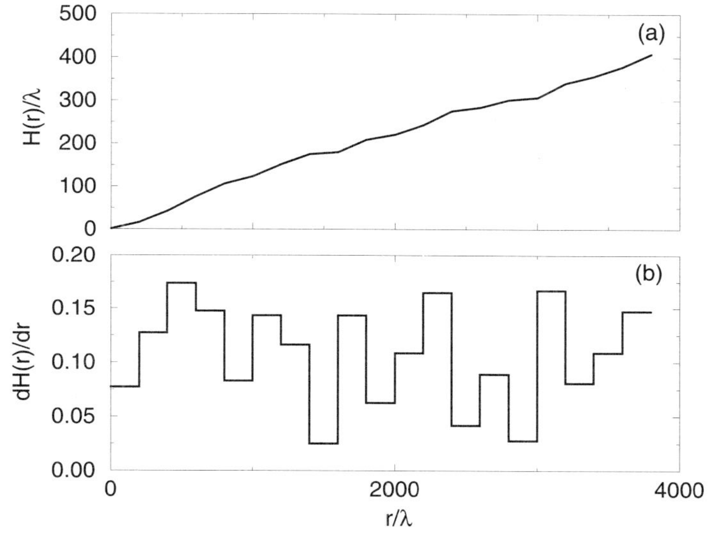

FIGURE 8. A segment of one realization of a numerically generated surface profile function $H(r)$ (a) and its derivative (b) for a two–dimensional surface that acts as a circular band–limited uniform diffuser. The parameters employed are $\theta_m = 20°$, $b = 200\lambda$ and $N = 20$.

generated by our approach indeed acts as a band-limited uniform diffuser. There is no scattered intensity for scattering angles larger than 20°, and the cutoff is very sharp. For values of θ_s in the interval $0 \leq \theta_s < 20°$ the scattered intensity is very closely uniform.

A comment needs to be made concerning the calculations that led to Fig. 9. If the coefficients $\{a_n\}$, and hence the surface profile function $H(r)$, are obtained by a straightforward application of the rejection method to the pdf $f(\gamma)$ given by Eq. (3.37), a fraction of the surfaces generated possess regions in which several consecutive a_n are very small, namely smaller than 10^{-2}. The presence of such practically horizontal regions on the surface gives rise to a peak in the mean differential reflection coefficient in the immediate vicinity of the specular direction $\theta_s = 0°$. To eliminate this peak, as a particular realization of the function $H(r)$ was being generated the differential reflection coefficient at $\theta_s = 0°$ was calculated, and if its value exceeded $2\max\langle \partial R/\partial \Omega_s \rangle$, that surface was discarded. Although this procedure alters the statistics of the resulting surface from the statistics defined by the pdf $f(\gamma)$, the altered statistics produces a surface with the desired scattering properties, and can be employed in fabricating surfaces on photoresist with these properties.

Surfaces of the type considered here can be fabricated on photoresist in the following way. Consider the function $h(\rho)$ that constitutes a single realization of the random process constructed according to Eqs. (3.30), (3.31), and (3.35), and satisfies the requirement that the corresponding differential reflection coefficient at

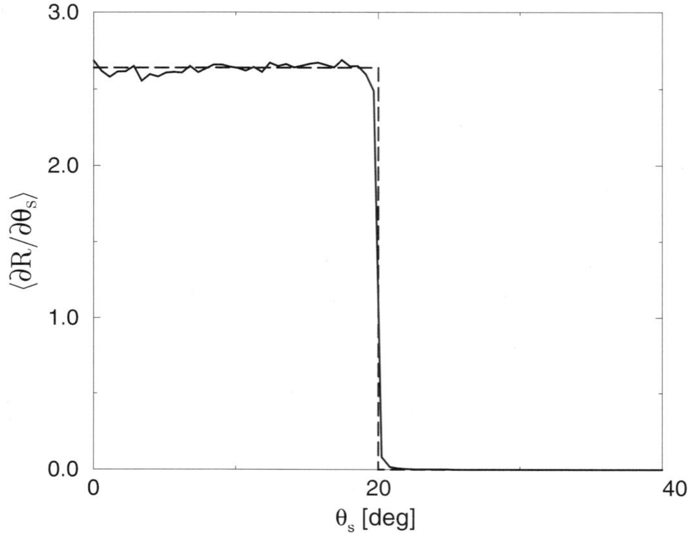

FIGURE 9. The mean differential reflection coefficient $\langle \partial R/\partial \Omega_s \rangle$ estimated from $N_p = 8,000$ realizations of the surface profile function for the case of normal incidence. The calculations were carried out for a two–dimensional Dirichlet surface that acts as a band–limited uniform diffuser within a circular region of scattering angles on the basis of the Kirchhoff approximation given by Eqs. (3.38) - (3.39). The parameters employed are $\theta_m = 20°$, $b = 200\lambda$ and $N = 200$. The dashed curve is a plot of the function given by Eq. (3.36).

$\theta_s = 0°$ does not exceed $2max\langle \partial R/\partial \Omega_s \rangle$. We define the function $\theta_h(\rho) = -\kappa h(\rho)$ that, with an appropriate choice of the constant κ can be interpreted as an angle. A disk-shaped mask is then constructed, whose transparent section is defined by the condition $\theta_h(\rho) < \theta < \theta_g$, where θ_g is a constant greater than $[\theta_h(\rho)]_{max}$.

A well-corrected optical system forms a demagnified image of the mask on a photoresist-coated plate, as shown in Fig. 10. The rotating photoresist plate is exposed for a time T_e, during which it executes a large integer number of revolutions. The total exposure of the plate will then be circularly symmetric, with a radial dependence of the form $E(r) = KT_e\,\theta_g - \theta_H(r)/2\pi$, where $\theta_H(r)$ is a scaled version of the function $\theta_h(\rho)$, and K is a constant related to the sensitivity of the photoresist and the intensity of the illumination. The exposure of the plate can be put in the final form $E(r) = E_0 + \alpha H(r)$, where E_0 and α are constants that can be adjusted by varying the shape of the mask and the exposure time. Assuming that the relation between exposure and resulting height on the plate is linear, the developed surface will have the desired properties. Moreover, characterized nonlinearities of the photoresist response can be taken into account by prescribed deformations of the mask in Fig. 10.

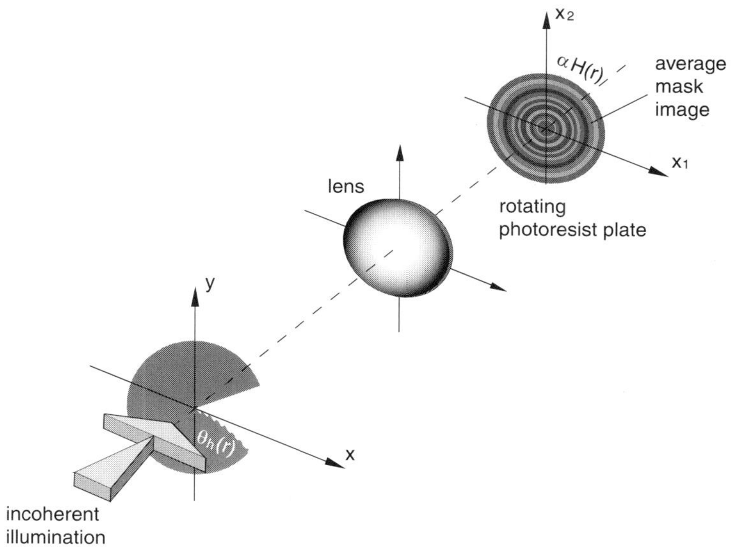

FIGURE 10. Schematic diagram of the setup proposed for the fabrication of a circulaly symmetric random surface.

4. Conclusions

In this paper we have presented methods for designing one- and two-dimensional random surfaces that produce a specified angular dependence of the intensity of the light scattered from them. These methods are based on the geometrical optics limit of the Kirchhoff approximation, a single-scattering approximation, for the scattering of scalar plane waves from impenetrable surfaces. However, the results obtained by these methods have a significantly wider range of applicability.

In the case of one-dimensional random surfaces we have presented a method for generating a perfectly conducting surface that produces a specified angular dependence of the intensity of the light scattered from it when it is illuminated at an arbitrary angle of incidence by $s-$polarized light. This method has been used to design a one-dimensional random surface that behaves as a band-limited uniform diffuser. Within the Kirchhoff approximation a perfectly conducting surface designed in this way behaves as a band-limited uniform diffuser when illuminated by $p-$polarized light as well as when illuminated by $s-$polarized light, and when multiple-scattering effects of all orders are taken into account. For a perfectly conducting surface it is an achromatic diffuser. The surface also acts as a band-limited uniform diffuser when it bounds a semi-infinite metal rather than a perfect conductor, and is also very nearly achromatic in this case as well. A method for fabricating such a surface on photoresist has also been described.

In the case of two-dimensional random surfaces we have presented methods for designing Dirichlet and Neumann surfaces that within the Kirchhoff approximation behave as band-limited uniform diffusers within a rectangular or circular region of scattering angles when illuminated at normal incidence by a scalar plane wave. These methods have been validated by computer simulation calculations, and approaches to fabricating such surfaces on photoresist have been described.

References

[1] C. N. Kurtz, *Transmittance characteristics of surface diffusers and the design of nearly band-limited uniform diffusers*, J. Opt. Soc. Am. **62** (1972) 929-989.

[2] C. N. Kurtz, H. O. Hoadley, and J. J. DePalma, *Design and synthesis of random phase diffusers*, J. Opt. Soc. Am. **63** (1973) 1080-1092.

[3] Y. Nakayama and M. Kato, *Linear recording of Fourier transform holograms using a pseudorandom diffuser*, Appl. Opt. **21** (1982) 1410-1418.

[4] M. Kowalczyk, *Spectral and imaging properties of uniform diffusers*, J. Opt. Soc. Am. A**1** (1984) 192-200.

[5] See, e.g. Thorlabs Inc. Newton, New Jersey 07860-0366 USA.

[6] Physical Optics Corp. Torrance, California 90501-1821 USA.

[7] T. A. Leskova, A. A. Maradudin, I. V. Novikov, A. V. Shchegrov, and E. R. Méndez, *Design of one-dimensional band-limited uniform diffusers of light*, Appl. Phys. Lett. **73** (1998) 1943-1945.

[8] E. R. Méndez, G. Martinez-Niconoff, A. A. Maradudin, and T. A. Leskova, *Design and synthesis of random uniform diffusers*, Proc. SPIE **3426** (1998) 2-13.

[9] A. A. Maradudin, I. Simonsen, T. A. Leskova, and E. R. Méndez, *Random surfaces that suppress single scattering*, Opt. Lett. **24** (1999) 1257-1259.

[10] A. A. Maradudin, T. A. Leskova, E. R. Méndez, and A. V. Shchegrov, *The design and fabrication of one-dimensional random surfaces with specified scattering properties*, Physics of the Solid State **41** (1999) 835-841.

[11] I. Simonsen, T. A. Leskova, A. A. Maradudin, and E. R. Méndez, *Design of one-dimensional Lambertian diffusers of light*, Waves in Random Media **11** (2001) 529-533.

[12] E. R. Méndez, E. E. García-Guerrero, H. M. Escamilla, A. A. Maradudin, T. A. Leskova, and A. V. Shchegrov, *Photofabrication of random achromatic optical diffusers for uniform illumination*, Appl. Opt. **40** (2001) 1098-1108.

[13] A. A. Maradudin, T. A. Leskova, E. R. Méndez, and I. Simonsen, *Band-limited uniform diffuser in transmission*, Proc. SPIE **4100** (2000) 113-123.

[14] A. A. Maradudin, E. R. Méndez, T. A. Leskova, and I. Simonsen, *A band-limited uniform diffuser in transmission. II. A multilayer system*, Proc. SPIE **4447**, (2001) 130-139.

[15] E. I. Chaikina, E. E. García-Guerrero, Zu-Han Gu, T. A. Leskova, A. A. Maradudin, E. R. Méndez, and A. V. Shchegrov,*Multiple-scattering phenomena in the scattering of light from randomly rough surfaces*, in *Frontiers of Laser Physics and Quantum Optics*, eds. Z. Xu, S. Xie, S. Y. Zhu, and M. O. Scully (Springer-Verlag, New York, 2000), pp. 225-259.

[16] A. E. Danese, *Advanced Calculus*, Vol. I (Allyn and Bacon, Boston, 1965), p. 123.

[17] A. A. Maradudin, T. Michel, A. R. McGurn, and E. R. Mendez, *Enhanced backscattering of light from a random grating*, Ann. Phys. (N.Y.) **203** (1990) 255-307.

[18] W. H. Press, S. A. Teukolsky, W. T. Vetterling, and B. P. Flannery, *Numerical Recipes in Fortran*, 2nd Edition (Cambridge University Press, New York, 1992), pp. 281-282.

DEPARTMENT OF PHYSICS AND ASTRONOMY, UNIVERSITY OF CALIFORNIA, IRVINE, CA 92697
E-mail address: `leskova@duey.ps.uci.edu`

DEPARTMENT OF PHYSICS AND ASTRONOMY, UNIVERSITY OF CALIFORNIA, IRVINE, CA 92697
E-mail address: `aamaradu@uci.edu`

DIVISIÓN DE FÍSICA APLICADA, CENTRO DE INVESTIGACIÓN CIENTIÍFICA Y DE EDUCACIÓN SUPERIOR DE ENSENADA, KM 107 CARRETERA TIJUANA-ENSENADA, ENSENADA, B.C. 22860, MÉXICO
E-mail address: `emendez@cisese.mx`

DEPARTMENT OF PHYSICS AND ASTRONOMY, UNIVERSITY OF CALIFORNIA, IRVINE, CA 92697
E-mail address: `javierml@duey.ps.uci.edu`

Contemporary Mathematics
Volume **339**, 2003

Resonance optical switch: calculation of resonance eigenvalues

Boris Pavlov and Kieran Robert

ABSTRACT. Mathematical model of a resonance triadic optical switch is suggested in form of a planar circular lens with one linear input fiber and three linear output fibers (terminals) attached to the boundary of the lens. Assuming that the refractive coefficient of the lens can be manipulated by electric field, we can choose the places of contacts such that the beam of light incoming from the input fiber is re-directed to one of three terminals, depending on electric field applied.The problems of choosing the geometric parameters of the switch and of its temperature stability are addressed. Both problems are reduced to the classical problem of spectral analysis of the wave equation on a circular domain.The techniques developed previously in a similar problem for Schrödinger equation obtained as an additive perturbation of Laplacian by the potential is not applicable to the wave equation

$$u_{tt} = \frac{1}{\rho^2(x)} \triangle u$$

defined by the similar multiplicative perturbation. To overcome the resulting analytic complications, a new perturbation technique is introduced, which also allows one to consider more general problems of large perturbations of operators with discrete asymptotically rare spectrum.

1991 *Mathematics Subject Classification.* Primary 47E05, 35P99, 35Q60; Secondary 35Q72, 78A48.

Key words and phrases. perturbation, eigenvalue, optical switch, quantum wire.

The first author is grateful to the Japanese Foundation JSPS and to the Russian Academy of Sciences' RFFI grant for the partial support of this work.

1. Introduction. A resonance quantum switch: a basic outline

Our approach to an optical switch is based on the previous study of resonance quantum switches. This is why in this section we provide a brief outline of the work concerning the quantum switch case.

In a standard mechanical switch, the input can be connected to and disconnected from any of the terminals by flipping a control knob. In the quantum situation of a device consisting of an input wire, a quantum well, and a few output terminals, a similar function can be performed by changing the resonance eigenfunction of the quantum well that corresponds to the resonance energy level equal to the energy of incoming particles. The idea of the resonance manipulation of the quantum current was developed in the series of papers [**7, 5, 20, 21, 22, 23, 26, 27**] and patented in [**24**]. It is based on the relation between the conductance and scattering discovered by R. Landauer [**15**] (see also [**10, 1**] and an observation in [**11**] concerning the role of the shape of the resonance eigenfunction in the scattering problem for the resonator with a sufficiently small opening) [1]. Many interesting observations concerning the resonance manipulation can be made on zero-range solvable models, see for instance [**3**].

There is a lot in common between the linear wave processes in quantum and optical systems. We use here not only this general connection, but also some technical details from the previous papers cited above.

Consider a quantum switch Ω constructed as a planar (contained in a plane L) network consisting of a circular domain (a quantum well, or a lens in the optical situation) Ω_0 with four quasi-one-dimensional wires $\Omega_j, j = 1,..,4$ of width δ attached to it. An incoming electron enters the network via the input wire Ω_1 and then leaves via one of three similar terminals $\Omega_s,\ s = 2,3,4$, passing the well Ω_0 on its way.

The working regime of the quantum switch is defined by the physical and geometrical parameters of the device (e.g., the Fermi level E_F and the effective electron mass in the wires, the radius of the well, the intensity $\mathcal{E}$ and direction ν ($|\nu| = 1,\ \nu \parallel L$) of the electric field $\mathcal{E}\nu$, and the minimal values of the potential on the well and on the wires). We demonstrate below the principle of the resonance manipulation of the quantum current across the circular quantum well. The dynamics of electrons in the device depends on spectral properties of the Schrodinger equation on the corresponding network. The structure

[1]Review of essential physical literature concerning the resonance manipulation of the quantum current can be found in [**9**]. None of the devices discussed there uses the shape of the wave-function.

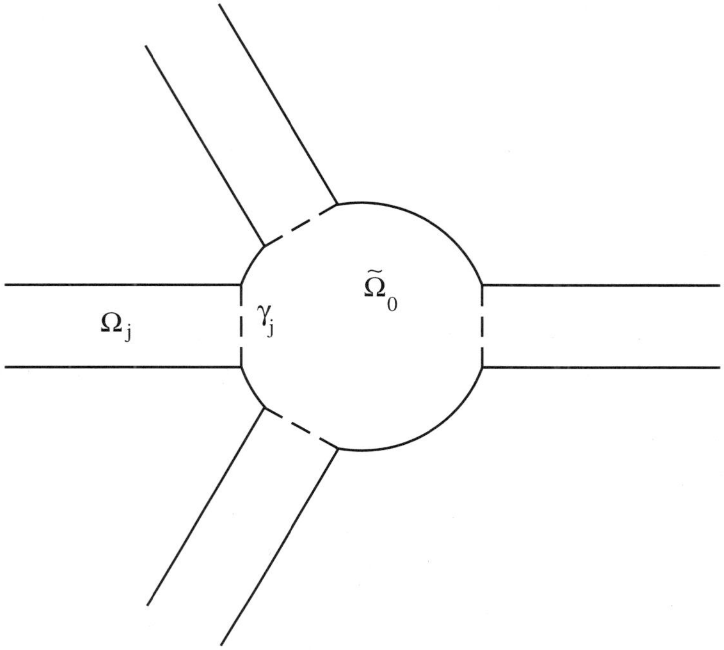

FIGURE 1. The circular lens Ω_0 with wires Ω_j attached. The modified lens $\widetilde{\Omega}_0$ has small circular arks of the circular boundary replaced by the flat cross-sections γ_j of the entering wires.

of the spectrum and the shape of the eigenfunctions is especially important for the values of energy E near the Fermi energy level $E \approx E_F$ [**17**], since only electrons with energy close to Fermi level can contribute to the conductance, see for instance [**6**].

Consider the scattering problem on the above network $\Omega = \cup_{s=0}^{4} \Omega_s$ for the Schrödinger equation

$$-\frac{\hbar^2}{2m_0} \Delta \psi + V\psi = E\psi \tag{1}$$

with Dirichlet boundary condition $\psi|_{\partial\Omega} = 0$. Assuming that the potential is constant $V = V_1$ along each wire, we can separate variables in the non-perturbed infinite wires $\psi = \psi^{\perp}\psi^{\|}$. Here $\psi^{\perp}$ and $\psi^{\|}$ are functions that depend only on the transversal variable y and the axial variable x correspondingly [2]. We choose $\psi^{\perp}$ on the cross-section $0 < y < \delta$ of the

[2]We will in general use the signs $\|$ and $\perp$ to indicate dependence on the axial and transversal variable respectively.

wire as a transversal mode, i.e. solution of the problem

$$-\frac{\hbar^2}{2m_0}\,\triangle^{\perp}\,\psi^{\perp}=E^{\perp}\psi^{\perp}$$

with the boundary condition $\psi|_{\partial\Omega_s}=0$, $s=1,2,3,4$. This implies quantization across the wire:

$$\psi^{\perp}_m=\sqrt{\frac{2}{\delta}}\sin\frac{m\pi y}{\delta}.$$

The wave function in a wire assumes the form of a linear combination of the exponential modes $\psi_m(x,y)=\psi^{\perp}_m(y)e^{\pm\frac{ipx}{\hbar}}$. Substituting this expression into the corresponding Schrodinger equation, one obtains a dispersion relation between the momentum p of the electron along the wire and the total energy E

$$\frac{\hbar^2}{2m_0}\,\frac{m^2\pi^2}{\delta^2}+\frac{p^2}{2m_0}+V_1=E.$$

The exponential modes in the wires can be oscillating or decreasing[3], depending on the sign of the expression

$$\frac{p^2}{2m_0}\;:=\;-\frac{\hbar^2}{2m_0}\left[\frac{n\pi}{\delta}\right]^2+E-V_1.$$

Namely,

$$\psi(y,x)=\psi^{\perp}(y)e^{\pm\frac{ipx}{\hbar}},\;p\in\mathbb{R}$$

on the "open channels" labelled with the index n, where $p^2>0$, and

$$\psi(y,x)=\psi^{\perp}(y)e^{-\frac{|p|x}{\hbar}},\;ip=|p|>0$$

on the "closed channels" with purely imaginary momentum p. The levels $E^{\perp}=\frac{n^2\pi^2\hbar^2}{2m\delta^2}$ play the role of thresholds separating the branches of the continuous spectrum of the wire with different spectral multiplicities. The resonance manipulation of the electron current is possible if only two oscillatory modes with n=1 are present in each wire, that is: if the Fermi energy E_F sits on the first spectral band, below the second spectral threshold,

$$V_1+\frac{\hbar^2\pi^2}{2m_0\delta^2}<E_F<V_1+\frac{4\hbar^2\pi^2}{2m_0\delta^2}.$$

If the width of the wires δ is chosen according to this condition, then all modes corresponding to the higher thresholds $E^{\perp}_m=\frac{\hbar^2\pi^2m^2}{2m_0\delta^2}$, $m\geq 2$

[3]We do not consider here the resonance modes which are exponentially growing.

are exponentially decaying [4]. Consider the space $L_2(\Gamma)$ of all square-integrable functions on the union $\Gamma = \cup_{s=1}^4 \Gamma_s$ of the flat "bottom" cross-sections Γ_s of the wires (see Fig. 1). The subspace $E_+ \subset L_2(\Gamma)$ of the functions corresponding to the "open" first channel with $n = 1$ is four-dimensional and the orthogonal subspace $E_- = L_2(\Gamma) \ominus E_+$ spanned by the cross-section functions of the higher channels with $n \geq 2$ is infinite-dimensional. Denote by P_+ the orthogonal projection onto E_+ in $L_2(\Gamma)$.

Scattered waves, i.e., solutions of the scattering problem on the whole network Ω can be obtained via matching solutions of the Schrödinger equation on the quantum well Ω_0 with the exponential modes on the wires. The corresponding technique is developed in [**22, 26**] where the scattering matrix is computed based on the Dirichlet-to-Neumann map (see e.g. [**8, 25, 31**] concerning this map). The approximate "one-pole" formula for the scattering matrix near the resonance eigenvalue λ_0 of the Schrödinger operator on the well Ω_0 can be presented in terms of the orthogonal projection onto an one-dimensional "entrance subspace". This subspace is spanned by the E_+ portion $P_+ \frac{\partial \varphi_0}{\partial n}\Big|_\Gamma = \phi_0$ of the resonance eigenfunction in the quantum well Ω_0. If the wires are "not too wide", then the scattering matrix can be presented in terms of the orthogonal projection onto the *entrance subspace* of E_+ spanned by the entrance vector ϕ_0

$$P_0 = \frac{\phi_0\rangle\ \langle\phi_0}{|\phi_0|^2} = \left\{ \frac{\langle \frac{\partial \Psi}{\partial n_l}, e_l^1 \rangle \langle \frac{\partial \Psi}{\partial n_s}, e_s^1 \rangle}{\sum_{t=1}^4 |\langle \frac{\partial \Psi}{\partial n_t}, e_t^1 \rangle|^2} \right\}_{l,s=1}^4 .$$

Denoting by $P_0^\perp = I_+ - P_0$ the complementary orthogonal projection, one can derive the following "one-pole approximation" [**26**] near the resonance eigenvalue λ_0:

$$\begin{aligned} \mathbf{S} &\approx -\frac{\frac{|\varphi_0|^2}{\lambda - \lambda_F - \Delta_0} + i\sqrt{\lambda - \frac{\pi^2}{\delta^2}}}{\frac{|\varphi_0|^2}{\lambda - \lambda_F - \Delta_0} - i\sqrt{\lambda - \frac{\pi^2}{\delta^2}}} P_0 + P_0^\perp \\ &= I - \frac{2|\phi_0|^2}{|\phi_0|^2 - i\sqrt{\lambda - \frac{\pi^2}{\delta^2}}(\lambda - \lambda_F - \Delta_0)} P_0 \\ &= \left\{ \delta_{st} - 2i \frac{(\phi_0)_s\ (\phi_0)_t}{|\phi_0|^2 - i\sqrt{\lambda - \frac{\pi^2}{\delta^2}}(\lambda - \lambda_F - \Delta_0)} \right\}_{s,t=1}^4 . \end{aligned} \tag{2}$$

Here Δ_0 is the shift of the resonance eigenvalue caused by the interaction of the electron with closed channels, $\lambda = \dfrac{2mE}{\hbar^2}$, and the resonance

[4]For a detailed analysis of spectra of waveguides see [**30**].

eigenvalue $\lambda_F = \dfrac{2mE_F}{\hbar^2}$. In dimensionless coordinates $\Delta_0 = -0.07$, so it may be neglected, compared with the distance 2.3 between the resonance eigenvalue 14.62 and the nearest (non-resonance) eigenvalue 12.32. One can see that the transmission coefficients T_{1t} are presented as

$$T_{1t} \approx -2i \frac{(\phi_0)_1\ (\phi_0)_t}{|\phi_0|^2 - i\sqrt{\lambda - \frac{\pi^2}{\delta^2}}(\lambda - \lambda_F - \Delta_0)}.$$

Hence for the resonance energy $\lambda = \lambda_F + \Delta_0$ and zero temperature the transmission coefficients are approximately equal to $-2i\dfrac{(\phi_0)_1\ (\phi_0)_t}{|\phi_0|^2}$. The resonance manipulation of the quantum current boils down, due to the Landauer formula for the quantum conductance [**6, 15**] [5]

$$\sigma_{1t} = \frac{e^2}{h}\ \frac{\left|T_{1t}\right|^2}{1 - \left|T_{1t}\right|^2},$$

to changing the transmission coefficients T_{1s}. It was proved in [**21, 22**] that the magnitude $\mathcal{E}$ of the electric field $\mathcal{E}\nu$ and the minimal value of the potential on the quantum well Ω_0 can be chosen such that for the resulting linear potential

$$V(x) = \mathcal{E}e\langle \nu,\, x\rangle + V_0$$

one of eigenvalues, namely the "resonance eigenvalue" [6] $\lambda_2 = \lambda_F + \Delta \approx \lambda_F$ coincides with the Fermi level E_F in the wires and the normal derivative of the corresponding "resonance" eigenfunction $\Psi = \Psi_2$ has two zeroes on the boundary at the points $\theta_1(\nu), \theta_2(\nu)$ such that the corresponding radii form the central angles $\pm\frac{\pi}{6}$ with the direction of the vector ν.

If the wires are sufficiently thin, the behavior of the resonance eigenfunction on the bottom sections γ_j of the wires is almost linear. Hence, the Fourier coefficients $\langle \frac{\partial \Psi}{\partial n}, e_l\rangle$ are close to zero if the axis of the wire enters the zero of the normal derivative of the resonance eigenfunction. It was shown in [**22**] that the second eigenfunction of the "even" series [7] of eigenfunctions of the Schrödinger operator on the quantum well Ω_0 with the above potential and Dirichlet boundary conditions satisfies this requirement if the radius of the well R, the Fermi level E_F, the

[5]This expression should be doubled if the spin is taken into account

[6]We borrow this term from [**14**].

[7]See the description of the even and odd series at the end of Section 2.

effective mass of the electron m and the magnitude $\mathcal{E}$ of the electric field satisfy the conditions

$$\frac{2m\mathcal{E}eR^3}{\hbar^2} = 18.86, \; 2mR^2\,\frac{E_F - V_0}{\hbar^2} = 14.62.$$

Here $\hbar$ is the conventional "divided" Planck constant and e is the electron charge.

A similar switching effect for the homogeneous Neumann conditions on the boundary of the well and zero-dimensional wires was described in [**21**]. The approximate expression for the scattering matrix in this case is

$$T^N_{ls} \approx -2i\frac{\langle \Psi, e^1_l\rangle\langle \Psi, e^1_s\rangle}{\sum_{t=1}^4 |\langle \Psi, e^1_t\rangle|^2},$$

and the switching condition can be formulated in terms of zeroes of the resonance eigenfunction on the boundary: the corresponding radii should form the angles $\pm\frac{\pi}{6}$ with the direction ν of the electric field. It was shown in [**21**] that the second eigenfunction of the even series of eigenfunctions of the Schrödinger operator on the circular well Ω fulfills this requirement if the radius of the well R, the Fermi level E_F, the effective mass of the electron m and the magnitude $\mathcal{E}$ of the electric field are chosen such that:

$$\frac{2m\mathcal{E}eR^3}{\hbar^2} = 3.558, \; 2mR^2\,\frac{E_F - V_0}{\hbar^2} = 3.79.$$

The temperature stability of the quantum switch is defined by the minimal distance of the resonance eigenvalue from the rest of the spectrum. In case of the Dirichlet boundary condition the nearest to 14.62 non-dimensional eigenvalue is 12.32, hence the condition of temperature stability is [**5**]

$$\kappa T < \frac{(14.62 - 12.32)\hbar^2}{2mR^2}.$$

In particular the resonance quantum switch self-assembled[8] on the Si matrix is stable at Helium temperature $4.2K$ if the radius of the corresponding quantum well is not larger than 8.5 nm, and the wires are 2 nm wide.

Similarly, for Neumann conditions the nearest non-dimensional eigenvalue is -0.79, hence the condition of the temperature stability is [**21**]:

$$\kappa T < \frac{(3.79 + 0.79)\hbar^2}{2mR^2}.$$

[8]via controlled diffusion of atoms [**4**]

2. A resonance optical switch: a basic outline

In the previous section we mentioned a close similarity between quantum and optical processes. We thus propose a model of a resonance optical switch that is manipulated via change in the refractive coefficient due to applied electric field. We do not discuss here the important issue of practical implementability of an optical switch. The reader can notice, for instance, that the spectra below should be rescaled in order for the refractive coefficient to satisfy the condition $\rho > 1$. This is easy to do, but it would however require the values of refractive index somewhat higher than the currently achievable in dielectric materials. We also neglect the practical limitation on how much the refractivity can be manipulated by the electric field. These important questions should and probably will be addressed elsewhere. Here we will just show that our previous results and techniques [**21**] concerning the resonance quantum switch can be used as an analytic base for the optical switch construction.

We assume that the optical switch is constructed as a network Ω consisting of a circular lens $\Omega_0 = \{x : |x| < 1\}$ (in non-dimensional coordinates) and attached to it linear optical fibers Ω_s, $(s = 1, 2, 3, 4)$ of width δ. We assume that the whole scattering problem on the network Ω can be considered as a planar one. In attempt to skip several long steps in analysis of the Resonance Optical Switch, we use our previous results on the Schrödinger equation (1) with a linear potential inside the quantum well radius 1 , $|x| \leq 1$ (in non-dimensional coordinates):

$$-\triangle u + \varepsilon |x| \cos\theta \; u = \lambda_2 u.$$

We will just replace this equation by the equivalent stationary wave equation

$$-\triangle u = \lambda_2 \left[1 - \frac{\varepsilon}{\lambda_2} |x| \cos\theta \right] u.$$

The we assume that the refractive index $\rho(x)$ of the lens is given by the formula

$$\rho^2(x) = 1 - \frac{\varepsilon}{\lambda_2} |x| \cos\theta, \tag{3}$$

where $\lambda_2 = 3.79$, $\varepsilon = 3.558$, and the angle θ measures the deviation of the vector x from some fixed direction ν. The bounded positive function ρ defines the metric of the Hilbert space $L_2(\Omega_0, \rho)$ on the lens where the spectral problem will be treated, with respect to the

Euclidean distance $|dx|$ or Lebesgue measure m:

$$dl = \rho\,|dx|,\ \ d\Omega = \rho^2 dm,\ \ \langle u,\, v\rangle = \int_{\Omega_0} u\bar{v}\rho^2 dm$$

We assume that the spectral problem for the Maxwell system inside the lens is reduced to the spectral problem for the stationary wave equation on the planar circular lens Ω_0 :

$$-\triangle\,\Psi = \mu\left[1 - \frac{\varepsilon}{\lambda_2}|x|\cos\theta\right]\Psi, \tag{4}$$

with the homogeneous Neumann boundary condition [9]

$$\left.\frac{\partial\Psi}{\partial n}\right|_{\partial\Omega} = 0.$$

The spectral parameter μ is the square of the frequency, $\mu = \omega^2$. Note that the equation (4) for the special value of the spectral parameter $\mu = \lambda_2$ coincides with the Schrödinger equation discussed in the previous section and thus the previously obtained results can be applied. The parameters ε, λ_2 are chosen above in such a way that the resonance eigenfunction of the Schrödinger equation fulfills all requirements that were needed in order to ensure the resonance manipulation of the quantum current. We will see that the same function satisfies the conditions for the resonance manipulation of the optical switch, see (2.1) below. This gives us for free a "tentative solution" [10] of the problem of choosing of the parameters of the switch ensuring the resonance manipulation. To solve the remaining problem of *stability* we will have to calculate the neighboring eigenvalues of the wave equation on the lens, which requires special perturbation techniques. This will be done in the next sections.

Our aim is to verify that methods of control of conductance based on perturbation theory developed in [**5, 21, 22, 23, 26, 27**] for the Schrödinger equation can be modified for the wave equation. This question is not entirely trivial, since the potential term in the order Schrödinger equation is a zero-order perturbation, while the change of the refractivity index is a second order perturbation. We will show in the next section that the classical analytical perturbation techniques can be modified to cover the case of the wave equation. We follow the idea of [**21**], but add also some new details. In this section we reduce the problem of manipulating transmission across the optical switch to

[9] Other self-adjoint boundary conditions can be handled as well.

[10] Up to the aforementioned problems of implementation.

the spectral problem for the wave equation which will be treated in the following section.

In the remaining part of this section we sketch, based on rigorous analysis presented in [**22, 26**], the derivation of an approximate expression for the scattering matrix of the resonance optical switch.

We assume now that the lens, i.e. the circular domain Ω_0 is modified in the natural manner to $\tilde{\Omega}$ to match the flat sections γ_s, $s = 1, 2, 3, 4$ of the entering fibers Ω_s (see Fig. 1). It is shown in [**26**] that the spectral characteristics of the Schrödinger equation on the modified domain are close in an appropriate sense to the corresponding characteristics in the original domain. A similar statement is valid for the wave equation. All characteristics of the modified domain, if not indicated specifically, will be denoted with tilde. For instance, the resonance eigenvalue will be denoted $\tilde{\lambda}_2$ and the resonance eigenfunction in the modified domain will be denoted $\tilde{\Psi}$.

We assume that the refractive index of the material used for fabrication of the fibers is equal to the minimal value

$$\rho_0 = \sqrt{\frac{3.79 - 3.558}{3.79}} \tag{5}$$

of the refractive index inside the lens [11] and that the matching conditions at the contact segments γ_s between the lens and the fibers imply the continuity of the solution and of its normal derivative. Then the wave equation in the fibers takes the form

$$- \triangle u = \mu \rho_0^2 u, \ 0 < x < \infty, \ 0 < y < \delta,$$

with the homogeneous Neumann conditions on the boundary $y = 0, \delta$:

$$\left. \frac{\partial u}{\partial n} \right|_{\partial \Omega_s} = 0, \ s = 1, 2, 3, 4.$$

The solutions have a form of exponential modes similar to ones in the previous section:

$$u^{\pm}(x, y) = \cos \frac{\pi m y}{\delta} e^{\pm i \frac{\sqrt{\mu - \frac{\pi^2 m^2}{\delta^2}}}{\rho_0} x}, \ n = 0, 1, , \ldots .$$

The bounded exponential modes $\Im \sqrt{\mu - \frac{\pi^2 m^2}{\delta^2}} > 0$ are oscillating or decreasing, depending on the sign of the expression under the square

[11]This assumption does not play any essential role and the case of any constant refractivity of the wires can be handled.

root. We consider the case

$$0 < \mu < \frac{\pi^2}{\delta^2},$$

when only two oscillating modes $e^{\pm i \frac{\sqrt{\mu}}{\rho_0} x}$ are present, which correspond to the open channel labelled with $m = 0$. All higher exponential modes

$$u^+_{m,s}(x,y) = \cos \frac{\pi m y}{\delta} e^{-\frac{\sqrt{\frac{4\pi^2 m^2}{\delta^2} - \mu}}{\rho_0} x}, \ m = 1, 2, \ldots; \ s = 1, 2, 3, 4.$$

are decreasing. They correspond to the closed channels. Scattered waves, i.e. the eigenfunctions of the wave equation on the network Ω, are constructed via matching the linear combinations of the above exponential modes on the fibers Ω_s, $s = 1, 2, 3, 4$:

$$u^-_{0,1}(x,y)\delta_{s1} + u^+_{0,s}(x,y) S^0_{s1} + \sum_{s=1,\, m=1}^{4,\infty} u^+_{m,s}(x,y) \mathbf{s}^m_{s1} \tag{6}$$

with the solution u_0 of the homogeneous wave equation inside the modified lens. The coefficients $\{S^0_{st}\}^4_{s,t=1}$ constitute the 4×4 scattering matrix in the open channel, while other coefficients are just amplitudes determined from the matching conditions. We impose homogeneous Neumann conditions on the complement of the union $\Gamma = \cup^4_{s=1} \gamma_s$ of the bottom sections of the fibers in the boundary $\partial \tilde{\Omega}_0$ of the modified lens. It is explained in the Appendix of [**26**] that only the open channel ($m = 0$ in our case) contributes essentially to the scattering process if the fibers are sufficiently narrow.

Solution of the non-homogeneous Neumann boundary problem for the modified domain with the boundary data u_s supported by the bottom sections γ_s of the fibers

$$- \triangle u - \lambda \rho^2 u = 0, \ \left. \frac{\partial u}{\partial n} \right|_{\gamma_s} = u_s$$

can be presented via the Green's function $G^N(x, y, \lambda)$ of the Neumann problem in the modified domain:

$$- \triangle_x G^N(x,y,\lambda) - \lambda \rho^2 G^N(x,y,\lambda) = \delta(x,y),$$

$$\left. \frac{\partial G^N(x,y,\lambda)}{\partial n} \right|_{x \in \partial \tilde{\Omega}} = 0, \ y \in \tilde{\Omega} :$$

$$u = \int_{\partial \tilde{\Omega}} G^N(x,y,\lambda) \left. \frac{\partial u}{\partial n} \right|_{\Gamma} (y) d\Gamma \tag{7}$$

Consider the space $L_2(\partial\tilde{\Omega})$ of all functions square integrable on the boundary $\partial\tilde{\Omega}$ of the modified domain. Its subspace $L_2(\Gamma)$ that consists of all functions supported in $\Gamma = \cup_{s=1}^4 \gamma_s$ plays an essential role in the scattering processes. We decompose this subspace into an orthogonal sum of two subspaces spanned by the eigenfunctions of the cross-sections of open and closed channels

$$E := L_2(\Gamma,\, \rho) = E_+ \oplus E_-. \tag{8}$$

In our case the subspace $E_+ = \bigvee_{s=1}^4 e_s^1$ is four-dimensional. We call it *the entrance subspace* of the scattering problem. The scattering matrix is an operator in E_+ (e.g., [**22, 26**]). We denote by P_+ the orthogonal projection of $L_2(\Gamma,\, \rho)$ onto E_+ and form the entrance vector ϕ_0 based on the resonance eigenfunction φ_0 as in previous section. Denoting by P_0 the orthogonal projection in $L_2(\Gamma, \rho_0)$ onto the entrance vector and neglecting as before the contribution of the closed channels, we obtain the following one-pole approximation for the scattering matrix:

THEOREM 2.1. *For sufficiently small width δ of the fibers, the scattering matrix $S(\mu)$ of the network Ω near the resonance eigenvalue $\mu_0 = \lambda_2$ can be approximated by the rational matrix-function of the frequency $\omega = \sqrt{\mu}$, :*

$$\mathbf{S}(\mu) \approx -\frac{\frac{|\varphi_0|^2}{\mu-\mu_0} + i\frac{\omega}{\rho_0}}{\frac{|\varphi_0|^2}{\mu-\mu_0} - i\frac{\omega}{\rho_0}} P_0 + P_0^{\perp} \tag{9}$$

REMARK 2.2. *One can show that this approximation is exact for a solvable model of the network with one-dimensional fibers (i.e., when $\delta = 0$), see the corresponding model for the Schrödinger equation in* [**23**].

PROOF. The theorem can be proven by matching the linear combinations of the exponential modes in the fibers with appropriate solutions of the wave equation inside the modified domain. The standard solution for the scattered waves on the fibers with incoming mode $e^{i\frac{\sqrt{\mu}}{\rho_0}x}e$, $e \in E$, $\| e \| = 1$ is presented in the form (6). The restrictions onto Γ of this expression and its derivatives along the wires

$$e + \mathcal{S}e,\ \mathcal{K}\left[-e + \mathcal{S}e\right]$$

with $\mathcal{K} = i\,\mathrm{diag}\left\{\rho_0^{-1}\sqrt{\mu - \frac{\pi^2 m^2}{\delta^2}}\right\}_{m=0}^{\infty}$ are related via the formula (7). The exact expression for the scattering matrix can be obtained from this relation [**26**]. For $\mu << \frac{\pi^2}{\delta^2}$ this expression can be simplified to obtain formula (9). We use the representation of the Green's function via the spectral series converging in the space of square-integrable functions on the lens:

$$G^N(x,y,\mu) = \sum_{l=1}^{\infty} \frac{\Psi_l(x)\Psi_l(y)}{\mu_l - \mu},$$

where μ_l are the eigenvalues and Ψ_l are the eigenfunctions of the Neumann problem on $\tilde{\Omega}$ ortho-normalized with respect to the weight ρ^2. Near the resonance eigenvalue λ_2 we can neglect all terms except the residue term $\frac{\Psi_0(x)\Psi_0(y)}{\mu_0-\mu}$ with the resonance eigenfunction Ψ_0. For sufficiently narrow fibers (see the estimates in [**26**]) we may consider only the modes in the open channel with $n = 0$. Then we obtain the approximate equation for the scattering matrix in the form

$$[I + \mathbf{S}]\, e + |P_E\Psi_0\rangle\langle P_E\Psi_0| \frac{i\sqrt{\mu}}{\rho_0}\, [I - \mathbf{S}]\, e = 0,$$

which implies the result. □

REMARK 2.3. *One can derive the transmission coefficients from (9).*

Indeed,

$$\mathbf{S}(\mu) \approx I - \frac{2 \parallel P_E\Psi_0 \parallel^2}{\parallel P_E\Psi_0 \parallel^2 - i\,(\mu - \lambda_2)\, \frac{\sqrt{\mu}}{\rho_0}} P_0.$$

Then for the resonance value of the spectral parameter $\mu = \lambda_2$ we obtain the approximate transmission coefficients from the input fiber Ω_1 to the output fiber Ω_s, $s \neq 1$:

$$\mathbf{S}_{s1} \approx -2\frac{\langle \Psi_0, e_s^1\rangle\langle \Psi, e_1^1\rangle}{\parallel P_E\Psi_0 \parallel^2} \tag{10}$$

as off-diagonal elements of the approximate scattering matrix.

Now transmission of light at the resonance frequency $\sqrt{\lambda_2}$ can be manipulated in the same way as the transmission of electrons for the quantum switch.

REMARK 2.4. *One can use the above approximate formula (9) for the scattering matrix on the network to estimate the switching time.*

Indeed, the speed of transition processes in scattering systems with an orthogonal pair of incoming and outgoing subspaces $D_{in,out}$ is estimated by the "life-time" of resonances, i.e. the complex zeroes of the scattering matrix [**16**]. The wave dynamics on the open channels in our network is approximately described by the one-dimensional wave equations $\rho^2 u_{tt} = u_{xx}$, hence the Cauchy data of the d'Alambert waves $u(x \pm \rho_0^{-1} t)$ on the fibers form a pair of incoming and outgoing subspaces. The pair of resonances (zeroes of the scattering matrix situated near the resonance eigenvalue, $\omega^2 = \mu \approx \mu_0 = \omega_0^2$) can be found from the approximate formula (9) for the scattering matrix:

$$\frac{2|\phi_0|^2}{\omega_0^2 - \omega^2} = i\omega\rho_0,$$

$$\omega = \omega_0 + i\frac{|\phi_0|^2}{2\rho_0\omega_0^2}.$$

Hence,

$$\tau = \frac{2\rho_0\omega_0^2}{||P_E\Psi_0||^2}$$

is the life-time of the resonances.

The estimation of stability of transmission with respect to the deviation of frequency from the resonance value or deviation of the geometric details of the optimal construction requires calculation not only of the resonance eigenvalue, but also of the nearest eigenvalues of the wave equation. Normally calculation of eigenvalues by conventional perturbation techniques requires imposing strong conditions on the magnitude of the perturbation compared with the distance between eigenvalues of the non-perturbed operator (see below). This condition is not fulfilled in our situation. We will suggest in the next section a modified perturbation technique for the nearest eigenvalues of the wave equation, based on a version of perturbation theory used previously in [**21**] for Schrödinger equation. We begin with the same non-perturbed operator as in [**21**], the Laplace operator A in the unit disc $\zeta < 1$ with homogeneous Neumann boundary conditions:

$$-\Delta\psi = \mu\psi, \ \left.\frac{\partial\psi}{\partial n}\right|_{\zeta=1} = 0.$$

Standard separation of variables gives the normalized eigenfunctions of the even series

$$\begin{aligned} \psi_0^s &= \frac{J_0(k_0^s\zeta)}{(2\pi\int_0^1(J_0(k_0^s\zeta))^2\zeta d\zeta)^{1/2}}, \\ \psi_n^t &= \frac{J_n(k_n^t\zeta)\cos(n\theta)}{(\pi\int_0^1(J_n(k_{n,}^t\zeta))^2\zeta d\zeta)^{1/2}}, \quad n = 1, 2, \ldots, \end{aligned} \tag{11}$$

and the eigenfunctions of the odd series

$$\psi_n^t = \frac{J_n(k_n^t\zeta)\sin(n\theta)}{(\pi\int_0^1(J_n(k_n^t\zeta))^2\zeta d\zeta)^{1/2}}, \quad n = 1, 2, \ldots. \tag{12}$$

The eigenvalues of the non-perturbed problem $\alpha_n^s = [k_n^s]^2 = [k_n^t]^2$, $s = t$ have multiplicity 2 if $n > 0$ and are equal to the squares of zeroes k_n^s of the derivatives of Bessel functions $J_n'(k_n^s) = 0$. We enumerate them in a monotonically increasing order $[k_n^t]^2 = \alpha_l$, $l = 0, 1, 2, \ldots$.

The spectral analysis of the non-perturbed operator can be done in explicit form. Note that this operator can be decomposed into the orthogonal sum of two operators acting in spectral subspaces constituted by even and odd eigenfunctions

$$A = A_{even} \oplus A_{odd}.$$

The spectral analysis of the perturbed problem

$$-\triangle u = \mu\rho^2 u, \left.\frac{\partial u}{\partial n}\right|_{\partial\Omega_0} = 0 \tag{13}$$

on the domains Ω_0 and $\tilde{\Omega}_0$ with $\rho \neq 1$ can be developed using a version of perturbation theory. In fact, we do not need a complete description of the spectrum of the corresponding perturbed operator A_ρ, but rather only calculation of its eigenvalues and eigenfunctions near the resonance ω^2. For additive perturbations $A \longrightarrow A+V$ this problem can be solved by conventional analytical techniques, see for instance [**13, 28**], if the perturbation is dominated by the spacing of the eigenvalues of the non-perturbed operator near the resonance eigenvalue α_0 :

$$\| V \| \leq \frac{1}{2} \min_l \left| \alpha_0 - \alpha_l \right|. \tag{14}$$

To adjust methods of the perturbation theory to our situation, when the above condition obviously is not fulfilled, we have to extend the field of applicability of the analytic perturbation techniques. In fact, the first step in this direction was made in the paper [**21**], where *additive* perturbation of an operator with asymptotically rare spectrum was considered when studying the resonance conductance of the quantum switch. The main idea of the new perturbation technique introduced in [**21**] was based on using an intermediate non-perturbed operator $\tilde{A}$ that admits exact spectral analysis via standard approaches. Then the calculation of the spectral characteristics of the perturbed operator $A+V$ can be done based on the comparison with the intermediate operator for which the basic condition (14) is satisfied in a neighborhood of the resonance:

$$\| (A+V) - \tilde{A} \| \leq \frac{1}{2} \min_l \left| \tilde{\alpha}_0 - \tilde{\alpha}_l \right|.$$

The role of the non-perturbed operator in [**21**] was played by a finite matrix - the restriction $[A+V]_N$ of the perturbed operator $A+V$ onto a certain spectral subspace E_N of the non-perturbed operator. The deviation of the restriction from the Schur complement of the perturbed operator in the same subspace [**2, 18**] happens to be small. This already allows one to apply the standard techniques of the perturbation theory to this pair, and thus the whole problem is reduced to a finite-dimensional one. Unfortunately, this is done at a price: the problem becomes non-linear with respect to the spectral parameter. In fact, in the case of for additive perturbations this difficulty can be eliminated [**19**], if one constructs a *global* intermediate non-perturbed operator as the orthogonal sum of the above finite matrix $[A+V]_N$ with the part $A_\perp$ of the non-perturbed operator in the orthogonal complement:

$$[A+V]_N \oplus A_\perp := \tilde{A}_N. \tag{15}$$

The global intermediate $\tilde{A}_N$ can be directly compared with the perturbed operator and so the Theorem 4.1 of [**21**] can be considered from this point of view as a *standard perturbation result for the pair* $\tilde{A}_N$ and $A+V$ [**19**].

Unfortunately, this approach based on existence of the global intermediate operator does not work in case of second order perturbations. That is why in this paper we apply the basic idea of [**21**] concerning a finite-dimensional intermediate operator to the case of second order perturbations. Our analysis will be done in general form: the basic structures are introduced axiomatically, and the required conditions are easily verified in our specific situation. We hope thus that the developed technique has a general value.

In our approach we use one more observation from [**21**]. For a given number α one can consider complementary invariant subspaces E_N and $E_\perp$ of the non-perturbed Laplace operator A in $E := L_2(\Omega_0)$ with $N = N(\alpha)$. These subspaces are obtained as linear hulls of eigenfunctions with eigenvalues satisfying respectively the inequalities $0 < \alpha_l < \alpha$ and $\alpha < \alpha_l$. The non-perturbed operator A can be decomposed into the orthogonal sum of two parts $A = A_N \oplus A_\perp$ acting in E_N and $E_\perp$ respectively. We denote by P^N, $P^\perp$ the orthogonal projections onto E_N and $E_\perp$. Another pair of complementary invariant subspaces can be obtained as linear hulls of *even* and *odd* eigenfunctions respectively:

$$E = E^{even} \oplus E^{odd}, \; A = A^{even} \oplus A^{odd}$$

We denote by P^{even}, P^{odd} the orthogonal projections onto E_{even}, E_{odd}. The projections P_N and P^{even} commute, so one can also consider the decompositions of even and odd components

$$A_N = A_N^{even} \oplus A_N^{odd}, \ A_\perp = A_\perp^{even} \oplus A_\perp^{odd},$$

where

$$A_N^{even} = P_N A^{even} P_N, \ A_\perp^{even} = P_\perp A^{even} P_\perp,$$

and

$$A_N^{odd} = P_N A^{odd} P_N, \ A_\perp^{odd} = P_\perp A^{odd} P_\perp.$$

The perturbation $V = \varepsilon\langle x, \nu\rangle$, where as before ν is a unit vector, is an even function on the unit disc with respect to the reflection in the line spanned by the vector ν. Hence, the perturbed operator $A + V$ can also be decomposed into the orthogonal sum of two operators

$$A + V = (A + V)^{even} \oplus (A + V)^{odd}.$$

This observation enables us to apply the perturbation procedure suggested in [**21**] to each pair $A^{even} \to (A+V)^{even}$ and $A^{odd} \to (A+V)^{odd}$ separately, thus reducing effectively the dimension of the space E_N by half. We will use this observation when applying the general technique developed in the next section to even and odd parts of the perturbed operator separately.

3. Perturbation procedure with an intermediate operator

Consider the spectral problem

$$-\Delta\psi = \mu(1 - \varepsilon\frac{\langle \nu, x\rangle}{\lambda_2})\psi,$$

$$\psi|_{\partial\Omega} = 0,$$

in the subspace $E_{even} = L_2(\Omega,\, even)$ of all square-integrable functions which are symmetric in the unit disc Ω with respect to the reflection in the line $x = s\nu$, $s \in \mathbb{R}$. We can assume that $\nu = (0, 1)$ (i.e., $\cos\theta = 1$). The above spectral problem can be rewritten as a generalized eigenvalue problem with infinite matrices:

$$(A - \mu K)\,\psi = 0.$$

Here $A = A_{even} = diag((k_m^s)^2)$ is the matrix of the Laplacian with respect to the basis ψ_m (11) and is arranged in non-decreasing order of the diagonal elements, while K is the Gram matrix of this basis with respect to the weighted dot product in $L_2(\rho, \Omega_0)$:

$$K_{mn} = \int_\Omega \rho^2 \psi_m^s \psi_n^t d\sigma,$$

where

$$\rho^2 = 1 - \frac{\varepsilon}{\lambda_2} r \cos\theta.$$

This matrix is related to the matrix V used in [**21**] as follows:

$$K_{ml} = I - \frac{\varepsilon}{\lambda_2} V^{st}_{ml}.$$

Here

$$V^{st}_{mn} = \frac{\int_0^{2\pi}\int_0^1 J_m(k^s_m\zeta) cos(m\theta) J_n(k^t_n\zeta) cos(l\theta)\zeta^2 cos(\theta) d\zeta d\theta}{(\pi\int_0^1 (J_m(k^s_m\zeta))^2\zeta d\zeta)^{1/2}(\pi\int_0^1 (J_n(k^t_n\zeta))^2\zeta d\zeta)^{1/2}}$$

The denominator in this formula is just the product of normalization constants and the factors π should be replaced by 2π when n or m (or both) equal to zero. The elementary calculation

$$\int_0^{2\pi} \cos(\theta)\ \cos(n\theta)\ \cos(m\theta)\, d\theta =$$

$$\frac{1}{2}\int_0^{2\pi} \cos\theta[\cos(m-n)\theta + \cos(m+n)\theta] d\theta$$

$$= \frac{1}{4}\int_0^{2\pi} [\cos(m-n-1)\theta + \cos(m-n+1)\theta]\ d\theta =$$

$$\begin{cases} \frac{\pi}{2} & \text{if } m-n = \pm 1, \\ 0 & \text{otherwise.} \end{cases},$$

shows that $\{V^{st}_{mn}\}$ is not zero only when $m - n = \pm 1$:

$$V^{st}_{mn} = \frac{\pi}{\sqrt{\pi}\sqrt{\pi}} \frac{\int_0^1 J_m(k^s_m\zeta) J_n(k^t_n\zeta)\zeta^2 d\zeta}{2(\int_0^1 (J_m(k^s_m\zeta))^2\zeta d\zeta)^{1/2}(\int_0^1 (J_n(k^t_n\zeta))^2\zeta d\zeta)^{1/2}}, \quad m, n \geq 1.$$

If at least one of the numbers m and n is equal to zero, then the corresponding factors $\sqrt{\pi}$ in the denominator are replaced by $\sqrt{2\pi}$. According to the scheme suggested in the previous section, we fix a number α, which we choose to be $\alpha = 100$, and construct the decomposition of the spaces E_{even}, E_{odd} into the orthogonal sum of subspaces

$$E_{even} = E^{\alpha}_{N,even} \oplus E^{\alpha}_{\perp,even}\ ;\ E_{odd} = E^{\alpha}_{N,odd} \oplus E^{\alpha}_{\perp,odd}.$$

The dimensions $N = N_{even,\,odd}$ of the corresponding spaces $E^{\alpha}_N = E^{\alpha}_{even,odd}$ can be estimated as

$$\dim E^{\alpha}_N \leq 17,$$

(see [**21**]). The operator K of multiplication by

$$1 - \frac{\epsilon}{\lambda_2}\langle x, \nu\rangle = 1 - \frac{3.558}{3.79}\langle x, \nu\rangle$$

is positive and can be estimated as

$$0.06\, I < K < 2\, I.$$

We consider the usual space $E := L_2(\Omega)$ on the lens Ω with the dot-product denoted by $\langle x, y\rangle$ and its weighted version $E_K := L_2(\Omega, K)$ that has the same elements but different dot-product $[x, y]_K := \langle Kx, y\rangle$. We represent the operators A, K as infinite matrices in $L_{2,even}(\Omega) = E_{even}$ (or $L_{2,odd}(\Omega) = E_{odd}$) with respect to the corresponding bases of even (odd) eigenfunctions of the (non-perturbed) Laplacian. For instance, in $E_{even} = L_{2,even}(\Omega)$ we have

$$A_{even} = A^{\alpha}_{N,even} \oplus A^{\alpha}_{\perp,even},$$

$$K = \begin{pmatrix} K_{NN} & K_{N\perp} \\ K_{\perp N} & K_{\perp\perp} \end{pmatrix},$$

where the matrix elements of K are obtained via framing K by the orthogonal projections P_N, $P_\perp$ in E_{even} onto E^{α}_N, $E^{\alpha}_\perp \subset E_{even}$. Aiming to the spectral analysis of the operator $K^{-1}A$ in the weighted space E_K, we proceed with the general spectral problem, assuming that the operator A^{-1} is positive and compact in the (non-weighted) Hilbert space E and the positive weight operator K is bounded and invertible: $\theta\, I \leq K \leq \Theta\, I$. According to the construction proposed in the previous section, we decompose the space E into the sum of two mutually orthogonal spectral subspaces $E = E_N \oplus E_\perp$ of the operator A, where the parts A_N and $A_\perp$ satisfy the estimates:

$$A_N \leq \alpha I \leq A_\perp, \text{where } N = N(\alpha),$$

with $N \to \infty$ when $\alpha \to \infty$. We will obtain estimates for eigenvalues of the above spectral problem $Au = \mu K u$ in the weighted space, considering them as singularities of the corresponding resolvent equation

$$Au - \mu K u = K f,\ u = R^K_\mu f. \tag{16}$$

A natural approximation of this infinitely-dimensional operator function can be obtained via framing it by projections P_N onto the finite-dimensional subspace E_N:

$$R^K_\mu \longrightarrow P_N R^K_\mu P_N. \tag{17}$$

The sequence (with respect to N) of matrix-functions obtained this way is strongly convergent to the resolvent when $P_N \to I$, $N \to \infty$, so all singularities of the resolvent eventually appear as singularities of this finite-dimensional approximation for sufficiently large N. Unfortunately, the structure of this finite-dimensional operator-function is still too complicated to calculate these singularities. In particular, the above "framed " resolvent is not a true resolvent, since it does not satisfy the Hilbert identity. Another natural candidate for a finite-dimensional approximation of the resolvent (16) of the infinitely-dimensional spectral problem is the resolvent of the finite-dimensional *reduced* operator A_N in E_N, which can be obtained as the solution of the corresponding *reduced* finite-dimensional equation

$$A_N u - \mu K_{NN} u = K_{NN} f. \tag{18}$$

A similar object was used in **[21]** as a convenient approximation for the framed resolvent. It is still too far from (17) in our optics case, though, due to the different structure of the perturbation. However, we can consider a modified equation containing the reduced weight-matrix with the entry $K_{\perp N}$ removed:

$$\begin{pmatrix} A_N & 0 \\ 0 & A_\perp \end{pmatrix} \begin{pmatrix} v_N \\ v_\perp \end{pmatrix} - \mu \begin{pmatrix} K_{NN} & K_{N\perp} \\ 0 & K_{\perp\perp}, \end{pmatrix} \begin{pmatrix} v_N \\ v_\perp \end{pmatrix}$$

$$= \begin{pmatrix} K_{NN} & K_{N\perp} \\ K_{\perp N} & K_{\perp\perp}, \end{pmatrix} \begin{pmatrix} f_N \\ 0 \end{pmatrix} := KF. \tag{19}$$

The component v_N of the solution of this equation can be used as a "link" between the framed resolvent and the resolvent of the "reduced" finite-dimensional problem (18). We assume now that the finite-dimensional problem (18) is already solved, that is its resolvent is found, and we know all its eigenvalues $\tilde{\alpha}_l^N$, $l = 1, 2, \ldots$ and eigenvectors $\tilde{\varphi}_l^N$, $l = 1, 2, \ldots$. We aim at the calculation of the eigenvalue α_0 of the infinitely-dimensional perturbed problem which is located near the eigenvalue $\tilde{\alpha}_0$ of the reduced problem. Denote by $\tilde{\delta}_0$ the "spacing" of the finite-dimensional reduced problem (18) at $\tilde{\alpha}_0$:

$$\tilde{\delta}_0 = \min_l | \tilde{\alpha}_0 - \tilde{\alpha}_l |.$$

We will also consider the reduced infinitely-dimensional problem in the complementary subspace $E_\perp$.

$$A_{\perp\perp} u - \mu K_{\perp\perp} u = K_{\perp\perp} f. \tag{20}$$

The spectrum of the problem $A_{\perp\perp} u - \mu K_{\perp\perp} u = 0$ is discrete and its resolvent can be found as the solution $u = u_\perp \in E_\perp$ of the above equation (20). We assume that the eigenvalues $\tilde{\alpha}_s^\perp$ and the eigenfunctions $\tilde{\varphi}_s^\perp$ of the reduced infinitely-dimensional problem are also known. In fact, we need only a naive estimate for them: $\tilde{\alpha_s^\perp} \geq \alpha\, \Theta^{-1}$, which shows together with a complementary estimate $\tilde{\alpha}_s^N \leq \alpha\,\theta^{-1}$ for eigenvalues of the finite-dimensional reduced operator that the spectra of both reduced problems can overlap only on the interval $(\alpha\Theta^{-1}, \alpha\theta^{-1})$. Generically, the two reduced problems do not have common eigenvalues, hence assuming that

$$\sigma(A_N) \cap \sigma(A_\perp) = \emptyset, \tag{21}$$

we can use the short hand notation $\tilde{\alpha}_s$ for all of these eigenvalues and similar notation $\tilde{\varphi}_s$ for the corresponding eigenfunctions. The following statement holds:

THEOREM 3.1. *Let $\tilde{\alpha}_0$ be an isolated simple eigenvalue of the reduced problem in E_N and the corresponding spacing $\tilde{\delta}_0$ is sufficiently large compared to the inverse lower bound α^{-1} of the spectrum of the non-perturbed operator A in the complementary subspace $E_\perp = E \ominus E_N$:*

$$\tilde{\Delta} := \frac{[\tilde{\alpha}_0 + 1/2\;\tilde{\delta}_0]^2 \parallel K \parallel^2}{\left(\alpha - [\tilde{\alpha}_0 + 1/2\;\tilde{\delta}_0] \parallel K \parallel\right)} < 1/2\;\tilde{\delta}_0. \tag{22}$$

Then there exists a single simple eigenvalue α_0 of the perturbed infinitely-dimensional problem in the disc $\tilde{D}_0$ of radius $0.5\tilde{\delta}_0$ centered at $\tilde{\alpha}_0$.

PROOF. This theorem can be proven using an operator version of the Rouche theorem discovered by I. Gohberg and E. Sigal [**12**]. We formulate and discuss this important fact in the Appendix below. The Gohberg-Sigal theorem is a tool of comparison of distributions of vector zeroes (see the Appendix) of two analytic operator functions. In our situation such functions appear while solving the infinite matrix equations in the above orthogonal decomposition of the space E. The original perturbed resolvent equation (16) looks in the matrix form as

$$\begin{pmatrix} A_N & 0 \\ 0 & A_\perp \end{pmatrix}\begin{pmatrix} u_N \\ u_\perp \end{pmatrix} - \mu\begin{pmatrix} K_{NN} & K_{N\perp} \\ K_{\perp N} & K_{\perp\perp}, \end{pmatrix}\begin{pmatrix} u_N \\ u_\perp \end{pmatrix}$$

$$= \begin{pmatrix} K_{NN} & K_{N\perp} \\ K_{\perp N} & K_{\perp\perp}, \end{pmatrix}\begin{pmatrix} f_N \\ 0 \end{pmatrix}. \tag{23}$$

If the conditions of the theorem are satisfied, the operator $A_\perp - \mu K_{\perp\perp}$ is invertible and hence we can eliminate the second component $u_\perp$ of the solution, obtaining an equation for the first component only:

$$\left[(A_N - \mu K_{NN}) - \mu^2 K_{N\perp}\frac{I}{A_\perp - \mu K_{\perp\perp}}K_{\perp N}\right] u_N$$

$$= \left[K_{NN} + K_{N\perp} \frac{\mu I}{A_\perp - \mu K_{\perp\perp}} K_{\perp N} \right] f_N.$$

Solution of the above equation exists, due to (22) for values of μ on the circle $\Sigma_0 = \partial D$ and can be presented as

$$u_N = P_N R_\mu P_N f_N = \frac{I}{\left[I - \frac{\mu^2}{(A_N - \mu K_{NN})} K_{N\perp} \frac{I}{A_\perp - \mu K_{\perp\perp}} K_{\perp N} \right]}$$

$$(24) \qquad (A_N - \mu K_{NN})^{-1} K_{NN} \left[I + \frac{\mu}{K_{NN}} K_{N\perp} \frac{I}{A_\perp - \mu K_{\perp\perp}} K_{\perp N} \right] f_N.$$

We are going to compare the above framed resolvent with the resolvent of the finite-dimensional reduced operator - the solution of the equation (18)

$$\tilde{R}^N_\mu f_N = \frac{I}{A_N - \mu K_{NN}} K_{NN} f_N.$$

The direct comparison of the above expressions for the reduced resolvent and the resolvent of the reduced operator is obstructed by the term $I + \frac{\mu}{K_{NN}} K_{N\perp} \frac{I}{A_\perp - \mu K_{\perp\perp}} K_{\perp N}$, which seemingly prevents us from straightforward applying of the above Rouche theorem to the formula (24). Consider one more intermediate equation suggested above (see (19)), which can be used as a missing link between the reduced resolvent of the perturbed problem and the resolvent of the reduced equation.One can check via straightforward solution of the resolvent equation (19) that the second factor in the right side of the above formula (24) for the framed resolvent of the complete problem coincides exactly with the projection of the solution of the equation (19) onto the space E_N:

(25)

$$v_N = (A_N - \mu K_{NN})^{-1} K_{NN} \left[I + \frac{\mu}{K_{NN}} K_{N\perp} \frac{I}{A_\perp - \mu K_{\perp\perp}} K_{\perp N} \right] f_N.$$

We will obtain further the spectral representation for v_N which enables one to accomplish the comparison of the reduced resolvent with the resolvent of the reduced operator. The infinite-dimensional spectral problem corresponding to the above system of equations (19)is

$$(26) \qquad (A_N - \mu K_{NN}) v_N = \mu K_{N\perp} v_\perp, \; (A_\perp - \mu K_{\perp\perp}) v_\perp = 0.$$

Note that the reduced weight is not positive. Hence, we will not attempt to formulate this spectral problem in any conventional operator form, but rather will consider it just as a homogeneous linear system. Nevertheless, this non-conventional spectral problem is “solvable” due to above generically satisfied *non-overlapping condition* (21): its eigenvalues coincide with the eigenvalues $\tilde{\alpha}_s$ of the reduced spectral problems for operators A_N and $A_\perp$ and the corresponding eigenvectors

$$\vec{\varphi} = \begin{pmatrix} \varphi_s^N \\ \varphi_s^\perp \end{pmatrix}$$

of the problem (26) have only first non-zero component coinciding with corresponding eigenvector of A_N:

$$\vec{\varphi}_s = \begin{pmatrix} \varphi_s^N \\ 0 \end{pmatrix}.$$

This in turn coincides with the eigenvector of the reduced operator A_N

$$A_N \varphi_s^N = \tilde{\alpha}_s K_{NN} \varphi_s^N,$$

while the second component is trivial $\varphi_s^\perp = 0$. The eigenvalues of the spectral problem (19) in the complementary subspace coincide with those of the reduced infinitely-dimensional operator $A_\perp$. The corresponding two-component eigenvectors have the second component equal to the eigenvector of the reduced problem

$$\left(A_\perp - \tilde{\alpha}_s K_{\perp\perp}\right) \varphi_s^\perp = 0$$

and the first component found from the equation

$$A_N \varphi_s^N - \tilde{\alpha}_s K_{N\perp} \varphi^\perp - \tilde{\alpha}_s K_{NN} \varphi_s^N = 0.$$

This equation has a unique solution due to the non-overlap assumption (21):

$$\varphi_s^N = \frac{1}{A_N - \tilde{\alpha}_s K_{NN}} \tilde{\alpha}_s K_{N\perp} \varphi_s^\perp.$$

The eigenvalues $\tilde{\alpha}$ of the hermitian-adjoint spectral problem

$$\begin{pmatrix} A_N & 0 \\ 0 & A_\perp \end{pmatrix} \begin{pmatrix} \psi_N \\ \psi_\perp \end{pmatrix} - \tilde{\alpha} \begin{pmatrix} K_{NN} & 0 \\ K_{\perp N} & K_{\perp\perp}, \end{pmatrix} \begin{pmatrix} \psi_N \\ \psi_\perp \end{pmatrix} = 0 \tag{27}$$

also coincide with those of the reduced problems. For the eigenvalues coinciding with those of $A_\perp$, the eigenvectors

$$\vec{\psi}_s = \begin{pmatrix} \psi_s^N \\ \psi_s^\perp \end{pmatrix}$$

have only second non-zero component

$$\vec{\psi}_s = \begin{pmatrix} 0 \\ \varphi_s^\perp \end{pmatrix},$$

where

$$A_\perp \psi_s^\perp - \tilde{\alpha}_s K_{\perp\perp} \psi_s^\perp = 0, \ \psi^\perp = \varphi_s^\perp.$$

For the eigenvalues coinciding with those of A_N, with the first component satisfying the reduced equation

$$A_N \psi_s^N - \tilde{\alpha}_s K_{NN} \varphi_s^N = 0,$$

and the second one found uniquely from the non-homogeneous equation

$$A_\perp \psi_s^\perp - \tilde{\alpha}_s K_{\perp\perp} \psi_s^\perp = \tilde{\alpha}_s K_{\perp N} \varphi_s^N,$$

$$\vec{\psi}_s = \begin{pmatrix} \varphi_s^N \\ \psi_s^\perp \end{pmatrix}.$$

Eigenvalues of both problems are real and coincide with the eigenvalues $\tilde{\alpha}_s$ of the reduced problems in E_N and $E_\perp$. The corresponding eigenvectors are bi-orthogonal with respect to the bi-linear form defined by the matrix

$$\tilde{K} = \begin{pmatrix} K_{NN} & K_{N\perp} \\ 0 & K_{\perp\perp}, \end{pmatrix}.$$

Indeed, for $s \neq t$ we have

$$0 = \langle A\vec{\varphi}_s, \vec{\psi}_t \rangle - \langle \vec{\varphi}_s, A\vec{\psi}_t \rangle = \tilde{\alpha}_s \langle \tilde{K}\vec{\varphi}_s, \vec{\psi}_t \rangle - \tilde{\alpha}_t \langle \vec{\varphi}_s, \tilde{K}^+ \vec{\psi}_t \rangle = [\tilde{\alpha}_s - \tilde{\alpha}_t] \langle \tilde{K}\vec{\varphi}_s, \vec{\psi}_t \rangle.$$

Hence, $\langle \tilde{K}\vec{\varphi}_s, \vec{\psi}_t \rangle = 0$ for $s \neq t$. For $s = t$ we have for $\varphi_s \in E_N$:

$$\langle \tilde{K}\vec{\varphi}_s, \vec{\psi}_s \rangle = \langle K_{NN}\varphi_s, \varphi_s \rangle = \langle K\varphi_s, \varphi_s \rangle > 0.$$

Similarly, for $\varphi_s \in E_\perp$

$$\langle \tilde{K}\vec{\varphi}_s, \vec{\psi}_s \rangle = \langle K_{\perp\perp}\varphi_s, \varphi_s \rangle = \langle K\varphi_s, \varphi_s \rangle > 0.$$

The properties of the constructed bi-orthogonal system are very convenient, due to the above representation formulae. Using these formulae, we can find now an explicit representation for the solution of the resolvent equation (19). Using the bi-orthogonality one can present it in form of the convergent spectral series:

$$\vec{v} = \begin{pmatrix} v_N \\ v_\perp \end{pmatrix} = \sum_s \frac{\vec{\phi}_s \langle \tilde{K}\vec{F}, \vec{\psi}_s \rangle}{\alpha_s - \mu} \frac{1}{\langle K\vec{\varphi}_s, \vec{\varphi}_s \rangle}.$$

Now we form the projection of the solution onto the subspace E_N:

$$v_N = P_N \vec{v} = \sum_{\varphi_s \in E_N} \frac{\langle K_{NN} f_N, \varphi_s^N \rangle}{\alpha_s - \mu} \varphi_s^N = [\frac{I}{K_{NN}} A_N - \mu I]^{-1} f_N.$$

Comparing with the projection v_N of the solution of the resolvent equation obtained earlier in (25), we see that

$$v_N = (A_N - \mu K_{NN})^{-1} K_{NN} \left[I + \frac{\mu}{K_{NN}} K_{N\perp} \frac{I}{A_\perp - \mu K_{\perp\perp}} K_{\perp N} \right] f_N =$$

$$[\frac{I}{K_{NN}} A_N - \mu I]^{-1} f_N. \tag{28}$$

Now we can apply the Rouche theorem (see Appendix) to the pair of matrix functions

$$m_0(\mu) = P_N R_\mu P_N, \; m_1 = [\frac{I}{K_{NN}} A_N - \mu I]^{-1},$$

and use the above formula (24) connecting them:

$$P_N R_\mu P_N \; [\frac{I}{K_{NN}} A_N - \mu I] = \frac{I}{\left[I - \frac{\mu^2}{(A_N - \mu K_{NN})} K_{N\perp} \frac{I}{A_\perp - \mu K_{\perp\perp}} K_{\perp N}\right]}.$$

Let us now μ belongs to the circle Σ_0 of radius $\delta_0/2$ centered at the eigenvalue α_0 of the reduced operator A_N. Due to the assumption of Theorem 3.1, we can estimate the second term in the square bracket of the denominator in the right hand side on as follows:

$$\| \frac{\mu^2}{(A_N - \mu K_{NN})} K_{N\perp} \frac{I}{A_\perp - \mu K_{\perp\perp}} K_{\perp N} \| <$$

$$\frac{2}{\delta_0} \frac{[\tilde{\alpha}_0 + 1/2\tilde{\delta}_0]^2 \, \| K \|^2}{\left(\alpha - [\tilde{\alpha}_0 + 1/2\tilde{\delta}_0] \, \| K \|\right)} < 1.$$

Hence, we can apply the Gohberg-Sigal theorem to the pair of operator-functions $\tilde{R}^N_\mu$ and $P_N R_\mu P_N$, having in mind that the first of them is invertible. This shows that the eigenvalues of the finite-dimensional reduced spectral problem approximate the eigenvalues of the original spectral problem $Au = \mu K u$. Vice versa, if α_0 is a simple isolated eigenvalue of the original spectral problem, then for sufficiently large α the reduced finite-dimensional spectral problem has a simple isolated eigenvalue $\tilde{\alpha}_0$ near to α. □

4. Appendix

Let E_0 be proper subspace of a finite-dimensional Hilbert space E, P_0 be an orthogonal projection onto E_0, and $P_0^\perp = I - P_0$ be the complementary projection onto $E_0^\perp = E \ominus E_0$. We say that the analytic matrix-function m with the domain D_m has an *simple isolated vector zero* at the point $\mu_0 \in D_m$ if it can be represented in a neighborhood $U_0 \subset D_m$ as a product

$$m(\mu) = m_0(\mu) \left[(\mu - \mu_0) P_0 + b P_0^\perp\right] \tag{29}$$

with some nonzero constant b and an invertible analytic matrix-function $m_0(\mu)$

$$m_0(\mu) = m_0(\mu_0) + \frac{\mu - \mu_0}{1!} m_0'(\mu_0) + \ldots, \quad \text{Ker } m_0(\mu_0) = 0.$$

Vectors $\{e_0\} \in E_0$ are called *root-vectors.* They satisfy $m(\mu_0)e_0 = 0$. For vectors $e_\perp$ from the complementary subspace $e_\perp \in E_{0\perp}$ we have $m(\mu_0)e_\perp \neq 0$. Similarly, a *simple isolated vector pole* can be defined: we say that function m has a simple isolated vector pole at the point μ_0 if it is represented as

$$m(\mu) = m_0(\mu) \left[\frac{P_0}{\mu - \mu_0} + b P_0^\perp \right]. \tag{30}$$

Here b is a non-zero constant, P_0 - the orthogonal projection onto a proper subspace $E_0 \subset E$, $P_0^\perp$ the complementary projection, m_0 an analytic invertible function in a neighborhood U_0 of the point μ_0 in D_m. Both isolated poles and zeroes of analytic matrix-functions are called in [**12**] *characteristic values* of the variable μ. The logarithmic residue of the function μ at a simple isolated zero or pole is defined as the anti-clockwise integral of the logarithmic derivative $m'(\mu)m^{-1}$ over a simple smooth curve $\Gamma_0 \subset U_0$ around the characteristic value m_0:

$$I_{m,\mu_0} = \frac{1}{2\pi i} \oint_{\Gamma_0} m'(\mu) m^{-1} d\mu.$$

In [**12**], the *period* of the logarithmic derivative $m'(\mu)m^{-1}$

$$M_{m,\mu_0} = \frac{1}{2\pi i} Trace \oint_{\Gamma_0} m'(\mu) m^{-1} d\mu.$$

is called the "multiplicity" of the characteristic value. The straightforward calculation of above integrals gives

$$I_{m,\mu_0} = \frac{1}{2\pi i} \oint_{\Gamma_0} m'(\mu) m^{-1} d\mu =$$

$$\frac{1}{2\pi i} \oint_{\Gamma_0} m_0(\mu) P_0 \left[(\mu_0 - \mu) P_0 + b P_0^\perp \right]^{-1} m_0^{-1}(\mu)(\mu) d\mu =$$

$$\frac{1}{2\pi i} m_0 \mu_0 \oint_{\Gamma_0} P_0 \left[(\mu_0 - \mu) P_0 + b P_0^\perp \right]^{-1} d\mu m_0^{-1}(\mu_0) = m_0 \mu_0 P_0 m_0^{-1}(\mu_0)$$

and

$$M_{m,\mu_0} = \pm \text{dim} P_0,$$

where the sign $\pm$ is defined by the type of the characteristic value: plus for a zero and minus is for a pole. We will use the "Hermitian" version of the Gohberg-Sigal theorem, assuming that all poles (zeroes) are simple (of first order):

THEOREM 4.1. *Let two finite square matrices* $m_s(\mu)$, $s = 0, 1$ *depend analytically on the parameter* μ *in the disc* D *of radius* $\tilde{\delta}$ *centered at the point* $\mu = \tilde{\alpha}_0$. *Let also* m_0 *have several characteristic points* μ_0^s *inside the disc*

$$D_0 = \left\{ \mu : |\mu - \tilde{\alpha}_0| \leq 0.5\tilde{\delta}_0 < 0.5\tilde{\delta} \right\}$$

with the total multiplicity

$$M_0 = \sum_s M_0^s.$$

If on the circle $\Sigma_0 = \{\mu : |\mu - \tilde{\alpha}_0| = 0.5tilde\delta_0\}$ *both functions have no characteristic values and the inequality*

$$max_{\mu\in\Sigma_0} \parallel m_0^{-1}(\mu)\,[M_1(\mu) - M_0(\mu)] \parallel < 1$$

is satisfied, then the total multiplicities M_s *of characteristic values of both functions* m_s *inside the circle* Σ_0 *are the same.*

PROOF. A much more general statement concerning analytic functions with multiple poles and zeroes is proven in [**12**].

In our paper (Section 3) we actually employ only a partial statement concerning the case when both functions are analytic and $M_0 = 1$. Then m_1 has a single simple isolated zero inside the disc Σ_0. The case of a single pole can be reduced to the case of a single zero via considering inverse functions.

5. Acknowledgements

The authors are grateful to A. Mikhailova for attracting their attention to the practical importance of respecting the symmetry of the problem when developing the perturbation techniques. B.P. is grateful to the Japanese Foundation JSPS for the support of his stay at the Solid State Computer Laboratory of Aizu University with Professor V. Ryzhii, where an essential part of this paper was written. B.P. is also thankful to the Russian Academy of Sciences for the support from RFFI grant during his visit to St. Petersburg in February 2003. The authors of the paper express their gratitude to the referee for important remarks and to Professor P. Kuchment for numerous essential suggestions.

References

[1] V. Adamyan, Scattering matrices for microschemes, in *Operator Theory: Adv. and Appl.*, **59** (1992), pp 1-10.

[2] V. Adamyan and H. Langer, Spectral properties of a class of rational operator-valued functions, *Journ. Operator Theory* **33**(1995), pp 259-277.

[3] S. Albeverio and P. Kurasov, *Singular Perturbations of Differential Operators*, London Math. Soc. Lect. Notes Series 271, Cambridge Univ. Press 2000.

[4] N. T. Bagraev, A. D. Bouravlev, A. M. Malyarenko, and S. A. Rykov, Room temperature single-hole memory, Low-Dim. Structures **9/10** (2000), pp 51–60.

[5] N. T. Bagraev, A. B. Mikhailova, B. S. Pavlov, L. V. Prokhorov, and A. M. Yafyasov, Temperature stability of the triadic resonance quatum switch, 10-th MEL-ARI/NID Workshop, Helsinki,1-3 July 2002, Academy of Finland.

[6] C. W. J. Beenakker and H. van-Houten, *Quantum Transport in semiconductor nanostructures*, Solid State Physics. Advances and Applications, v. 44, Acad. Press, San Diego 1991, pp 1-228.

[7] V. Bogevolnov, A. Mikhailova, B. Pavlov, and A. Yafyasov, About Scattering on the ring, in *Operator Theory: Advances and Applications*, v. 124 (The Israel Gohberg Anniversary Volume : International Workshop in Gronningen, June 1988) ed. A. Dijksma, A.M. Kaashoek, A.C.M. Ran). Birkhauser, Basel 2001, pp 155-187.

[8] A.P.Calderón, On an inverse boundary value problem, in Seminar on Numerical Analysis and its Applications to Continuum Physics, *Soc. Brasiliera de Mathematica, Rio de Janeiro* (1980), pp 65-73.

[9] European Comission IST programme: Future and emerging Technologies. *Technology Roadmap for Nanoelectronics*, R. Compano (ed.), Second edition, Nov. 2000. Luxemburg: Office for Official Publications of the European Communities. 105 p.

[10] P. Exner and P. Sheba, A new type of quantum interference transistor, *Phys. Lett. A* 129:8,9 (1988), pp 477-480.

[11] M. Faddeev and B. Pavlov, Scattering by resonator with the small opening, Proc. LOMI (**126**), 1983. (English transl. in J. of Sov. Math. vol. 27 (1984) pp 2527-2533)

[12] I. T. Gohberg and E. I. Sigal, Operator extension of the theorem about logarithmic residue and Rouchet theorem, *Mat. Sbornik* **84**(1971), p 607

[13] T. Kato, *Perturbation theory for linear operators*, Springer Verlag, Berlin-Heidelberg-NY, 2nd edition 1984.

[14] V. Kostrykin and R. Schrader, Kirchhoff's rule for quantum wires, *J. Phys. A: Math. Gen.* **32** (1999) pp 595-630.

[15] R. Landauer, Spatial variation of currents and fields due to localized scatterers in metallic constructions, *IBM Jornal for Research and Development,* 1 (1957), pp 223-231.

[16] P. Lax and R. Phillips, *Scattering theory*, Academic Press, New York 1967.

[17] O. Madelung, *Introduction to solid-state theory*, Springer Series in Solid-State Sciences, v. 2, Springer-Verlag, Berlin, New York 1978.

[18] R. Mennicken and A. Shkalikov, Spectral decomposition of symmetric operator matrices, *Math. Nachrichten* **179** (1996), pp 259-273.

[19] A. Mikhailova, Private communication, Nov. 2002.

[20] A. Mikhailova and B. Pavlov, Quantum Domain as a triadic relay, April 2000, Department of Mathematics Report series No.420, ISSN 1173-0889, the University of Auckland, Auckland, NZ,16p. New technologies for narrow-gap semiconductors. Esprit project N 28890 ESPRIT NTCONGS Progress Reports (July 1, 1999 - December 31,1999).

[21] A. Mikhaylova and B. Pavlov, Quantum domain as a triadic relay, in *Unconventional Models of Computations UMC'2K (Proceedings of the UMC'2K Conference, Brussels, Dec 2000) eds. I. Antoniou, C. Calude, M.J. Dinneen*, Springer Verlag Series for Discrete Mathematics and Theoretical Computer Science, Springer Verlag, Berlin 2001, pp 167-186.

[22] A. Mikhailova and B. Pavlov, Resonance Quantum Switch, in *S.Albeverio, N.Elander, W.N.Everitt and P.Kurasov (eds.), Operator Methods in Ordinary and Partial Differential Equations, (S.Kovalevski Symposium, Univ. of Stockholm, June 2000)*, Operator Methods: Advances and Applications, v. 132, Birkhuser, Basel-Boston-Berlin, 2002.

[23] A. Mikhailova, B. Pavlov, I. Popov, T. Rudakova, and A. M. Yafyasov, Scattering on a Compact Domain with Few Semi-Infinite Wires Attached: Resonance Case, *Math. Nachr.* **235** (2002), pp 101-128.

[24] B.Pavlov, Provisional Patent: *A System and Method for Resonance manipulation of Quantum Currents Through Splitting* Auckland University Limited, 504590, 17 Can 2000, New Zealand.

[25] B. Pavlov, S-Matrix and Dirichlet-to-Neumann Operators, in *Scattering, ed. R. Pike, P. Sabatier*, Academic Press, Harcourt Science and Tech. Company 2001, pp 1678-1688.

[26] B. Pavlov, Resonance quantum switch: Matching domains, in *Proceedings of the center of Mathematics of the Australian national University*, Canberra, Australia, 2002, v. 40.

[27] B. Pavlov, I. Popov, V. Geyler, and O. Pershenko, Possible construction of a quantum multiplexer, *Europhyics Letters*, **52**(2000), no. 2, pp 196-202.

[28] M. Reed and B. Simon, *Methods of modern mathematical physics*, Acad. Press, New York, London, 1972.

[29] K. Robert, Resonance Optical Switch, Talk at the New-Zealand mathematical Simposium, December 2002.

[30] A. Silbergleit and Yu. Kopilevich, *Spectral theory of guided waves*, Institute of Physics publishing, Bristol 1996.

[31] J. Sylvester and G. Uhlmann, The Dirichlet to Neumann map and applications, in *Proceedings of the Conference "Inverse problems in partial differential equations (Arcata, 1989)"*, SIAM, Philadelphia (1990), pp 101–139.

Department of Mathematics, University of Auckland, New Zealand

E-mail address: pavlov@math.auckland.ac.nz, krob049@ec.auckland.ac.nz

Contemporary Mathematics
Volume **339**, 2003

High energy asymptotics of the density of states for certain periodic pseudo-differential operators in dimension one

Alexander V. Sobolev

Abstract. We consider a pseudo-differential operator on the real line, which is a lower-order perturbation of an elliptic operator with a homogeneous symbol and constant coefficients. It is proved that the density of states of such an operator admits a complete asymptotic expansion at large energies. A few first terms of this expansion are calculated explicitly.

1. Introduction

Spectral theory of periodic operators offers an abundant array of problems with regard to all aspects of spectral structure. One of such problems is the asymptotics of the (integrated) density of states $D(\lambda)$ as the spectral parameter (or energy) λ tends to infinity. Although the study of this object presents an independent interest, it also helps to obtain other important spectral properties. For instance, the high energy asymptotics of $D(\lambda)$ provides information on the size of spectral gaps, or, it can help to find out whether or not the number of gaps is finite. In the recent years a number of results was obtained for the density of states for the Schrödinger operator

$$H = -\Delta + V, \tag{1.1}$$

in $\mathsf{L}^2(\mathbb{R}^d), d \geq 1$, with a real-valued function V, periodic with respect to a d-dimensional lattice $\Gamma \subset \mathbb{R}^d$. Below we denote by $\mathcal{O} \subset \mathbb{R}^d$ a fundamental domain of the lattice Γ. For this operator, as well as for any other elliptic self-adjoint differential operator the density of states is defined by the formula

$$D(\lambda) = \lim_{L\to\infty} \frac{N(\lambda; H_D^{(L)})}{L^d}. \tag{1.2}$$

Here $H_D^{(L)}$ is the restriction of H to the cube $[0, L)^d$ with the Dirichlet boundary conditions, and $N(\lambda;\ \cdot\ )$ is the counting function of the discrete spectrum of $H_D^{(L)}$. To be precise, the quantity $D(\lambda)$ is called the *integrated density of states*, but for the sake of brevity we call it simply the density of states. This will not cause

2000 *Mathematics Subject Classification.* Primary 35P20, 47G30, 47A55; Secondary 81Q10

The author was supported in part by EPSRC Grant RV513700, AMS and NSF.

any confusion. It is an elementary exercise to check that for the unperturbed operator $H_0 = -\Delta$, the density of states is proportional to the volume of the ball $\{\boldsymbol{\xi} \in \mathbb{R}^d : |\boldsymbol{\xi}|^2 \le \lambda\}$, that is $D_0(\lambda) = (2\pi)^{-d}\, \mathrm{w}_d\, \lambda^{d/2}$, where we have denoted by w_d the volume of the unit ball in $\mathbb{R}^d$.

The first asymptotic formulas for the operator (1.1) were obtained in the case $d = 1$ (see [**6**] [**9**]). It was shown that for C^∞-smooth 2π-periodic functions V the density of states admits a complete expansion in the powers of λ^{-1}:

$$\begin{cases} D(\rho^2) = \dfrac{1}{\pi}\rho + \sum_{j=1}^N d_j \rho^{1-2j} + O\big(\rho^{1-2(N+1)}\big), \ \rho \to \infty, \ \ \forall N; \\ d_1 = -\dfrac{1}{4\pi^2}\int_0^{2\pi} V(x)dx, \ \ d_2 = -\dfrac{1}{16\pi^2}\int_0^{2\pi} |V(x)|^2 dx. \end{cases} \tag{1.3}$$

The remaining coefficients are found via certain explicit and simple recursion relations. Note also that in addition to this formula, an asymptotics for the spectral function of H was derived in [**9**].

The first results on the multidimensional Schrödinger operator were established in [**10**] (see also [**11**] for an elementary proof) for arbitrary bounded periodic V. It was proved that

$$D(\rho^2) = D_0(\rho^2) + O(\rho^{d-2}), \ \rho \to \infty.$$

This formula was improved in [**3**] with the help of a sophisticated microlocal technique: for C^∞-smooth V it was shown that

$$D(\rho^2) = D_0(\rho^2) + c_2 \rho^{d-2} + O(\rho^{d-3+\epsilon}), \ \forall \epsilon > 0,$$
$$c_2 = -\frac{d\, \mathrm{w}_d}{2(2\pi)^d |\mathcal{O}|}\int_{\mathcal{O}} V(\mathbf{x})d\mathbf{x}.$$

A further improvement was achieved in [**4**] through the use of a suitable perturbation argument:

$$D(\rho^2) = D_0(\rho^2) + c_2 \rho^{d-2} + O(\rho^{d-3}\ln \rho). \tag{1.4}$$

To be precise, the asymptotics of this type was also established in [**4**] for the perturbed polyharmonic operator.

A number of results on the density of states is known for almost periodic operators as well, but this subject is beyond the scope of this paper.

The difference between the one- and multi-dimensional cases is not surprising. For $d = 1$ one can carry out more or less complete perturbation analysis of the Floquet eigenvalues. On the contrary, for $d \ge 2$ the eigenvalues split in two groups behaving differently under the perturbation V. One group of eigenvalues behaves essentially as in the case $d = 1$, and their contribution can be relatively easily evaluated by means of the perturbation theory. The analysis of the other group of eigenvalues presents a major obstacle. This group consists of "anomalous" eigenvalues which move by a quantity of order $\|V\|$ under the perturbation V, and therefore cannot be considered as small perturbations of the unperturbed eigenvalues (see e.g. [**12**], Ch. 4). In the relevant literature they are sometimes called resonant, unstable or singular eigenvalues. Various spectral properties of the operator H can be established by checking that the quantity of these "anomalous" eigenvalues is negligible in some sense. For instance, this strategy was successfully applied in [**12**] for the justification of the Bethe-Sommerfeld conjecture, and in [**4**] for the proof of (1.4). On the other hand, this approach does not allow one to study further terms (if they exist) in the asymptotics of $D(\lambda)$. In spite of these difficulties at least a partial analysis of the resonant eigenvalues is possible. As was shown in [**2**] (see also [**5**], Section 4.6), they can be described by a suitable effective one-dimensional Schrödinger operator.

The aim of the present paper is to study the density of states of a periodic elliptic pseudo-differential operator in $\mathsf{L}^2(\mathbb{R})$, of a form more general than the Schrödinger operator. For the operator under consideration it is shown that the density of states admits a complete asymptotic expansion, a few first terms of which being calculated explicitly (see Theorem 2.2). Applying this result to the Schrödinger operator one recovers the first coefficients in the decomposition (1.3) found previously in [**6**] and [**9**]. Although the study of such a one-dimensional problem presents an independent interest, the main motivation for the present work comes from the link with the multi-dimensional case, see above. When writing this paper, the author intended to outline the main ideas and indicate the principal steps of the proof, avoiding tedious technical details: the proofs are either omitted altogether or just sketched. An exposition of the results with full proofs will be published elsewhere.

2. Main result

2.1. Periodic pseudo-differential operators. We start with introducing the classes of relevant symbols. For any function $u \in \mathsf{L}^2([0, 2\pi])$ we denote by $\hat{u}(n)$ its Fourier coefficients:

$$\hat{u}(\theta) = \frac{1}{\sqrt{2\pi}} \int_0^{2\pi} e^{-i\theta x} u(x)dx, \ \theta \in \mathbb{Z}.$$

Let $b = b(x, \xi), x, \xi \in \mathbb{R}$ be a 2π-periodic complex-valued function, that is

$$b(x + 2\pi n, \xi) = b(x, \xi), \ n \in \mathbb{Z}.$$

We say that the symbol b belongs to the class $\mathbf{S}_\alpha$ if its Fourier coefficient $\hat{b}(\theta, \xi)$ with respect to the variable x satisfy the condition

$$\text{(2.1)} \qquad \big| b \big|_{l,s}^{(\alpha)} := \max_{p \le l} \max_{r \le s} \sup_{\xi, \theta} \langle \theta \rangle^p \ \langle \boldsymbol{\xi} \rangle^{-\alpha + r} |\partial_{\boldsymbol{\xi}}^r \hat{b}(\theta, \boldsymbol{\xi})| < \infty, \ \ \langle t \rangle = \sqrt{1 + t^2}.$$

We call α the order of the symbol b. Under the condition $\big| b \big|_{l,0}^{(\alpha)} < \infty$, $l > 1$, the integral

$$\mathrm{Op}(b)u(x) := \frac{1}{2\pi} \int_{\mathbb{R}} \int_{\mathbb{R}} b(x, \xi) e^{i\xi(x-y)} u(y) dy d\xi,$$

is finite for all u from the Schwarz class $\mathsf{S}(\mathbb{R})$, so that the pseudo-differential operator (PDO) $\mathrm{Op}(b)$ is properly defined on $\mathsf{S}(\mathbb{R})$. Below this class is always taken as a default domain for all PDO's unless otherwise stated. It is natural to call such PDO's periodic. It is easy to show that for $\alpha = 0$ the operator $\mathrm{Op}(b)$ is bounded and

$$\text{(2.2)} \qquad \| \mathrm{Op}(b) \| \le C_l \big| b \big|_{l,0}^{(0)}, \ \forall l > 1.$$

We study a PDO in $\mathsf{L}^2(\mathbb{R})$ of the form

$$\text{(2.3)} \qquad H = H_0 + \mathrm{Op}(b), \quad H_0 = |D|^m, \ \ D = -i\frac{d}{dx},$$

where $m > 0$ and $b \in \mathbf{S}_\alpha$ with $\alpha < m$. The symbol of H is $h(x, \xi) = h_0(\xi) + b(x, \xi)$, $h_0(\xi) = |\xi|^m$. To ensure that the operator under consideration is self-adjoint we suppose that the symbol b is *symmetric* in the sense that

$$\hat{b}(\theta, \xi) = \overline{\hat{b}(-\theta, \xi + \theta)}, \quad \xi \in \mathbb{R}, \theta \in \mathbb{Z}. \tag{2.4}$$

This guarantees that $\operatorname{Op}(b)$ is symmetric on its natural domain. Since $\alpha < m$, the perturbation $\operatorname{Op}(b)$ is H_0-bounded with a bound < 1 and thus H is self-adjoint on the domain $\mathcal{D}(H_0) = \mathsf{H}^m(\mathbb{R})$.

The definition of the density of states (1.2) is not applicable as the operator H in general is not local. In order to introduce $D(\lambda)$ appropriately, we need to describe first the *Floquet decomposition* for H.

2.2. The Floquet decomposition. We identify the space $\mathcal{H} = \mathsf{L}^2(\mathbb{R})$ with the direct integral

$$\mathfrak{G} = \int_0^1 \mathfrak{H} dk, \ \mathfrak{H} = \mathsf{L}^2(0, 2\pi).$$

This identification is implemented by the Gelfand transform

$$(Uu)(x, k) = e^{-ikx} \sum_{n \in \mathbb{Z}} e^{-i2\pi kn} u(x + 2\pi n), \ k \in [0, 1),$$

which is initially defined on $u \in \mathsf{S}(\mathbb{R})$ and extends by continuity to a unitary mapping from $\mathcal{H}$ onto $\mathfrak{G}$. For each $k \in [0, 1)$ define the pseudo-differential operator on the circle $\mathbb{T} = \mathbb{R}/(2\pi\mathbb{Z})$:

$$(H(k)u)(x) = \frac{1}{\sqrt{2\pi}} \sum_{n \in Z} h(x, n + k) e^{inx} \hat{u}(n), \quad u \in \mathsf{C}^\infty(\mathbb{T}).$$

One can show that the family $H(k)$ represents the decomposition of the operator H in the direct integral:

$$UHU^{-1} = \int_0^1 H(k) dk,$$

the spectrum of each $H(k)$ being discrete. Denote by $\lambda_j(k)$ the eigenvalues of $H(k)$ labelled in ascending order, and by

$$N(\lambda, k) = \sharp\{j : \lambda_j(k) \le \lambda\}, \ \lambda \in \mathbb{R},$$

their counting function. Now we are in a position to define the density of states:

$$D(\lambda) = \frac{1}{2\pi} \int_0^1 N(\lambda, k) dk.$$

One can show (see e.g. [**11**]), that for differential elliptic operators, and in particular for the Schrödinger operator, this definition is equivalent to (1.2). Sometimes we need to reflect the dependence of the counting function and density of states on the operator. In this case we use the notation $N(\lambda, k; H)$, $D(\lambda; H)$. Note that the density of states is preserved for a unitarily equivalent operator. Precisely, let W be a 2π-periodic unitary operator, that is an operator, commuting with shifts by $2\pi n, n \in \mathbb{Z}$. Then $D(\lambda; H) = D(\lambda; W^* H W)$. This simple property will be used in the proof of the main result.

The quantity $D(\lambda, H)$ has a transparent geometrical meaning if the symbol h does not depend on x. We state the appropriate formulas in the form of a lemma for later reference.

LEMMA 2.1. *Let $A = \operatorname{Op}(a)$ be a PDO with a real-valued symbol $a \in \mathsf{L}^\infty_{\mathrm{loc}}(\mathbb{R})$ such that $a(\xi) \to \infty$ as $|\xi| \to \infty$. Then*

$$D(\lambda; A) = \frac{1}{2\pi} \operatorname{vol}(\xi \in \mathbb{R} : a(\xi) \le \lambda).$$

PROOF. Denote

$$\Theta(t) = \begin{cases} 0, & t < 0; \\ 1, & t \ge 0. \end{cases}$$

Observe that the eigenvalues of $A(k)$ equal $a(n+k), n \in \mathbb{Z}$, so that

$$2\pi D(\lambda) = \sum_{n\in\mathbb{Z}} \int_0^1 \Theta(\lambda - a(n+k))dk = \int_{a(\xi)\le\lambda} d\xi = \operatorname{vol}(\xi \in \mathbb{R} : a(\xi) \le \lambda),$$

as required. □

2.3. Main result. The main result consists in an asymptotic decomposition of $D(\lambda)$ as $\lambda \to \infty$ for the operator (2.3). We prove the existence of such a decomposition and compute explicitly a few first terms thereof. Denote

$$\mathcal{B}_1(\xi) = \int_0^{2\pi} b(x,\xi)dx, \quad \mathcal{B}_2(\xi) = \int_0^{2\pi} |b(x,\xi)|^2 dx.$$

Similarly to the class $\mathbf{S}_\alpha$ of PDO's we also introduce a class $\mathcal{S}_j^\nu \subset \mathsf{C}^\infty(\mathbb{R})$, $\nu \in \mathbb{R}, j \in \mathbb{Z}$, of functions $\mu = \mu(\xi)$ such that

$$\partial_\xi^s \mu(\xi) \le C_s \langle\xi\rangle^{-j\kappa+\nu-s}, \quad \kappa = m - \alpha > 0,$$

for any non-negative integer s. We say that the function $a(\xi)$ is represented by an asymptotic series $\sum_{j=1}^\infty \mu_j(\xi)$ with $\mu_j \in \mathcal{S}_j^\nu$ if

$$u^{(\nu)}(\xi; a) := a(\xi) - \sum_{j=1}^M \mu_j(\xi) = O(\langle\xi\rangle^{-(M+1)\kappa+\nu}),$$

for all $M \in \mathbb{N}$. We write this down as follows: $a(\xi) = \sum_{j=1}^\infty \mu_j(\xi)$. The main result is contained in the following Theorem:

THEOREM 2.2. *Let $m > 0$. Let $b \in \mathbf{S}_\alpha$ be a symmetric symbol with $\alpha < m$. Then there exists a sequence of real-valued functions $\nu_j, j = 1, 2, \dots,$ that belong to $\mathcal{S}_j^1$, such that the sum*

$$\xi_M(\eta) = \eta + \sum_{j=1}^M \nu_j(\eta)$$

describes the asymptotics of the density of states $D(\rho^m)$ for all $\rho \ge \rho_0$, any $M \in \mathbb{N}$ by the formula

$$D(\rho^m) = \frac{1}{2\pi}\big[\xi_M(\rho) - \xi_M(-\rho)\big] + O(\langle\rho\rangle^{(M+1)(\alpha-m)+1})$$

The first two functions ν_j are given by the formulas

$$(2.5) \quad \begin{cases} \nu_1(\eta) = & -\dfrac{1}{2\pi m}\eta|\eta|^{-m}\mathcal{B}_1(\eta), \\ \nu_2(\eta) = & -\dfrac{1}{4\pi m^2}\eta^{-2m+1}\big[(m-1)\mathcal{B}_2(\eta) - \eta\mathcal{B}_2'(\eta)\big] + O(\langle\eta\rangle^{2\alpha-2m-1}). \end{cases}$$

Remark that for the Schrödinger operator $H = -d^2/dx^2 + V$ with a periodic potential $V \in \mathsf{C}^\infty$ the asymptotics obtained in Theorem 2.2 gives

$$D(\rho^2) = \frac{1}{\pi}\rho - \frac{1}{4\pi^2\rho}\int_0^{2\pi} V(x)dx - \frac{1}{16\pi^2\rho^3}\int_0^{2\pi} |V(x)|^2 dx + O(\rho^{-5}).$$

This formula agrees with the first three terms of the complete asymptotic expansion (1.3). Note also that the term with $O(\rho^{-4})$ is absent, as in (1.3).

3. Partition of the perturbation

Before proceeding to the proof we need to make some preparations. We split the perturbation $\mathrm{Op}(b)$ into components that will play different roles in the proof.

Let $\eta \in \mathsf{C}_0^\infty(\mathbb{R})$ be a non-negative function such that

$$0 \le \eta \le 1, \quad \eta(t) = \begin{cases} 1, & |t| \le 1/4; \\ 0, & |t| \ge 1/2. \end{cases}$$

Assume also for convenience that η is even, i.e. $\eta(t) = \eta(-t)$. Set for $L \ge 1$ and $\theta \ne 0$

$$\begin{cases} \zeta_\theta(\xi; L) = & \eta\left(\dfrac{\xi + \theta/2}{L}\right), \\ \varphi_\theta(\xi; L) = & 1 - \zeta_\theta(\xi; L). \end{cases}$$

Point out that

$$\varphi_\theta(\xi; L) = \varphi_{-\theta}(\xi + \theta; L), \quad \zeta_\theta(\xi; L) = \zeta_{-\theta}(\xi + \theta; L). \tag{3.1}$$

Note that both functions φ_θ and ζ_θ, viewed as symbols, belong to the class $\mathbf{S}_0$. Indeed, suppose that $|\theta| \le L/8$. Then on the support of $\partial_\xi \varphi_\theta$ and $\partial_\xi \zeta_\theta$ one has $3L/16 \le |\xi| \le 9L/16$, and thus

$$|\partial_\xi^s \varphi_\theta(\xi; L)| + |\partial_\xi^s \zeta_\theta(\xi; L)| \le C_s L^{-s} \le \tilde{C}_s \langle \xi \rangle^{-s}, \quad |\theta| \le L/8. \tag{3.2}$$

Using $\varphi_\theta, \zeta_\theta$, we shall introduce for a symbol $b \in \mathbf{S}_\alpha$ five new symbols $b^\uparrow, b^o, b^\sharp, b^\natural, b^\flat$ depending on the parameter $\rho \ge 1$ in the following way:

$$b^\uparrow(x, \xi; \rho) = \frac{1}{\sqrt{2\pi}} \sum_\theta \big(1 - \eta(4\theta\rho^{-1})\big) \hat{b}(\theta, \xi) e^{i\theta x}, \tag{3.3}$$

$$b^\sharp(x, \xi; \rho) = \frac{1}{\sqrt{2\pi}} \sum_{\theta \ne 0} \hat{b}(\theta, \xi) \varphi_\theta(\xi; 16\rho) \eta(4\theta\rho^{-1}) e^{i\theta x}, \tag{3.4}$$

$$b^\natural(x, \xi; \rho) = \frac{1}{\sqrt{2\pi}} \sum_{\theta \ne 0} \hat{b}(\theta, \xi) \varphi_\theta(\xi; \rho) \zeta_\theta(\xi; 16\rho) \eta(4\theta\rho^{-1}) e^{i\theta x}, \tag{3.5}$$

$$b^\flat(x, \xi; \rho) = \frac{1}{\sqrt{2\pi}} \sum_{\theta \ne 0} \hat{b}(\theta, \xi) \zeta_\theta(\xi; \rho) \eta(4\theta\rho^{-1}) e^{i\theta x}, \tag{3.6}$$

$$b^o(x, \xi) = b^o(\xi) = \frac{1}{\sqrt{2\pi}} \hat{b}(0, \xi). \tag{3.7}$$

The corresponding operators are denoted by $B^\dagger = \mathrm{Op}(b^\dagger)$ where $\dagger$ takes any of the values $\uparrow, \sharp, \flat, \natural$ or o. By definition of η

$$b = b^o + b^\flat + b^\natural + b^\sharp + b^\uparrow.$$

The role of each of these operator is easy to explain. The symbol $b^{\uparrow}$ contains only Fourier coefficients with $|\theta| \geq \rho/16$, and the remaining symbols contain the Fourier coefficients with $|\theta| \leq \rho/8$. The symbol $b^{\flat}$ is supported in the region where ξ is small compared to ρ, and thus, as it will be shown later, it does not contribute to the asymptotics of the density of states $D(\rho^m)$. The symbol $b^{\sharp}$, on the contrary, contains a cut-off on the "large" values of ξ and hence does not affect $D(\rho^m)$ either. The only significant components of b are the symbols $b^{\natural}$ and b^{o}. The symbol b^{o} will remain as it is, and the symbol $b^{\natural}$ will be transformed in the next Section to another symbol, independent of x.

Often we combine $B^{\flat}$, $B^{\sharp}$ and $B^{\uparrow}$: for instance $B^{\flat,\sharp} = B^{\flat} + B^{\sharp}$, $B^{\flat,\sharp,\uparrow} = B^{\flat,\sharp} + B^{\uparrow}$. A similar convention applies to the symbols. In view of (3.2) under the condition $b \in \mathbf{S}_\alpha$ the above symbols belong to class $\mathbf{S}_\alpha$ and the following bounds hold:

$$\left\| b^{\flat} \right\|_{l,s}^{(\alpha)} + \left\| b^{\natural} \right\|_{l,s}^{(\alpha)} + \left\| b^{\sharp} \right\|_{l,s}^{(\alpha)} + \left\| b^{o} \right\|_{l,s}^{(\alpha)} + \left\| b^{\uparrow} \right\|_{l,s}^{(\alpha)} \leq C_{l,s} \left\| b \right\|_{l,s}^{(\alpha)}, \tag{3.8}$$

see (2.1) for definition of $\left\| \cdot \right\|_{l,s}^{(\alpha)}$. To check that the introduced operations preserve symmetry, calculate using (3.1)

$$\begin{aligned}\overline{\hat{b}^{\flat}(-\theta, \xi+\theta; \rho)} &= \overline{\hat{b}(-\theta, \xi+\theta)\zeta_{-\theta}(\xi+\theta; \rho)}\eta(4\theta\rho^{-1}) \\ &= \hat{b}(\theta, \xi)\zeta_{\theta}(\xi; \rho)\eta(4\theta\rho^{-1}) = \hat{b}^{\flat}(\theta, \xi; \rho).\end{aligned}$$

Therefore, by (2.4) the operator $B^{\flat}$ is symmetric if so is B. Similarly for $B^{\sharp}$, $B^{\natural}$ and $B^{\uparrow}$.

Let us list some other elementary properties of the introduced operators. Let $\chi(\,\cdot\,, \mathcal{C})$ be the indicator of a measurable set $\mathcal{C} \subset \mathbb{R}$.

LEMMA 3.1. *Let $b \in \mathbf{S}_\alpha$, with some $\alpha \in \mathbb{R}$. Denote for some $R \geq 0$*

$$\Omega_R = \{\xi \in \mathbb{R} : |\xi| \geq R\}, \quad \Lambda_R = \mathbb{R} \setminus \Omega_R. \tag{3.9}$$

Then the following statements hold:

(i) *The operator* $\operatorname{Op}(b^{\flat})$ *is bounded and*

$$\| \operatorname{Op}(b^{\flat}) \| \leq C_l \left\| b \right\|_{l,0}^{(\alpha)} \rho^{\alpha_1},$$

with $\alpha_1 = \max(\alpha, 0)$. *Moreover, for any* $R \geq 11\rho/16$

$$\chi(D, \Omega_R)\operatorname{Op}(b^{\flat}) = \operatorname{Op}(b^{\flat})\chi(D, \Omega_R) = 0.$$

(ii) *For any* $\beta \in \mathbb{R}$ *one has* $b^{\natural} \in \mathbf{S}_\beta$ *and*

$$\left\| b^{\natural} \right\|_{l,s}^{(\beta)} \leq C \left\| b \right\|_{l,s}^{(\alpha)} \rho^{\alpha-\beta},$$

with a constant C depending only on l, s, α, β. In particular, the operator $B^{\natural}$ is bounded and

$$\|B^{\natural}\| \leq C_l \left\| b \right\|_{l,0}^{(\alpha)} \rho^{\alpha},$$

for any $l > 1$.

(iii) *Let* $R \leq 3\rho$. *Then*

$$\chi(D, \Lambda_R)B^{\sharp} = B^{\sharp}\chi(D, \Lambda_R) = 0.$$

(iv) *If $R \le C\rho$, then*

$$\|\chi(D, \Lambda_R)B^{\uparrow}\| + \|B^{\uparrow}\chi(D, \Lambda_R)\| \le C_{l,p} \big| b \big|_{l,0}^{(\alpha)} \rho^{p-l}.$$

for some fixed p and any sufficiently big $l \ge p$.

The proof of this lemma is elementary: it uses only simple facts of the PDO calculus. Properties (i), (iii), (iv) will ensure that the operators $B^{\flat}$, $B^{\sharp}$, $B^{\uparrow}$ do not contribute to the sought asymptotics.

4. The "gauge transformation"

4.1. Our strategy will be to find a unitary operator which reduces $H = H_0 + \mathrm{Op}(b)$ to another PDO, whose symbol, up to some controllable small errors, depends only on ξ. The approach described in this section is similar to the method developed in [**8**] to study the asymptotics of the eigenvalues for pseudo-differential operators on the circle (see also [**1**] for matrix-valued symbols). The method of [**8**] consisted in reducing the given elliptic PDO of non-zero order to another PDO having constant coefficient, up to an error of any prescribed order. In [**8**] this is implemented with the help of a similarity transformation using a suitably chosen Fourier Integral Operator.

In the present paper, in order to perform the sought reduction we use a unitary operator of the form $U = e^{i\Psi}$ with a suitable bounded self-adjoint periodic PDO Ψ. Somewhat abusing the terminology one can call this operator a "gauge transformation". It is useful to consider U as an element of the group

$$U(t) = \exp\{i\Psi t\}, \quad \forall t \in \mathbb{R}.$$

Introduce the notation for multiple commutators: $\mathrm{ad}(H, \Psi) = i[H, \Psi]$, $\mathrm{ad}^N(H; \Psi) = \mathrm{ad}\big(\mathrm{ad}^{N-1}(H; \Psi); \Psi\big)$. **We assume that the operator $\mathrm{ad}(H_0, \Psi)$ is bounded, so that** $U(t)D(H_0) = D(H_0)$. This assumption will be justified later on. Let us express the operator

$$A_t = U(-t)HU(t)$$

via its (weak) derivative w.r.t. t:

$$A_t = H + \int_0^t U(-\tau)\,\mathrm{ad}(H; \Psi)U(\tau)d\tau.$$

By induction it is easy to show that

$$A_1 = H + \sum_{j=1}^{M} \frac{1}{j!}\,\mathrm{ad}^j(H; \Psi) + R_{M+1}^{(1)}, \tag{4.1}$$

$$R_{M+1}^{(1)} = \int_0^1 d\tau_1 \int_0^{\tau_1} d\tau_2 \dots \int_0^{\tau_M} U(-\tau_{M+1})\,\mathrm{ad}^{M+1}(H; \Psi)U(\tau_{M+1})d\tau_{M+1}.$$

The operator Ψ is sought in the form

$$\Psi = \sum_{k=1}^{M} \Psi_k, \ \Psi_k = \mathrm{Op}(\psi_k), \ \psi_k \in \mathbf{S}_{k(\alpha-m)+1}. \tag{4.2}$$

Substitute this formula in (4.1) and rewrite, regrouping the terms:

$$A_1 = H_0 + B + \sum_{l=1}^{M} \mathrm{ad}(H_0; \Psi_l) + \sum_{l=2}^{M}\sum_{j=2}^{l} \frac{1}{j!} \sum_{k_1+k_2+\cdots+k_j=l} \mathrm{ad}(H_0; \Psi_{k_1}, \Psi_{k_2}, \dots, \Psi_{k_j})$$

$$+ \sum_{l=2}^{M+1}\sum_{j=1}^{l-1} \frac{1}{j!} \sum_{k_1+k_2+\cdots+k_j=l-1} \mathrm{ad}(B; \Psi_{k_1}, \Psi_{k_2}, \dots, \Psi_{k_j}) + R^{(1)}_{M+1} + R^{(2)}_{M+1}.$$

$$R^{(2)}_{M+1} = \sum_{j=1}^{M} \frac{1}{j!} \sum_{k_1+k_2+\cdots+k_j \geq M+1} \mathrm{ad}(H; \Psi_{k_1}, \Psi_{k_2}, \dots, \Psi_{k_j}).$$

Introduce the notation

$$B_1 = B,$$

(4.3) $$B_l = \sum_{j=1}^{l-1} \frac{1}{j!} \sum_{k_1+k_2+\cdots+k_j=l-1} \mathrm{ad}(B; \Psi_{k_1}, \Psi_{k_2}, \dots, \Psi_{k_j}),\ l \geq 2,$$

(4.4) $$T_l = \sum_{j=2}^{l} \frac{1}{j!} \sum_{k_1+k_2+\cdots+k_j=l} \mathrm{ad}(H_0; \Psi_{k_1}, \Psi_{k_2}, \dots, \Psi_{k_j}),\ l \geq 2.$$

We emphasise that the operators B_l and T_l depend only on $\Psi_1, \Psi_2, \dots, \Psi_{l-1}$. Let us make one more rearrangement:

$$A_1 = H_0 + B + \sum_{l=1}^{M} \mathrm{ad}(H_0, \Psi_l) + \sum_{l=2}^{M} B_l + \sum_{l=2}^{M} T_l + R_{M+1},$$

$$R_{M+1} = B_{M+1} + R^{(1)}_{M+1} + R^{(2)}_{M+1}.$$

Now we can specify our algorithm for finding Ψ_k's. Recall that the operators with superscripts $\uparrow, \sharp, \natural, \flat, o$ are defined in (3.3), (3.4), (3.5), (3.6), (3.7). The symbols ψ_k will be found from the following system of commutator equations:

(4.5) $$\mathrm{ad}(H_0; \Psi_1) + B_1^{\natural} = 0,$$

(4.6) $$\mathrm{ad}(H_0; \Psi_l) + B_l^{\natural} + T_l^{\natural} = 0,\ l \geq 2,$$

and hence

(4.7) $$A_1 = A_0 + L_M^{\flat,\sharp,\uparrow}(\Psi) + R_{M+1},$$

(4.8) $$L_M = \sum_{l=1}^{M} B_l + \sum_{l=2}^{M} T_l,$$

(4.9) $$A_0 = H_0 + \sum_{l=1}^{M} B_l^o + \sum_{l=2}^{M} T_l^o.$$

Recall that by Lemma 3.1(ii), the operators $B_l^{\natural}, T_l^{\natural}$ are bounded, and therefore, in view of (4.5), (4.6), so is the commutator $\mathrm{ad}(H_0; \Psi)$. This justifies the assumption made in the beginning of the formal calculations in this Section.

4.2. Commutator equations. The next Lemma is central for the study of the commutator equations (4.5), (4.6).

LEMMA 4.1. *Let $A = \mathrm{Op}(a)$ be a symmetric PDO such that $a \in \mathbf{S}_\alpha$. Then the PDO Ψ with the Fourier coefficients of the symbol $\psi(x,\xi)$ given by*

$$\begin{cases} \hat\psi(\theta,\xi) = i\dfrac{\hat a^\natural(\theta,\xi;\rho)}{\tau_\theta(\xi)}, \ \tau_\theta(\xi) = h_0(\xi+\theta) - h_0(\xi), \ \theta \neq 0; \\ \hat\psi(0,\xi) = 0, \end{cases} \tag{4.10}$$

solves the equation

$$\mathrm{ad}(H_0;\Psi) + \mathrm{Op}(a^\natural) = 0. \tag{4.11}$$

Moreover, the operator Ψ is bounded and self-adjoint, $\psi \in \mathbf{S}_\beta$ for all $\beta \in \mathbb{R}$, and

$$\big|\psi\big|^{(\beta)}_{l,s} \leq C\big|a\big|^{(\alpha)}_{p,s}\rho^{\alpha-m+1-\beta},$$

with some $p = p(l,s)$. The constant C is independent of the parameter $\rho \geq 1$ and the symbol a.

PROOF. Let t be the symbol of $\mathrm{ad}(H_0;\Psi)$. The Fourier transform $\hat t(\theta,\xi)$ is easy to find:

$$\hat t(\theta,\xi) = i\big(h_0(\xi+\theta) - h_0(\xi)\big)\hat\psi(\theta,\xi) = i\tau_\theta(\xi)\hat\psi(\theta,\xi).$$

Therefore the equation (4.11) amounts to

$$i\tau_\theta(\xi)\hat\psi(\theta,\xi) = -\hat a^\natural(\theta,\xi;\rho) = -\hat a(\theta,\xi;\rho)\varphi_\theta(\xi;\rho)\zeta_\theta(\xi;16\rho)\eta\big(4|\theta|\rho^{-1}\big).$$

By definition of the function φ_θ, a solution $\hat\psi$ exists and is given by (4.10). This symbol satisfies the condition (2.4), so that Ψ is a symmetric operator. Note also that by (3.2), (3.8) and due to the inequality

$$|\partial^s_\xi \tau_\theta^{-1}| \leq C_s|\theta|^{-1}\langle\xi\rangle^{-m+1-s}, \ \xi \in \operatorname{supp}\varphi_\theta(\,\cdot\,;\rho)\eta(4|\theta|\rho^{-1}),$$

the symbol ψ belongs to $\mathbf{S}_{\alpha-m+1}$ and one easily shows that

$$\big|\psi\big|^{(\alpha-m+1)}_{l,s} \leq C\big|a\big|^{(\alpha)}_{p,s},$$

with some p. Remembering that $\hat\psi(\theta,\xi) \neq 0$ only for $3\rho \leq |\xi| \leq 9\rho$, one concludes that $\psi \in \mathbf{S}_\beta$ for any $\beta \in \mathbb{R}$ and the required bound for the norm $\big|\psi\big|^{(\beta)}_{l,s}$ holds. This estimate with $\beta = 0$ and (2.2) ensure the boundedness of Ψ. □

Applying this lemma to equations (4.5) and (4.6), we obtain the following lemma.

LEMMA 4.2. *Let $b \in \mathbf{S}_\alpha$ be a symmetric symbol. Then there exists a sequence of self-adjoint bounded PDO's Ψ_k, $k = 1, 2, \dots$ with the symbols $\psi_k \in \mathbf{S}_\beta$ for any $\beta \in \mathbb{R}$, such that (4.5) and (4.6) hold, and*

(i)

$$\big|\psi_l\big|^{(\beta)}_{r,s} \leq C\big(\big|b\big|^{(\alpha)}_{p,n}\big)^l\rho^{l(\alpha-m)+1-\beta}, \ l \geq 1; \tag{4.12}$$

(ii) *The symbols b_l, t_l of the corresponding operators B_l, T_l belong to $\mathbf{S}_\beta$ with arbitrary $\beta \in \mathbb{R}$ and*

$$\big|b_l\big|^{(\beta)}_{r,s} + \big|t_l\big|^{(\beta)}_{r,s} \leq C\big(\big|b\big|^{(\alpha)}_{p,n}\big)^l\rho^{l(\alpha-m)+m-\beta}, \ l \geq 2; \tag{4.13}$$

(iii) *The bound holds: $\|R_{M+1}\| \leq C\big(\big|b\big|^{(\alpha)}_{p,n}\big)^{M+1}\rho^{(M+1)(\alpha-m)+m}$;*

(iv) *The symbols* $b_l^o(\xi;\rho)$ *and* $t_l^o(\xi;\rho)$ *belong to the class* $\mathcal{S}_l^m$, *and*

$$b_2^o(\xi;\rho) + t_2^o(\xi;\rho) = -\frac{1}{2\pi}\sum_{\theta\neq 0}\frac{|\hat{b}(\theta,\xi)|^2}{\tau_\theta(\xi)}\eta(4|\theta|\rho^{-1})\Big[2-\eta(4|\theta|\rho^{-1})\Big] \tag{4.14}$$

for all $3\rho/4 \le |\xi| \le 2\rho$.

The integer-valued parameters p, n *in* (4.12) *and* (4.13) *depend on* $l, r, s, \alpha, \beta, m$. *The constants in the above estimates do not depend on* b, *but depend on* α, β, r, s, l.

The proof of (4.12), (4.13) uses induction and Lemma 4.1. It relies on the fact that b_l, t_l depend on the symbols $\psi_1, \psi_2, \dots, \psi_{l-1}$ only, see (4.3) and (4.4). The identity (4.14) follows by a direct computation based on the explicit formula for ψ_1, and the relations

$$B_2 = \operatorname{ad}(B;\Psi_1),\ T_2 = -\frac{1}{2}\operatorname{ad}(B^\natural;\Psi_1).$$

Let us now summarize the results of this section in the following theorem:

THEOREM 4.3. *Let* $b \in \mathbf{S}_\alpha, \alpha < m$, *be a symmetric symbol. Then for any positive integer* M *there exists a symbol* $\psi \in \mathbf{S}_{\alpha-m+1}$ *of the form* (4.2) *such that*

(i) *the operator* $A_1 = e^{-i\Psi}He^{i\Psi}$ *has the form* (4.7), *and*
(ii) *the conclusions of Lemma 4.2 hold for the symbols* ψ_l, b_l, t_l *and the remainder* R_{M+1}.

This theorem reduces the operator H to a form convenient for further calculation of the density of states. The next step is to carry out such a calculation for the operator A_0 with constant coefficients. After that it will be shown that the remaining components of A_1, that is $L^\uparrow$, $L^\flat$ and $L^\sharp$ do not affect the asymptotics.

5. Sketch of the proof of Theorem 2.2

5.1. Operator A_0. We begin our study of the density of states with the operator A_0 defined in (4.9). According to Lemma 4.2(iv) we have $b_l^o, t_l^o \in \mathcal{S}_l^m$, so that the symbol $a_0(\xi;\rho)$ of A_0 can be written as

$$a_0(\xi;\rho) = |\xi|^m + \sum_{j=1}^{M}\mu_j(\xi;\rho),\ \rho\ge 1, \tag{5.1}$$

where $\mu_j \in \mathcal{S}_j^m$ and $\mu_1 = b^o$, $\mu_j = b_j^o + t_j^o$, $j\ge 2$. Since A_0 has a symbol independent of x, in order to find its density of states $D(\lambda; A_0)$, we can immediately use Lemma 2.1. To this end we need to identify the set where $a_0(\xi;\rho) < \rho^m$. This is the objective of the next lemma.

LEMMA 5.1. *Let* a_0 *be as above. Then there exists an* $\eta_0 > 0$ *such that for all* $\eta : |\eta| \ge \eta_0$ *the equation* $a(\xi;\rho) = |\eta|^m$ *has exactly one root* $\xi(\eta;\rho)$ *having the same sign as* η. *The function* $\xi(\eta;\rho)$ *can be represented as an asymptotic series*

$$\xi(\eta;\rho) = \eta + \sum_{j=1}^{\infty}\omega_j(\eta;\rho),$$

with functions $\omega_j \in \mathcal{S}^1_j$ *that are defined by the following recursion relations:*

$$m|\eta|^m \eta^{-1}\omega_1 + \mu_1 = 0,$$

$$\omega_l = -\frac{\eta}{m|\eta|^m}\mu_l - \sum_{s=2}^{l} \frac{(m-1)\cdots(m-s+1)}{s!}\eta^{-s+1} \sum_{p_1+p_2+\dots+p_s=l} \omega_{p_1}\cdots\omega_{p_s}$$
$$-\frac{\eta}{m|\eta|^m}\sum_{s=1}^{l-1}\frac{1}{s!}\sum_{j+p_1+p_2+\dots+p_s=l} \partial^s\mu_j(\eta)\omega_{p_1}\cdots\omega_{p_s},\ l\ge 2.$$

The first two terms, i.e. $\omega_1(\eta;\rho)$, $\omega_2(\eta;\rho)$ *are defined by the formulas* (2.5) *for all* $3\rho/4 \le |\eta| \le 2\rho$. *Moreover, the functions* ω_l *satisfy the estimates*

$$|\partial_\rho^n \partial_\xi^s \omega_l(\xi;\rho)| \le C_{l,n,s}\langle\xi\rangle^{-l\kappa+1-s}\rho^{-n}, \ \kappa = m-\alpha. \tag{5.2}$$

The proof of this Lemma is quite elementary: it uses only Taylor's formula and a fixed point theorem. The calculation of the coefficients ω_1, ω_2 is based on the explicit formula (4.14).

This Lemma enables us to find the asymptotics of the density of states for A_0. Note that in contrast to Lemma 4.2 in the following statements we do not establish the precise dependence of the remainders on the perturbation b.

THEOREM 5.2 (Operator A_0). *Let* $\rho \ge 1$, *and let* $\omega_j \in \mathcal{S}^1_j$ *be the symbols from Lemma 5.1 constructed for the symbol* $a_0(\xi;\rho)$ *of the operator* A_0 *given by* (5.1) *with some* M. *Let*

$$\xi_M(\eta;\rho) = \eta + \sum_{j=1}^{M}\omega_j(\eta;\rho).$$

Then for sufficiently large ρ *the integrated density of states* $D(\eta^m; A_0), \eta > 0$ *satisfies the formula*

$$D(\eta^m; A_0) = \frac{1}{2\pi}\big[\xi_M(\eta;\rho) - \xi_M(-\eta;\rho)\big] + O(\langle\rho\rangle^{(M+1)(\alpha-m)+1}), \tag{5.3}$$

for all $3\rho/4 \le \eta \le 2\rho$.

PROOF. Let $\xi(\eta)$ be the function constructed in Lemma 5.1. According to Lemma 2.1

$$2\pi D(\eta^m; A_0) = \xi(\eta;\rho) - \xi(-\eta;\rho).$$

In view of Lemma 5.1 we have $\xi(\eta;\rho) = \xi_M(\eta;\rho) + O(\rho^{-(M+1)\kappa+1})$ for $3\rho/4 \le \eta \le 2\rho$, which leads to (5.3). □

5.2. Operator A_1. Relying on Theorem 5.2 we shall now establish the asymptotics of the density of states for the operator A_1, see (4.7). The first step is to establish some bounds relating the counting functions of the fibre operators $A_0(k)$ and $A_1(k)$ for $k \in (0,1)$.

LEMMA 5.3. *For any number* $M > 0$, *one has*

$$N(\lambda_-, k; A_0) \le N(\rho^m, k; A_1) \le N(\lambda_+, k; A_0),$$
$$\lambda_\pm = \rho^m \pm c\rho^{-(M+1)\kappa+m},$$

with $\kappa = m - \alpha$, *for all* $k \in [0,1]$. *The constant* c *is uniform in* b *and depends on the parameter* M.

PROOF. In view of the bound $\|R_{M+1}(H)\| \le C\rho^{-(M+1)\kappa+m}$ found in Lemma 4.2(iii), we have

$$N(\lambda_-, k; A_0 + L^{\flat,\sharp,\uparrow}) \le N(\rho^m, k; A_1) \le N(\lambda_+, k; A_0 + L^{\flat,\sharp,\uparrow}), \ L = L_M. \tag{5.4}$$

To get rid of $L^{\flat,\sharp,\uparrow}$ in this inequality, represent $A_0+L^{\flat,\sharp,\uparrow}$ in a matrix form by decoupling the space $\mathcal{H} = \mathsf{L}^2(\mathbb{R})$ in the orthogonal sum of three subspaces corresponding to the following three disjoint subsets of $\mathbb{R}$: Λ_r, Ω_R (see (3.9) for definition) and $\Upsilon_{r,R} = \Omega_r \cap \Lambda_R$, where $0 < r < R < \infty$ with $r = 11\rho/16$ and $R = 2\rho$. Denote

$$\mathcal{P}_1 = \chi(D, \Lambda_r), \mathcal{P}_2 = \chi(D, \Upsilon_{r,R}), \ \mathcal{P}_3 = \chi(D, \Omega_R),$$

so that $I = \mathcal{P}_1 + \mathcal{P}_2 + \mathcal{P}_3$. By Lemma 3.1 we have

$$(\mathcal{P}_2 + \mathcal{P}_3)L^\flat = L^\flat(\mathcal{P}_2 + \mathcal{P}_3) = (\mathcal{P}_1 + \mathcal{P}_2)L^\sharp = L^\sharp(\mathcal{P}_1 + \mathcal{P}_2) = 0.$$

Therefore $A_0 + L^{\flat,\sharp,\uparrow}$ takes the form $\tilde{A}_0 + \tilde{L}$ with

$$\tilde{A}_0 = \begin{bmatrix} \mathcal{P}_1(A_0 + L^\flat)\mathcal{P}_1 & 0 & 0 \\ 0 & A_0\mathcal{P}_2 & 0 \\ 0 & 0 & \mathcal{P}_3(A_0 + L^{\sharp,\uparrow})\mathcal{P}_3 \end{bmatrix},$$

$$\tilde{L} = \begin{bmatrix} \mathcal{P}_1 L^\uparrow \mathcal{P}_1 & \mathcal{P}_1 L^\uparrow \mathcal{P}_2 & \mathcal{P}_1 L^\uparrow \mathcal{P}_3 \\ \mathcal{P}_2 L^\uparrow \mathcal{P}_1 & \mathcal{P}_2 L^\uparrow \mathcal{P}_2 & \mathcal{P}_2 L^\uparrow \mathcal{P}_3 \\ \mathcal{P}_3 L^\uparrow \mathcal{P}_1 & \mathcal{P}_3 L^\uparrow \mathcal{P}_2 & 0 \end{bmatrix}.$$

Here and below, slightly abusing the notation the operators of the form $\mathcal{P}_j T \mathcal{P}_l$, $j, l = 1, 2, 3$, are considered as operators from $\mathcal{P}_l\mathcal{H}$ to $\mathcal{P}_j\mathcal{H}$. By (4.8), Lemma 4.2 and Lemma 3.1(iv), the norm of the operator $\tilde{L}$ is bounded by $C_r\rho^{-r}$ with arbitrary $r > 0$. Therefore the estimate (5.4) transforms into the two-sided bound

$$N(\lambda_-, k; \tilde{A}_0) \le N(\rho^m, k; A_1) \le N(\lambda_+, k; \tilde{A}_0).$$

Notice that $\lambda_+ \le (5\rho/4)^m$, $\lambda_- \ge (4\rho/5)^m$. Thus the proof of the Lemma will be completed if we show that

$$N(\mu^m, k; \mathcal{P}_1(A_0 + L^\flat)\mathcal{P}_1) = N(\mu^m, k; \mathcal{P}_1 A_0), \tag{5.5}$$

$$N(\mu^m, k; \mathcal{P}_3(A_0 + L^{\sharp,\uparrow})\mathcal{P}_3) = N(\mu^m, k; \mathcal{P}_3 A_0 \mathcal{P}_3) = 0, \tag{5.6}$$

for all $\mu \in [4\rho/5, 5\rho/4]$. According to (4.13), the symbols $\sum_{l=1}^M b_l$ and $\sum_{l=2}^M t_l$ belong to the class $\mathbf{S}_\alpha$. In view of (3.8) the same applies to the sums involving $\sharp$- and $\uparrow$-operators. Using this fact and definition (4.8) it is now a matter of routine to show that

$$|((L^\sharp + L^\uparrow)u, u)| \le C(\Xi^\alpha u, u), \ \Xi = \mathrm{Op}(\langle\xi\rangle),$$

and hence

$$((A_0 + L^\sharp + L^\uparrow)\mathcal{P}_3 u, \mathcal{P}_3 u) \ge (2^m(\rho^m - C\rho^\alpha)\mathcal{P}_3 u, \mathcal{P}_3 u) \ge (3\rho/2)^m \|\mathcal{P}_3 u\|^2.$$

Therefore the fibre $\big(\mathcal{P}_3(A_0 + L^\sharp + L^\uparrow)\mathcal{P}_3\big)(k)$ satisfies the same bound for a.a. k. This ensures (5.6). Similarly, by Lemma 3.1(i)

$$\|L^\flat\| \le C\rho^{\alpha_1}, \ \alpha_1 = \max(\alpha, 0).$$

The operator $A_0\mathcal{P}_1$ satisfies $A_0\mathcal{P}_1 \le (3\rho/4)^m$ for large ρ. Since $\alpha_1 < m$, the norm $\|L^\flat\|$ is strictly less than $(4\rho/5)^m - (3\rho/4)^m$. Now a straightforward perturbation argument leads to (5.5). This completes the proof of the Lemma. □

Thus the question is reduced to the study of the operator A_0.

5.3. Proof of Theorem 2.2: conclusion. By Theorem 5.2

$$D(\lambda_\pm; A_0) = D(\rho^m; A_0) + O\big(\rho^{-\kappa(M+1)+1}\big).$$

Now Lemma 5.3 yields that

$$D(\rho^m; A_1) = D(\rho^m; A_0) + O\big(\rho^{-\kappa(M+1)+1}\big).$$

Since the unitary operator $U = e^{i\Psi}$ is periodic, we have $D(\lambda; A_1) = D(\lambda; H)$. It remains to use Theorem 5.2 for $\eta = \rho$ and denote $\nu_j(\xi) = \omega_j(\xi, \rho)$, $\xi = \pm\rho$. The property $\nu_j \in \mathcal{S}_j^1$ follows from the fact that $\omega(\ \cdot\ , \rho) \in \mathcal{S}_j^1$ and the estimate (5.2). The formulas (2.5) follow from Lemma 5.1.

□

References

1. M. S. Agranovich, *Elliptic pseudodifferential operators on a closed curve*, Trudy Moskov. Mat. Obshch. **47** (1984), 22–67 (Russian).
2. J. Feldman, H. Knörrer, E. Trubowitz, *Perturbatively unstable eigenvalues of periodic Schrödinger operator*, Comment. Math. Helvetici, **66** (1991), 557-579.
3. B. Helffer, A. Mohamed, *Asymptotics of the density of states for the Schrödinger operator with periodic electric potential*, Duke Math. J. **92** (1998), 1–60.
4. Yu. Karpeshina, *On the density of states for the periodic Schrödinger operator*, Ark. Mat. **38**(2000), 111–137.
5. ———, *Perturbation theory for the Schrödinger operator with a periodic potential*, Lecture Notes in Math. vol 1663, Springer Berlin 1997.
6. V. A. Marchenko, I.V. Ostrovskii, *A characterization of the spectrum of Hill's operator*, Mat. Sb. **97** (1975), 540–606 (Russian); Eng. transl. in Math USSR Sb. **26** (1975).
7. M. Reed, B. Simon, *Methods of modern mathematical physics, IV*, Academic Press, New York, 1975.
8. G. V. Rozenbljum, *Near-similarity of operators and the spectral asymptotic behavior of pseudodifferential operators on the circle*, Trudy Moskov. Mat. Obshch. **36** (1978), 59–84 (Russian).
9. D. Schenk, M.A. Shubin, *Asymptotic expansion of the state density and the spectral function of a Hill operator*, Math. USSR Sbornik, **56** (1987), no. 2, 473–490.
10. M.A. Shubin, *Weyl's theorem for the Schrödinger operator with an almost periodic potential*, Vestnik Moskov. Univ. Ser. I Mat. Mekh. **31** (1976), no 2, 84-88 (Russian). Engl. transl.: Moscow Univ. Math. Bull. **31** (1976), 133-137.
11. ———, *The spectral theory and the index of elliptic operators with almost periodic coefficients*, Russian Math Surveys **34** (1979), no 2, 109–157.
12. M. Skriganov, *Geometrical and arithmetical methods in the spectral theory of the multidimensional periodic operators*, Proc. Steklov Math. Inst. Vol. 171, 1984.

School of Mathematical Sciences, University of Sussex, Falmer, Brighton, BN1 9QH, UK

E-mail address: a.v.sobolev@sussex.ac.uk

Contemporary Mathematics
Volume **339**, 2003

On discrete spectrum in the gaps of a two-dimensional periodic elliptic operator perturbed by a decaying potential

T. A. Suslina

ABSTRACT. We study the discrete spectrum in the gaps of a two-dimensional periodic elliptic second order operator perturbed by a decaying potential. We are interested in the asymptotics (for large coupling constant) of the number of eigenvalues that have been "born" (or have "died") at the edges of the gap. The high-energy (Weyl) asymptotics and the threshold asymptotics are distinguished. At the right edge of the gap the competition between the Weyl contribution and the threshold contribution to the asymptotics is possible.

§ 0. Introduction

1. Let A be a selfadjoint elliptic periodic second order operator in $L_2(\mathbb{R}^d)$, $d \geq 2$, given by the expression $A = -\operatorname{div} g(\mathbf{x})\nabla + p(\mathbf{x})$, and let V be an operator of multiplication by a function $V(\mathbf{x}) \geq 0$ decaying at infinity. Let the interval (λ_-, λ_+) be a gap in the spectrum of A. We put $A_\pm(\alpha) = A \mp \alpha V(\mathbf{x})$, $\alpha > 0$. By $\mathfrak{N}_+(\alpha, \lambda_+)$ we denote the number of eigenvalues of the operator $A_+(t)$ that have been "born" at the point λ_+ as the coupling constant t has been growing from 0 to α. The function $\mathfrak{N}_-(\alpha, \lambda_-)$ is defined similarly for the operator A_-. We are interested in the asymptotics of these functions as $\alpha \to \infty$ (in the large coupling constant limit). The corresponding asymptotics may be rather diverse, depending both on A and V. They were studied in a number of papers. First of all, we mention [B3, B4, BL] and especially the survey [B2] and the references therein. Another approach to the class of problems under discussion was proposed in [Iv].

The asymptotic behavior of $\mathfrak{N}_\pm(\alpha, \lambda_\pm)$ depends on dimension d and on the character of decay of V. The situation for $d \geq 3$ differs substantially from that for $d = 2$. If $d \geq 3$ and $V \in L_{d/2}(\mathbb{R}^d)$, the function $\mathfrak{N}_+(\alpha, \lambda_+)$ has the Weyl asymptotics

$$\mathfrak{N}_+(\alpha, \lambda_+) \sim (2\pi)^{-d}\omega_d \alpha^{d/2} \int V^{d/2} (\det g)^{-1/2}\, d\mathbf{x}, \quad \alpha \to \infty; \qquad (0.1)$$

here ω_d is the volume of the unit ball in $\mathbb{R}^d$.

2000 *Mathematics Subject Classification.* Primary 35P20.
Key words and phrases. Periodic operator, perturbation, discrete spectrum, threshold effect.
Supported by RFBR (grant no. 02-01-00798)

If $V \notin L_{d/2}(\mathbb{R}^d)$, then the estimate $\mathfrak{N}_+(\alpha, \lambda_+) = O(\alpha^{d/2})$ is violated, and $\mathfrak{N}_+(\alpha, \lambda_+)$ can have an arbitrary order of growth greater than $d/2$. Essentially, the asymptotics (0.1) has a "high-energy" origin, while the behavior of $\mathfrak{N}_+(\alpha, \lambda_+)$ with $V \notin L_{d/2}(\mathbb{R}^d)$ is determined by the "threshold" effect near the edge of the gap of the unperturbed operator A. (See discussion in [B2, §2].)

2. If $d = 2$, the situation is much more complicated. Let $V \in L_1(\mathbb{R}^2)$. Already for $A = -\Delta$ in the case of the semi-infinite gap $(-\infty, 0)$, this condition is insufficient for (0.1). Due to the threshold effects, $\mathfrak{N}_+(\alpha, 0)$ may have an arbitrary order of growth greater than $d/2$. Moreover, it may happen that $\mathfrak{N}_+(\alpha, 0) = O(\alpha)$, but the asymptotics is not of Weyl type. In this case the asymptotic coefficient is the sum of the Weyl term and the "threshold" term. A *"special channel"* is responsible for the threshold effect. We mean the problem on the semi-axis that is obtained by restriction of $-\Delta - \alpha V$ to the subspace of functions depending only on $|\mathbf{x}|$. At the same time, the potential V is averaged over the polar angle. These effects were investigated in [BL] in detail. At the level of estimates this channel was discovered before in [S].

In [BLSu], the same effects were studied in the case where A is a periodic elliptic operator of the form $A = -\operatorname{div} g(\mathbf{x})\nabla + p(\mathbf{x})$. Adding an appropriate constant to p allows us to assume that the lower edge of the spectrum is the point $\lambda = 0$. In [BLSu], the *negative* discrete spectrum of the operator $A - \alpha V$, i. e., the case of the semi-bounded gap $(-\infty, 0)$ was studied. The description of the special channel was given in terms of the Floquet-Bloch decomposition for the unperturbed operator A. The answer involves the so-called *tensor of effective masses* at the edge of the spectrum and a *positive periodic solution* φ *of the equation* $A\varphi = 0$. The function φ can be eliminated from the answer under an additional "regularity" condition imposed on V.

The present paper is a continuation of [BLSu], but now we study the case of an internal gap in the spectrum of A. For this, we need to change the technique of investigation substantially.

3. For the study of the functions $\mathfrak{N}_\pm(\alpha, \lambda_\pm)$ in an *internal gap* of A, we impose certain restrictions on the structure of the edges of the gap (see Condition 1.3($\pm$) below). For the lower edge of the spectrum $\lambda = 0$ this condition is fulfilled automatically. The answers are given in terms of the *model operators*, which are simpler than $A_\pm(\alpha)$. The model operators involve the tensors of effective masses at the edges of the gap and the corresponding eigenfunctions. As well as for the semi-infinite gap, it is possible to eliminate the eigenfunctions from the answer under an additional "regularity" condition imposed on V. The main results are formulated in Theorems 2.2($\pm$), 2.5($\pm$). On the right (but not on the left) edge of the gap a competition between the Weyl contribution and the threshold contribution to the asymptotics is possible.

4. Acknowledgments. The author is grateful to M. Sh. Birman for useful discussions and constant attention to work.

5. Notation. In what follows, $\mathbb{Q}^2$ is an open unit square in $\mathbb{R}^2$. The symbol $\langle\cdot,\cdot\rangle$ stands for the standard inner product in $\mathbb{C}^m$; $\mathbf{1}$ is the unit (2×2)-matrix. Any integral without indication of the integration domain is over $\mathbb{R}^2$. Further, $\nabla = \operatorname{grad}$, $\nabla^* = -\operatorname{div}$. We denote by H^s, $s \geq 0$, the Sobolev classes.

Many statements and formulas contain the double indices "±". Unless otherwise explicitly stated, the upper and the lower versions should be read independently.

§ 1. Setting of the problem. Preliminaries

1. Differential operators. The *unperturbed* operator A is formally given by the expression $\mathcal{A}u = \nabla^* g\nabla u + pu$. Here g is a (2×2)-matrix-valued function, p is a real-valued function. We assume that

$$\left.\begin{aligned} &\overline{g} = g > 0,\quad \overline{p} = p,\quad g + g^{-1} \in L_\infty(\mathbb{R}^2),\quad p \in L_\infty(\mathbb{R}^2),\\ &g(\mathbf{x}+\mathbf{n}) = g(\mathbf{x}),\quad p(\mathbf{x}+\mathbf{n}) = p(\mathbf{x}),\quad \mathbf{x}\in\mathbb{R}^2,\quad \mathbf{n}\in\mathbb{Z}^2.\end{aligned}\right\} \tag{1.1}$$

There is no loss of generality in assuming that the lattice of periods is $\mathbb{Z}^2$. The precise definition of A as a selfadjoint operator in the Hilbert space $L_2(\mathbb{R}^2)$ is given via the closed lower semi-bounded quadratic form

$$a[u,u] = \int (\langle g\nabla u, \nabla u\rangle + p|u|^2)\, d\mathbf{x},\quad u \in H^1(\mathbb{R}^2). \tag{1.2}$$

Adding an appropriate constant to p allows us to assume that $\inf\operatorname{spec} A = 0$. Under this condition, in $H^1(\mathbb{R}^2)$ the form $a[u,u] + \gamma\int |u|^2\, d\mathbf{x}$, $\gamma > 0$, determines a metric equivalent to the standard one.

A *perturbation* is introduced as the operator of multiplication by a function $V(\mathbf{x})$ such that $V(\mathbf{x}) \geq 0$, $\mathbf{x}\in\mathbb{R}^2$. We impose the following condition on V (cf., e. g., [BL]).

CONDITION 1.1. For some $\sigma > 1$,

$$\Bigg(\int\limits_{|\mathbf{x}|\leq 1} |V|^\sigma\, d\mathbf{x}\Bigg)^{1/\sigma} + \sum_{k\geq 1}\Bigg(\int\limits_{e^{k-1}\leq|\mathbf{x}|\leq e^k} |V|^\sigma |\mathbf{x}|^{2(\sigma-1)}\, d\mathbf{x}\Bigg)^{1/\sigma} < \infty. \tag{1.3}$$

We mention at once that (1.3) implies that

$$(\!(V)\!)_\sigma := \sum_{\mathbf{n}\in\mathbb{Z}^2}\Bigg(\int\limits_{\mathbb{Q}^2+\mathbf{n}} V^\sigma\, d\mathbf{x}\Bigg)^{1/\sigma} < \infty,\quad \sigma > 1, \tag{1.4}$$

and, moreover, $V\in L_1(\mathbb{R}^2)$. Consider the quadratic form

$$v[u,u] = \int V|u|^2\, d\mathbf{x}.$$

Under condition (1.4) (and, moreover, under condition (1.3)), this form is compact in $H^1(\mathbb{R}^2)$. Consequently, the *form*

$$a_\pm(\alpha)[u,u] := a[u,u] \mp \alpha v[u,u],\quad u\in H^1(\mathbb{R}^2),\quad \alpha > 0,$$

is lower semi-bounded and closed in $L_2(\mathbb{R}^2)$. The form $a_\pm(\alpha)$ generates a selfadjoint operator $A_\pm(\alpha)$ in $L_2(\mathbb{R}^2)$. Formally, the operator $A_\pm(\alpha)$ corresponds to the differential expression $\mathcal{A}_\pm(\alpha)u = \nabla^* g\nabla u + pu \mp \alpha V u$. The *spectrum of* $A_\pm(\alpha)$ *in the spectral gaps of* A *is discrete.*

First, we recall the result for the semi-infinite gap. Let $\mathfrak{N}_+(\alpha,\lambda;A,V)$, $\alpha>0$, $\lambda\le 0$, denote the *number of eigenvalues of the operator* $A_+(\alpha)$, *lying to the left of the point* λ.

For the Weyl asymptotic coefficient we introduce the notation

$$J(V,g) := (4\pi)^{-1}\int V(\det g)^{-1/2}\,d\mathbf{x}. \tag{1.5}$$

PROPOSITION 1.2. *Under condition* (1.4), *we have*

$$\mathfrak{N}_+(\alpha,\lambda;A,V)\le C\alpha(\!(V)\!)_\sigma,\quad C=C(g,p,\sigma,\lambda),\quad \sigma>1,\quad \lambda<0,$$

$$\lim_{\alpha\to\infty}\alpha^{-1}\mathfrak{N}_+(\alpha,\lambda;A,V)=J(V,g),\quad \lambda<0. \tag{1.6}$$

Comments on Proposition 1.2 and necessary references can be found in [BLSu]. For $\lambda=0$ the Weyl asymptotics (1.6) may fail to occur even under condition (1.3) because of spectral "threshold" effects. These phenomena were studied in the paper [BL] for the operator $-\Delta-\alpha V$ and in [BLSu] in the general case of a periodic operator A. Below we shall impose one more condition on V (see Condition 2.1(q)), which ensures that $\mathfrak{N}_+(\alpha,0;A,V)=O(\alpha^q)$, $\alpha\to\infty$, $q\ge 1$. In the present paper the discrete spectrum of the operators $A_\pm(\alpha)$ in internal gaps of A is studied.

2. The Floquet decomposition. As usual, for the periodic operators we employ the Floquet-Bloch theory. Let $\widetilde{H}^1(\mathbb{Q}^2)$ be the subspace formed by the functions in $H^1(\mathbb{Q}^2)$ whose $\mathbb{Z}^2$-periodic extensions belong to the class $H^1_{\mathrm{loc}}(\mathbb{R}^2)$. Next, we denote by $\widetilde{H}^1_{\boldsymbol{\xi}}(\mathbb{Q}^2)$, $\boldsymbol{\xi}\in\mathbb{R}^2$, the subspace of functions of the form $u(\mathbf{x})=e^{i\langle\mathbf{x},\boldsymbol{\xi}\rangle}v(\mathbf{x})$, $v\in\widetilde{H}^1(\mathbb{Q}^2)$. Consider the following family of quadratic forms in $L_2(\mathbb{Q}^2)$:

$$a_{\boldsymbol{\xi}}[u,u]=\int_{\mathbb{Q}^2}(\langle g\nabla u,\nabla u\rangle+p|u|^2)\,d\mathbf{x},\quad u\in\widetilde{H}^1_{\boldsymbol{\xi}}(\mathbb{Q}^2),\quad \boldsymbol{\xi}\in\mathbb{R}^2. \tag{1.7}$$

The selfadjoint operator in $L_2(\mathbb{Q}^2)$ generated by the form (1.7) is denoted by $A(\boldsymbol{\xi})$. The operator $A(\boldsymbol{\xi})$ corresponds to the expression $\mathcal{A}$ with ($\boldsymbol{\xi}$)-quasi-periodic boundary conditions. Usually it is sufficient to consider $\boldsymbol{\xi}\in\mathbb{T}^2=\mathbb{R}^2/(2\pi\mathbb{Z})^2$. The parameter $\boldsymbol{\xi}$ is called the *quasi-momentum*. All operators $A(\boldsymbol{\xi})$ have discrete spectrum. Let $E_s(\boldsymbol{\xi})$, $s\in\mathbb{N}$, be consecutive eigenvalues (counted with multiplicity) of the operator $A(\boldsymbol{\xi})$; and let $\psi_s(\mathbf{x},\boldsymbol{\xi})$ be the corresponding eigenfunctions normalized in $L_2(\mathbb{Q}^2)$. The functions E_s are continuous and $(2\pi\mathbb{Z})^2$-periodic. The spectrum of A coincides with the union of the intervals (bands) that are the ranges of the functions E_s. The eigenfunctions ψ_s admit representation of the form $\psi_s(\mathbf{x},\boldsymbol{\xi})=e^{i\langle\mathbf{x},\boldsymbol{\xi}\rangle}\varphi_s(\mathbf{x},\boldsymbol{\xi})$, $\varphi_s(\cdot,\boldsymbol{\xi})\in\widetilde{H}^1(\mathbb{Q}^2)$. The functions ψ_s, φ_s are Hölder continuous in $\mathbf{x}$.

We consider the integral operators

$$(\Psi_s u)(\boldsymbol{\xi})=(2\pi)^{-1}\int\overline{\psi_s(\mathbf{x},\boldsymbol{\xi})}u(\mathbf{x})\,d\mathbf{x},\quad s\in\mathbb{N}.$$

The mappings $\Psi_s:L_2(\mathbb{R}^2)\to L_2(\mathbb{T}^2)$ *are partially isometric and surjective.* The operators $\Psi_s^*\Psi_s$, $s\in\mathbb{N}$, are orthoprojections in $L_2(\mathbb{R}^2)$. They are orthogonal to one another and $\sum_{s\in\mathbb{N}}\Psi_s^*\Psi_s=I$. Denoting by $[E_s]$ the operator of multiplication by the function $E_s(\boldsymbol{\xi})$ in $L_2(\mathbb{T}^2)$, we have $A=\sum_{s\in\mathbb{N}}\Psi_s^*[E_s]\Psi_s$.

3. A gap. The spectrum of A may have gaps other than the semi-infinite gap $(-\infty, 0)$. Let $\Lambda = (\lambda_-, \lambda_+)$ be a gap. Clearly,

$$\lambda_+ = \min_{\boldsymbol{\xi} \in \mathbb{T}^2} E_l(\boldsymbol{\xi}), \tag{1.8+}$$

$$\lambda_- = \max_{\boldsymbol{\xi} \in \mathbb{T}^2} E_{l-1}(\boldsymbol{\xi}), \tag{1.8−}$$

for some $l \in \mathbb{N}$. We impose certain restrictions on the "structure" of the edges of the gap. Let us formulate condition for λ_+.

CONDITION 1.3(+). a) $\min_{\boldsymbol{\xi} \in \mathbb{T}^2} E_{l+1}(\boldsymbol{\xi}) > \lambda_+$; b) *The minimum* (1.8+) *is attained only at finitely many points* $\boldsymbol{\xi}_j^{(+)} \in \mathbb{T}^2$, $j = 1, \dots, m_+$, *all of which are non-degenerate minimum points for* $E_l(\cdot)$.

REMARK 1.4. For the semi-infinite gap $(-\infty, 0)$ Condition 1.3(+) at $\lambda_+ = 0$ is fulfilled automatically with $l = 1$, $m_+ = 1$, $\boldsymbol{\xi}_1^{(+)} = 0$. This fact was used in [BLSu].

By Condition 1.3(+), λ_+ is a simple eigenvalue of the operator $A(\boldsymbol{\xi})$ with $\boldsymbol{\xi} = \boldsymbol{\xi}_j^{(+)}$, $j = 1, \dots, m_+$. Then, for some (sufficiently small) $\delta > 0$ the *eigenvalue* $E_l(\boldsymbol{\xi})$, $|\boldsymbol{\xi} - \boldsymbol{\xi}_j^{(+)}| \le \delta$, *is simple*. This implies the *real analyticity* of $E_l(\boldsymbol{\xi})$ in these neighborhoods of the points $\boldsymbol{\xi}_j^{(+)}$, $j = 1, \dots, m_+$. Then Condition 1.3(+), b) means that

$$E_l(\boldsymbol{\xi}) - \lambda_+ = \mathfrak{b}_j^{(+)}(\boldsymbol{\xi} - \boldsymbol{\xi}_j^{(+)}) + O(|\boldsymbol{\xi} - \boldsymbol{\xi}_j^{(+)}|^3), \quad |\boldsymbol{\xi} - \boldsymbol{\xi}_j^{(+)}| \le \delta, \quad j = 1, \dots, m_+, \tag{1.9+}$$

where $\mathfrak{b}_j^{(+)}$ is a positive definite quadratic form.

Let us formulate condition on λ_-.

CONDITION 1.3(−). a) $\max_{\boldsymbol{\xi} \in \mathbb{T}^2} E_{l-2}(\boldsymbol{\xi}) < \lambda_-$; b) *The maximum in* (1.8−) *is attained only at finitely many points* $\boldsymbol{\xi}_j^{(-)} \in \mathbb{T}^2$, $j = 1, \dots, m_-$, *all of which are non-degenerate maximum points for* $E_{l-1}(\cdot)$.

Similarly to (1.9+), Condition 1.3(−), b) means that

$$\lambda_- - E_{l-1}(\boldsymbol{\xi}) = \mathfrak{b}_j^{(-)}(\boldsymbol{\xi} - \boldsymbol{\xi}_j^{(-)}) + O(|\boldsymbol{\xi} - \boldsymbol{\xi}_j^{(-)}|^3), \quad |\boldsymbol{\xi} - \boldsymbol{\xi}_j^{(-)}| \le \delta, \quad j = 1, \dots, m_-, \tag{1.9−}$$

where $\mathfrak{b}_j^{(-)}$ is a positive definite quadratic form.

REMARK 1.5. We agree that the points $\boldsymbol{\xi}_j^{(\pm)} \in \mathbb{T}^2$ are represented as points of the semi-open cube: $\boldsymbol{\xi}_j^{(\pm)} \in [-\pi, \pi)^2$, $j = 1, \dots, m_\pm$. Accordingly, small neighborhoods of points $\boldsymbol{\xi}_j^{(\pm)} \in \mathbb{T}^2$ are understood as $\mathbb{R}^2$-neighborhoods of points $\boldsymbol{\xi}_j^{(\pm)} \in [-\pi, \pi)^2$.

The form $\mathfrak{b}_j^{(\pm)}(\boldsymbol{\xi} - \boldsymbol{\xi}_j^{(\pm)})$ can be written as

$$\begin{aligned} \mathfrak{b}_j^{(\pm)}(\boldsymbol{\xi} - \boldsymbol{\xi}_j^{(\pm)}) = \langle b_j^{(\pm)}(\boldsymbol{\xi} - \boldsymbol{\xi}_j^{(\pm)}), \boldsymbol{\xi} - \boldsymbol{\xi}_j^{(\pm)} \rangle &= |\beta_j^{(\pm)}(\boldsymbol{\xi} - \boldsymbol{\xi}_j^{(\pm)})|^2, \\ \beta_j^{(\pm)} &= (b_j^{(\pm)})^{1/2}, \quad j = 1, \dots, m_\pm, \end{aligned} \tag{1.10±}$$

where $b_j^{(\pm)}$ is a constant positive definite matrix. (The matrix $(b_j^{(\pm)})^{-1}$ determines the *tensor of effective masses* for the point $\boldsymbol{\xi}_j^{(\pm)}$.)

We put $E_+ := E_l$, $E_- := E_{l-1}$, $\psi^{(+)} := \psi_l$, $\varphi^{(+)} := \varphi_l$, $\psi^{(-)} := \psi_{l-1}$, $\varphi^{(-)} := \varphi_{l-1}$. The functions $\psi^{(\pm)}$, $\varphi^{(\pm)}$ can be chosen as real-analytic $H^1(\mathbb{Q}^2)$-valued functions of $\boldsymbol{\xi}$ for $|\boldsymbol{\xi} - \boldsymbol{\xi}_j^{(\pm)}| \le \delta$, $j = 1, \dots, m_\pm$. The functions $\psi^{(\pm)}$, $\varphi^{(\pm)}$ are Hölder continuous in $\mathbf{x}$. We introduce the notation

$$\psi_j^{(\pm)}(\mathbf{x}) := \psi^{(\pm)}(\mathbf{x}, \boldsymbol{\xi}_j^{(\pm)}), \quad \varphi_j^{(\pm)}(\mathbf{x}) := \varphi^{(\pm)}(\mathbf{x}, \boldsymbol{\xi}_j^{(\pm)}), \quad j = 1, \dots, m_\pm. \qquad (1.11\pm)$$

4. Let $\lambda \in (\lambda_-, \lambda_+) = \Lambda$. We denote by

$$\mathfrak{N}_\pm(\alpha, \lambda; A, V), \quad \alpha > 0, \quad \lambda \in \Lambda, \qquad (1.12)$$

the number of eigenvalues of $A_\pm(t)$ that have crossed λ as the coupling constant t has been growing from 0 to α. The following statement was proved in [B1] (see also [B2, Theorem 3.2]).

PROPOSITION 1.6. *Under condition* (1.4), *we have*

$$\lim_{\alpha \to \infty} \alpha^{-1} \mathfrak{N}_+(\alpha, \lambda; A, V) = J(V, g), \quad \lambda \in \Lambda, \qquad (1.13)$$

$$\lim_{\alpha \to \infty} \alpha^{-1} \mathfrak{N}_-(\alpha, \lambda; A, V) = 0, \quad \lambda \in \Lambda.$$

Thus, if the "observation point" λ lies inside the gap, then $\mathfrak{N}_+$ has the Weyl asymptotics (1.13). For $\lambda = \lambda_+$ the asymptotics (1.13) may fail to occur even under condition (1.3). Below we impose an additional condition (Condition 2.1(q)) on V that ensures the finiteness of the following limits:

$$\mathfrak{N}_\pm(\alpha, \lambda_+; A, V) = \lim_{\lambda \to \lambda_+ - 0} \mathfrak{N}_\pm(\alpha, \lambda; A, V), \qquad (1.14+)$$

$$\mathfrak{N}_\pm(\alpha, \lambda_-; A, V) = \lim_{\lambda \to \lambda_- + 0} \mathfrak{N}_\pm(\alpha, \lambda; A, V), \qquad (1.14-)$$

We are interested in the behavior of functions (1.14) as $\alpha \to \infty$.

5. On compact operators. Let $\mathfrak{H}$ be a separable Hilbert space. The space of continuous linear operators is denoted by $\mathfrak{R}$, and that of compact operators by $\mathfrak{S}_\infty$. Let $T \in \mathfrak{S}_\infty$, and let $s_k(T)$ be the singular numbers of T, i. e., the consecutive eigenvalues (counted with multiplicity) of the operator $(T^*T)^{1/2}$. We denote

$$n(s, T) := \operatorname{card}\{k : s_k(T) > s\}, \quad s > 0.$$

For $T = T^*$, we put $2T_\pm = |T| \pm T$ and $n_\pm(s, T) := n(s, T_\pm)$, $s > 0$. Clearly, $n_+(\cdot, T)$ is the counting function for the sequence $\{\lambda_k^{(+)}(T)\}$ of positive eigenvalues of T. For the sequence $\{\lambda_k^{(-)}(T)\}$ a similar role is played by $n_-(\cdot, T)$ (here $\lambda_k^{(-)}(T) = \lambda_k^{(+)}(-T)$). We have $n(s, T) = n_+(s, T) + n_-(s, T)$, $s > 0$.

We denote by Σ_q, $0 < q < \infty$, the space (ideal) of compact operators distinguished by the condition

$$|T|_q^q := \sup_{s > 0} s^q n(s, T) < \infty, \quad q > 0.$$

The space Σ_q is complete in the quasi-norm $|\cdot|_q$ and non-separable. On the space Σ_q we consider the functionals

$$\begin{aligned} \Delta_q(T) &:= \limsup_{s \to 0} s^q n(s, T), \\ \delta_q(T) &:= \liminf_{s \to 0} s^q n(s, T). \end{aligned} \qquad (1.15)$$

The condition $\Delta_q(T) = 0$ determines the separable subspace Σ_q^0 in Σ_q. For $T = T^* \in \Sigma_q$, we put

$$\Delta_q^{(\pm)}(T) := \Delta_q(T_\pm), \quad \delta_q^{(\pm)}(T) := \delta_q(T_\pm). \tag{1.16}$$

Below D_q stands for any of the functionals (1.15), (1.16).

If $T^* = T \in \mathfrak{S}_\infty(\mathfrak{H})$, then the numbers $\lambda_k^{(+)}(T)$ (the numbers $(-\lambda_k^{(-)}(T))$) coincide with the consecutive positive maxima (the negative minima) of the ratio of quadratic forms

$$(Tu, u)_{\mathfrak{H}} \,/ \|u\|^2_{\mathfrak{H}}, \quad u \in \mathfrak{H}. \tag{1.17}$$

Passage from T to the ratio (1.17) facilitates using variational arguments. Therefore, *we shall use the simpler notation* $n_\pm(s, (1.17))$ in place of $n_\pm(s, T)$, $\big|(1.17)\big|_q$ in place of $\big|T\big|_q$, $D_q(1.17)$ in place of $D_q(T)$, etc.

6. An auxiliary problem on the semi-axis. Let $\overline{f} = f \in L_{1,\mathrm{loc}}(\mathbb{R}_+)$. For some $R \geq 1$, consider the ratio of quadratic forms

$$\int_R^\infty f(r)|z(r)|^2 r\, dr \Big/ \int_R^\infty |z'(r)|^2 r\, dr, \quad z(R) = 0, \quad R \geq 1. \tag{1.18}$$

Here z runs through all functions absolutely continuous on $\mathbb{R}_+$ and such that the integral in the denominator is finite. On f we impose the following "implicit" condition: for some $q \geq 1$,

$$\big|(1.18)\big|_q < \infty, \quad q \geq 1. \tag*{$(1.19)_q$}$$

This condition is fulfilled (or not fulfilled) simultaneously for all $R \geq 1$. Moreover, under condition $(1.19)_q$ all six functionals $D_q(1.18)$ do not depend on $R \geq 1$.

We can give an elementary *sufficient* condition for $(1.19)_q$ (see [BL], and also [BS], [BLSu]). This condition becomes *necessary* for the non-negative f. Namely, we put $\boldsymbol{\zeta}(f) := \{\zeta_n(f)\}$, $n \in \mathbb{Z}_+$,

$$\zeta_0(f) := \int_0^1 |f(e^t)|e^{2t}\, dt, \quad \zeta_n(f) := \int_{e^{n-1}}^{e^n} t|f(e^t)|e^{2t}\, dt, \quad n \in \mathbb{N};$$

$$\|\boldsymbol{\zeta}(f)\|^q_{q,\infty} := \sup_{s>0} s^q \mathrm{card}\{n : \zeta_n(f) > s\}, \quad q \geq 1,$$

$$\Delta_q(\boldsymbol{\zeta}(f)) := \limsup_{s\to 0} s^q \mathrm{card}\{n : \zeta_n(f) > s\}, \quad q \geq 1,$$

$$\delta_q(\boldsymbol{\zeta}(f)) := \liminf_{s\to 0} s^q \mathrm{card}\{n : \zeta_n(f) > s\}, \quad q \geq 1.$$

PROPOSITION 1.7. a) *Assume that*

$$\|\boldsymbol{\zeta}(f)\|_{q,\infty} < \infty, \quad q \geq 1. \tag*{$(1.20)_q$}$$

Then $(1.19)_q$ *is true, and*

$$\Delta_q(1.18) \leq C(q)\, \Delta_q(\boldsymbol{\zeta}(f)).$$

b) *Assume that* $f(r) \geq 0$, $r \geq R_0$, *for some* $R_0 \geq 1$. *Then* $(1.19)_q$ *implies condition* $(1.20)_q$ *and also the inequalities*

$$\partial_q(1.18) \geq c(q)\, \partial_q(\boldsymbol{\zeta}(f)), \quad \partial = \Delta, \, \delta.$$

§ 2. Formulation of the main results

1. Our goal is to study the asymptotics of the functions $\mathfrak{N}_\pm(\alpha,\lambda_+;A,V)$, $\mathfrak{N}_\pm(\alpha,\lambda_-;A,V)$ (see (1.14)) as $\alpha\to\infty$. We introduce the following quantities:

$$\Delta_q^{(\pm)}(\lambda_\tau;A,V) := \limsup_{\alpha\to\infty}\alpha^{-q}\mathfrak{N}_\pm(\alpha,\lambda_\tau;A,V),\quad \tau=\pm,\quad q\geq 1, \tag{2.1τ}$$

$$\delta_q^{(\pm)}(\lambda_\tau;A,V) := \liminf_{\alpha\to\infty}\alpha^{-q}\mathfrak{N}_\pm(\alpha,\lambda_\tau;A,V),\quad \tau=\pm,\quad q\geq 1. \tag{2.2τ}$$

For a function $F(\mathbf{x})$, $\mathbf{x}\in\mathbb{R}^2$, we put $F_\beta(\mathbf{x})=F(\beta\mathbf{x})$; here β is a positive matrix. Let (r,θ) be the polar coordinates of a point $\mathbf{x}\in\mathbb{R}^2$; we write $F(\mathbf{x})=F(r,\theta)$. We put

$$\langle F\rangle(r)=(2\pi)^{-1}\int_{-\pi}^{\pi}F(r,\theta)\,d\theta.$$

Along with Condition 1.1, we impose the following condition on V.

CONDITION 2.1(q). *For some* $q\geq 1$, $\big|(1.18)\big|_q<\infty$ *with* $f=\langle V\rangle$.

Since $V\geq 0$, Condition 2.1(q) is equivalent (see Proposition 1.7) to the following relation: $\|\boldsymbol{\zeta}(\langle V\rangle)\|_{q,\infty}<\infty$. Examples (for any $q\geq 1$) demonstrating the compatibility of Conditions 1.1 and 2.1(q) can be found in [BL] and [BLSu, §8].

For $\varphi\in L_\infty(\mathbb{R}^2)$, we introduce the notation $f_{\beta,\varphi}:=\langle(|\varphi|^2V)_\beta\rangle$. Let

$$\Delta_q^{(+)}(V,\beta,\varphi),\quad \delta_q^{(+)}(V,\beta,\varphi),\quad q\geq 1, \tag{2.3}$$

denote the functionals $\Delta_q^{(+)}(1.18)$, $\delta_q^{(+)}(1.18)$ for $f=f_{\beta,\varphi}$. We mention that *Condition* 2.1(q) *with* $f=\langle V\rangle$ *is equivalent to the same condition with* $f=f_{\beta,\varphi}$.

Also, we note that the functionals (2.3) coincide for potentials asymptotically close as $|\mathbf{x}|\to\infty$ (see Proposition 2.2 from [BLSu]).

2. In [BLSu] it was shown that, if V satisfies Conditions 1.1 and 2.1(q), then

$$\mathfrak{N}_+(\alpha,0;A,V)=O(\alpha^q),\quad \alpha\to\infty, \tag{2.4}$$

and the corresponding asymptotic formulas were established. Earlier the same was established in [BL] in the case where $A=-\Delta$.

Below we formulate two theorems (Theorems 2.2(τ) and 2.4(τ)) about the asymptotics of $\mathfrak{N}_\pm(\alpha,\lambda_\tau;A,V)$, $\tau=\pm$. In Theorems 2.2($\pm$) the answers are formulated in terms of the corresponding *model Schrödinger operators with*, generally speaking, *matrix-valued potentials.* In Theorems 2.4($\pm$) the answers are formulated in terms of the auxiliary problem on the semi-axis, but V is subject to the additional restriction.

The description of the model operators involves the quadratic forms $\mathfrak{b}_j^{(\pm)}$ (see (1.9)) and the corresponding eigenfunctions $\psi_j^{(\pm)}$ (see (1.11)). In the Hilbert space $\mathfrak{H}_\pm=L_2(\mathbb{R}^2;\mathbb{C}^{m_\pm})$, we consider the following diagonal second order elliptic operator with constant coefficients:

$$\mathcal{B}_\pm(\mathbf{D})=\operatorname{diag}\big(\mathfrak{b}_1^{(\pm)}(\mathbf{D}),\dots,\mathfrak{b}_{m_\pm}^{(\pm)}(\mathbf{D})\big),\quad \mathbf{D}=-i\nabla. \tag{2.5$\pm$}$$

The expression (2.5$\pm$) generates a positive selfadjoint operator $B_\pm$ in $\mathfrak{H}_\pm$. Now, we introduce the matrix-row and the matrix-column

$$\Pi_\pm(\mathbf{x}):=\{\psi_j^{(\pm)}(\mathbf{x})\}_{j=1}^{m_\pm},\quad \Pi_\pm^*(\mathbf{x}):=\operatorname{col}\overline{\{\psi_j^{(\pm)}(\mathbf{x})\}}_{j=1}^{m_\pm}.$$

We denote $W(\mathbf{x}) = (V(\mathbf{x}))^{1/2}$ and define the non-negative matrix potential $\mathcal{U}_\pm(\mathbf{x}) := (W(\mathbf{x})\Pi_\pm(\mathbf{x}))^* W(\mathbf{x})\Pi_\pm(\mathbf{x})$. The functions $\psi_j^{(\pm)}(\mathbf{x})$ are bounded; therefore, the potential $\mathcal{U}_\pm(\mathbf{x})$ admits a point-wise estimate in terms of $V(\mathbf{x})$. Now, we introduce the model operators

$$B_\pm(\alpha) := B_\pm - \alpha\mathcal{U}_\pm(\mathbf{x}), \quad \alpha > 0. \tag{2.6$\pm$}$$

By $\mathfrak{N}_+(\alpha, \lambda; B_\pm, \mathcal{U}_\pm)$, $\alpha > 0$, $\lambda \le 0$, we denote the number of eigenvalues of the operator $B_\pm(\alpha)$ lying to the left of the point λ. The estimate (2.4) for $A = -\Delta$ is carried over to the operator (2.6): $\mathfrak{N}_+(\alpha, 0; B_\pm, \mathcal{U}_\pm) = O(\alpha^q)$, $\alpha \to \infty$. Consider the ratio of (finite-dimensional) forms

$$\frac{\langle \mathcal{U}_\pm(\mathbf{x})\mathbf{c}, \mathbf{c}\rangle}{\langle \mathcal{B}_\pm(\boldsymbol{\eta})\mathbf{c}, \mathbf{c}\rangle}, \quad \mathbf{c} \in \mathbb{C}^{m_\pm}; \quad \mathbf{x} \in \mathbb{R}^2, \quad \boldsymbol{\eta} \in \mathbb{R}^2. \tag{2.7$\pm$}$$

For $\mu > 0$, by $n^{(\pm)}(\mu; \mathbf{x}, \boldsymbol{\eta})$ we denote the number of eigenvalues of the ratio (2.7±) that are greater than μ.

For $\lambda < 0$, the function $\mathfrak{N}_+(\alpha, \lambda; B_\pm, \mathcal{U}_\pm)$ has the Weyl asymptotics

$$\lim_{\alpha\to\infty} \alpha^{-1}\mathfrak{N}_+(\alpha, \lambda; B_\pm, \mathcal{U}_\pm) = \widetilde{J}(\mathcal{B}_\pm, \mathcal{U}_\pm) := (2\pi)^{-2} \int\int n^{(\pm)}(1; \mathbf{x}, \boldsymbol{\eta})\, d\mathbf{x}\, d\boldsymbol{\eta}, \quad \lambda < 0. \tag{2.8$\pm$}$$

We introduce the notation

$$\Delta_q(B_\pm, \mathcal{U}_\pm) := \limsup_{\alpha\to\infty} \alpha^{-q}\mathfrak{N}_+(\alpha, 0; B_\pm, \mathcal{U}_\pm), \quad q \ge 1, \tag{2.9$\pm$}$$

$$\delta_q(B_\pm, \mathcal{U}_\pm) := \liminf_{\alpha\to\infty} \alpha^{-q}\mathfrak{N}_+(\alpha, 0; B_\pm, \mathcal{U}_\pm), \quad q \ge 1, \tag{2.10$\pm$}$$

$$\widetilde{\Delta}_1(B_\pm, \mathcal{U}_\pm) := \Delta_1(B_\pm, \mathcal{U}_\pm) - \widetilde{J}(\mathcal{B}_\pm, \mathcal{U}_\pm), \tag{2.11$\pm$}$$

$$\widetilde{\delta}_1(B_\pm, \mathcal{U}_\pm) := \delta_1(B_\pm, \mathcal{U}_\pm) - \widetilde{J}(\mathcal{B}_\pm, \mathcal{U}_\pm). \tag{2.12$\pm$}$$

THEOREM 2.2(+). *Let the operator A be generated by the form* (1.2) *under conditions* (1.1). *Let (λ_-, λ_+) be a gap in the spectrum of A. Suppose that Condition* 1.3(+) *is satisfied. Suppose that the potential $V \ge 0$ satisfies Conditions* 1.1 *and* 2.1(q). *Then the following is true for the quantities* (2.1+), (2.2+):

(a) *If $q = 1$, then*

$$\partial_1^{(+)}(\lambda_+; A, V) = J(V, g) + \widetilde{\partial}_1(B_+, \mathcal{U}_+), \qquad \partial = \Delta, \delta, \tag{2.13}$$

$$\Delta_1^{(-)}(\lambda_+; A, V) = 0. \tag{2.14}$$

Here $J(V, g)$ is as in (1.5), *and $\widetilde{\Delta}_1(B_+, \mathcal{U}_+)$, $\widetilde{\delta}_1(B_+, \mathcal{U}_+)$ are defined in* (2.11+), (2.12+).

For the validity of the Weyl asymptotics

$$\Delta_1^{(+)}(\lambda_+; A, V) = \delta_1^{(+)}(\lambda_+; A, V) = J(V, g) \tag{2.15}$$

it suffices that $\Delta_1^{(+)}(V, \mathbf{1}, 1) = 0$. Here $\Delta_1^{(+)}(V, \mathbf{1}, 1)$ is defined in (2.3).

(b) *If $q > 1$, then* (2.14) *is fulfilled and*

$$\partial_q^{(+)}(\lambda_+; A, V) = \partial_q(B_+, \mathcal{U}_+), \qquad \partial = \Delta, \delta. \tag{2.16}$$

Here $\Delta_q(B_+, \mathcal{U}_+)$, $\delta_q(B_+, \mathcal{U}_+)$ are defined in (2.9+), (2.10+).

THEOREM 2.2(−). *Let the operator A be generated by the form* (1.2) *under conditions* (1.1). *Let (λ_-, λ_+) be a gap in the spectrum of A. Suppose that Condition* 1.3(−) *is satisfied. Suppose that the potential $V \geq 0$ satisfies Conditions* 1.1 *and* 2.1(q). *Then the following is true for the quantities* (2.1−), (2.2−):

(a) *If $q = 1$, then*

$$\Delta_1^{(+)}(\lambda_-; A, V) = \delta_1^{(+)}(\lambda_-; A, V) = J(V, g), \tag{2.17}$$

$$\partial_1^{(-)}(\lambda_-; A, V) = \widetilde{\partial}_1(B_-, \mathcal{U}_-), \quad \partial = \Delta, \delta. \tag{2.18}$$

Here $J(V, g)$ is as in (1.5), *and $\widetilde{\Delta}_1(B_-, \mathcal{U}_-)$, $\widetilde{\delta}_1(B_-, \mathcal{U}_-)$ are defined in* (2.11−), (2.12−).

(b) *If $q > 1$, then*

$$\Delta_1^{(+)}(\lambda_-; A, V) \leq J(V, g), \tag{2.19}$$

$$\partial_q^{(-)}(\lambda_-; A, V) = \partial_q(B_-, \mathcal{U}_-), \quad \partial = \Delta, \delta. \tag{2.20}$$

Here $\Delta_q(B_-, \mathcal{U}_-)$, $\delta_q(B_-, \mathcal{U}_-)$ are defined in (2.9−), (2.10−).

3. The model operator (2.6±) involves the forms $\mathfrak{b}_j^{(\pm)}$ or, equivalently, the matrices $\beta_j^{(\pm)}$ (see (1.10)), and also the eigenfunctions $\psi_j^{(\pm)}$. It is impossible to avoid the dependence on $\beta_j^{(\pm)}$ in formulas (2.13), (2.16), (2.18), (2.20). Concerning the (more unpleasant) dependence on the functions $\psi_j^{(\pm)}$, the things are different. It is possible to eliminate these functions from the asymptotic formulas under some supplementary conditions of "regular" behavior of the perturbation V. We proceed to the formulation of Theorem 2.4 that solves this problem.

In addition to Conditions 1.1, 2.1(q), we impose the following condition on V.

CONDITION 2.3. *There exists a function $\mathcal{S} = \overline{\mathcal{S}}$ satisfying Condition* 1.1 *(with V replaced by S) and such that $V(\mathbf{x}) = \mathcal{S}(\mathbf{x})(1 + o(1))$ as $|\mathbf{x}| \to \infty$. Suppose that the Fourier-image $\Phi\mathcal{S}$ of $\mathcal{S}$ satisfies the following condition: for some $\varkappa > 1$, $\Phi\mathcal{S} \in H^{\varkappa}(\mathbb{R}^2 \setminus B_\varepsilon)$, $\forall \varepsilon > 0$, where $B_\varepsilon = \{\boldsymbol{\xi} \in \mathbb{R}^2 : |\boldsymbol{\xi}| \leq \varepsilon\}$.*

THEOREM 2.4(+). *Let the operator A be generated by the form* (1.2) *under conditions* (1.1). *Let (λ_-, λ_+) be a gap in the spectrum of A. Suppose that Condition* 1.3(+) *is satisfied. Suppose that the potential $V \geq 0$ satisfies Conditions* 1.1, 2.1(q), 2.3. *Then the following is true for the quantities* (2.1+), (2.2+):

(a) *If $q = 1$, then* (2.14) *is satisfied and*

$$\partial_1^{(+)}(\lambda_+; A, V) = J(V, g) + \sum_{j=1}^{m_+} \partial_1^{(+)}(V, \beta_j^{(+)}, 1), \quad \partial = \Delta, \delta. \tag{2.21}$$

Here $J(V, g)$ is as in (1.5), *and $\partial_1^{(+)}(V, \beta_j^{(+)}, 1)$, $\partial = \Delta, \delta$, are defined in accordance with* (2.3).

(b) *If $q > 1$, then* (2.14) *is satisfied and*

$$\partial_q^{(+)}(\lambda_+; A, V) = \sum_{j=1}^{m_+} \partial_q^{(+)}(V, \beta_j^{(+)}, 1), \quad \partial = \Delta, \delta. \tag{2.22}$$

THEOREM 2.4(−). *Let the operator A be generated by the form* (1.2) *under conditions* (1.1). *Let* (λ_-, λ_+) *be a gap in the spectrum of* A. *Suppose that Condition* 1.3(−) *is satisfied. Suppose that the potential* $V \geq 0$ *satisfies Conditions* 1.1, 2.1(q), 2.3. *Then the following is true for the quantities* (2.1−), (2.2−):

(a) *If* $q = 1$, *then* (2.17) *is satisfied and*

$$\partial_1^{(-)}(\lambda_-; A, V) = \sum_{j=1}^{m_-} \partial_1^{(+)}(V, \beta_j^{(-)}, 1), \quad \partial = \Delta, \delta. \tag{2.23}$$

(b) *If* $q > 1$, *then* (2.19) *is satisfied and*

$$\partial_q^{(-)}(\lambda_-; A, V) = \sum_{j=1}^{m_-} \partial_q^{(+)}(V, \beta_j^{(-)}, 1), \quad \partial = \Delta, \delta. \tag{2.24}$$

§ 3. Sketch of the proof

1. Reduction to compact operators. We denote

$$X(\lambda) := W(A - \lambda I)^{-1} W, \quad \lambda \in \Lambda. \tag{3.1}$$

The functions (1.12) are related to the counting functions of the spectrum of the operator $X(\lambda)$:

$$\mathfrak{N}_\pm(\alpha, \lambda; A, V) = n_\pm(t, X(\lambda)), \quad t\alpha = 1, \quad \lambda \in \Lambda. \tag{3.2}$$

In (3.2) we cannot pass to the limit as $\lambda \to \lambda_\pm$, because the operators (3.1) do not have limits. Therefore, we need an appropriate regularization.

PROPOSITION 3.1(±). *Suppose that near* $\lambda_\pm$ *the operator* $X(\lambda)$ *is represented in the form*

$$X(\lambda) = \Gamma_\pm(\lambda) + Y_\pm(\lambda), \tag{3.3±}$$

where $\Gamma_\pm(\lambda) = \Gamma_\pm(\lambda)^*$, *the limit (with respect to the operator norm)*

$$\text{(u)-}\lim_{\lambda \to \lambda_\pm} \Gamma_\pm(\lambda) =: \Gamma_\pm \in \Sigma_q, \quad q \geq 1,$$

exists, and (uniformly in λ*)* $\operatorname{rank} Y_\pm(\lambda) \leq r_\pm < \infty$. *Then*

$$\begin{aligned} \partial_q^{(+)}(\lambda_\pm; A, V) &= \partial_q^{(+)}(\Gamma_\pm), \quad \partial = \Delta, \delta, \\ \partial_q^{(-)}(\lambda_\pm; A, V) &= \partial_q^{(-)}(\Gamma_\pm), \quad \partial = \Delta, \delta. \end{aligned} \tag{3.4±}$$

2. For $N > 0$, let $\zeta_N(\mathbf{x})$ be the characteristic function of the disc $\{|\mathbf{x}| \leq N\}$, and let $\widetilde{\zeta}_N(\mathbf{x}) := 1 - \zeta_N(\mathbf{x})$. We denote $W_N := \zeta_N W$, $\widetilde{W}_N := \widetilde{\zeta}_N W$, $V_N := \zeta_N V$, $\widetilde{V}_N := \widetilde{\zeta}_N V$. The operator $X(\lambda)$ is represented in the form

$$X(\lambda) = L_N(\lambda) + K_N(\lambda) + 2\mathrm{Re}\, M_N(\lambda),$$

where $L_N(\lambda) := W_N(A - \lambda I)^{-1} W_N$, $K_N(\lambda) := \widetilde{W}_N(A - \lambda I)^{-1} \widetilde{W}_N$, $M_N(\lambda) := \widetilde{W}_N(A - \lambda I)^{-1} W_N$. We are going to regularize these operators separately and examine the contribution of each of them to the limit quantities (2.1), (2.2).

Below we suppose that *Conditions* 1.1, 2.1(q) *and* 1.3(±) *are satisfied.*

3. The operator $L_N(\lambda)$ is "responsible" for the Weyl contribution to the asymptotics.

PROPOSITION 3.2($\pm$). *We have*

$$L_N(\lambda) = \mathcal{L}_N^{(\pm)}(\lambda) + \widehat{L}_N^{(\pm)}(\lambda),$$

where $\operatorname{rank} \widehat{L}_N^{(\pm)}(\lambda) \le 2m_\pm$, *the limit*

$$(\Sigma_1)\text{-}\lim_{\lambda\to\lambda_\pm} \mathcal{L}_N^{(\pm)}(\lambda) =: \mathcal{L}_N^{(\pm)}(\lambda_\pm) \in \Sigma_1$$

exists, and

$$\Delta_1^{(+)}(\mathcal{L}_N^{(\pm)}(\lambda_\pm)) = \delta_1^{(+)}(\mathcal{L}_N^{(\pm)}(\lambda_\pm)) = J(V_N, g), \qquad (3.5\pm)$$
$$\Delta_1^{(-)}(\mathcal{L}_N^{(\pm)}(\lambda_\pm)) = 0.$$

We have $\mathcal{L}_N^{(\pm)}(\lambda_\pm) = \zeta_N \mathcal{L}_N^{(\pm)}(\lambda_\pm)\zeta_N$.

4. The operator $K_N(\lambda)$ is responsible for the threshold contribution to the asymptotics. We put $\mathcal{E}_j^{(\pm)} := \{\boldsymbol{\xi} : \ |\beta_j^{(\pm)}(\boldsymbol{\xi} - \boldsymbol{\xi}_j^{(\pm)})| \le \delta\}$, $j = 1, \dots, m_\pm$. Choose a number $\delta > 0$ so small that $E_\pm(\boldsymbol{\xi})$ is a simple eigenvalue of the operator $A(\boldsymbol{\xi})$ for $\boldsymbol{\xi} \in \mathcal{E}_j^{(\pm)}$, $j = 1, \dots, m_\pm$, and $\mathcal{E}_j^{(\pm)} \cap \mathcal{E}_k^{(\pm)} = \emptyset$ for $j \ne k$. Let $\chi_j^{(\pm)}$ denote the characteristic function of the ellipse $\mathcal{E}_j^{(\pm)}$.

We introduce the projections

$$\mathcal{X}^{(\pm)} := \sum_{j=1}^{m_\pm} \Psi_\pm^*[\chi_j^{(\pm)}]\Psi_\pm, \quad \widetilde{\mathcal{X}}^{(\pm)} := I - \mathcal{X}^{(\pm)},$$

which commute with A. Here $\Psi_+ := \Psi_l$, $\Psi_- := \Psi_{l-1}$. The operator $K_N(\lambda)$ is represented in the form

$$K_N(\lambda) = Q_N^{(\pm)}(\lambda) + P_N^{(\pm)}(\lambda), \qquad (3.6\pm)$$

where $Q_N^{(\pm)}(\lambda) := \widetilde{W}_N(A - \lambda I)^{-1}\mathcal{X}^{(\pm)}\widetilde{W}_N$, $P_N^{(\pm)}(\lambda) := \widetilde{W}_N(A - \lambda I)^{-1}\widetilde{\mathcal{X}}^{(\pm)}\widetilde{W}_N$.

The operator $P_N^{(\pm)}(\lambda)$ in the limit $N \to \infty$ gives no contribution to the asymptotics.

PROPOSITION 3.3($\pm$). *The limit*

$$(\Sigma_1)\text{-}\lim_{\lambda\to\lambda_\pm} P_N^{(\pm)}(\lambda) =: P_N^{(\pm)}(\lambda_\pm) \in \Sigma_1$$

exists, and

$$\Delta_1^{(+)}(P_N^{(\pm)}(\lambda_\pm)) = \delta_1^{(+)}(P_N^{(\pm)}(\lambda_\pm)) = J(\widetilde{V}_N, g),$$
$$\Delta_1^{(-)}(P_N^{(\pm)}(\lambda_\pm)) = 0.$$

The operator $Q_N^{(\pm)}(\lambda)$ can be written as follows:

$$Q_N^{(\pm)}(\lambda) = \sum_{j,k=1}^{m_\pm} (T_{jN}^{(\pm)}(\lambda)\Psi_\pm)\operatorname{sign}(A - \lambda I)(T_{kN}^{(\pm)}(\lambda)\Psi_\pm)^*, \qquad (3.7\pm)$$

$$T_{jN}^{(\pm)}(\lambda) := \widetilde{W}_N\Psi_\pm^*[\chi_j^{(\pm)}|E_\pm - \lambda|^{-1/2}], \quad j = 1, \dots, m_\pm.$$

The integral operator $T_{jN}^{(\pm)}(\lambda)$ has the following kernel:

$$(2\pi)^{-1}\widetilde{W}_N(\mathbf{x})\chi_j^{(\pm)}(\boldsymbol{\xi})\,|E_\pm(\boldsymbol{\xi}) - \lambda|^{-1/2}\, e^{i\langle \mathbf{x},\boldsymbol{\xi}\rangle}\varphi^{(\pm)}(\mathbf{x},\boldsymbol{\xi}).$$

Under regularization, the operator $T_{jN}^{(\pm)}(\lambda)$ can be replaced by the integral operator $\widehat{T}_{jN}^{(\pm)}(\lambda)$ with the following simpler kernel:

$$(2\pi)^{-1}\widetilde{W}_N(\mathbf{x})\chi_j^{(\pm)}(\boldsymbol{\xi})\left|\pm\mathfrak{b}_j^{(\pm)}(\boldsymbol{\xi}-\boldsymbol{\xi}_j^{(\pm)})+\lambda_\pm-\lambda\right|^{-1/2}e^{i\langle\mathbf{x},\boldsymbol{\xi}\rangle}\varphi_j^{(\pm)}(\mathbf{x}).$$

Recall that $\mathfrak{b}_j^{(\pm)}$ are defined in (1.9), and $\varphi_j^{(\pm)}$ are introduced in (1.11). It is elementary to reduce the operator $\widehat{T}_{jN}^{(\pm)}(\lambda)$ to the model integral operator studied in [BLSu, §3]. The results of [BLSu] imply the following statement.

PROPOSITION 3.4(±). *The following representations are valid:*

$$\widehat{T}_{jN}^{(\pm)}(\lambda)=\mathcal{T}_{jN}^{(\pm)}(\lambda)+\mathcal{Y}_{jN}^{(\pm)}(\lambda),$$

where $\operatorname{rank}\mathcal{Y}_{jN}^{(\pm)}(\lambda)=1$ *and the limit*

$$(\mathrm{u})\text{-}\lim_{\lambda\to\lambda_\pm}\mathcal{T}_{jN}^{(\pm)}(\lambda)=:\mathcal{T}_{jN}^{(\pm)}(\lambda_\pm)\in\Sigma_{2q}$$

exists.

As a result, we obtain the following statement for the operator (3.7±).

PROPOSITION 3.5(±). *We have*

$$Q_N^{(\pm)}(\lambda)=\mathcal{Q}_N^{(\pm)}(\lambda)+\widehat{Q}_N^{(\pm)}(\lambda),$$

where $\operatorname{rank}\widehat{Q}_N^{(\pm)}(\lambda)\le 2m_\pm$, *the limit*

$$(\mathrm{u})\text{-}\lim_{\lambda\to\lambda_\pm}\mathcal{Q}_N^{(\pm)}(\lambda)=:\mathcal{Q}_N^{(\pm)}(\lambda_\pm)\in\Sigma_q$$

exists, and $\mathcal{Q}_N^{(\pm)}(\lambda_\pm)=R_N^{(\pm)}\pmod{\Sigma_q^0}$, *where*

$$R_N^{(\pm)}:=\pm\sum_{j,k=1}^{m_\pm}(\mathcal{T}_{jN}^{(\pm)}(\lambda_\pm)\Psi_\pm)(\mathcal{T}_{kN}^{(\pm)}(\lambda_\pm)\Psi_\pm)^*.\tag{3.8±}$$

The quantities $\partial_q^{(+)}(R_N^{(+)})=:\partial_q^{(+)}(*)$ *and* $\partial_q^{(-)}(R_N^{(-)})=:\partial_q^{(-)}(*)$, $\partial=\Delta,\delta$, *do not depend on* N.

Relation (3.6±) and Propositions 3.3(±), 3.5(±) imply the following statement.

PROPOSITION 3.6(±). *We have*

$$K_N(\lambda)=\mathcal{K}_N^{(\pm)}(\lambda)+\widehat{K}_N^{(\pm)}(\lambda),$$

where $\operatorname{rank}\widehat{K}_N^{(\pm)}(\lambda)\le 2m_\pm$ *and the limit*

$$(\mathrm{u})\text{-}\lim_{\lambda\to\lambda_\pm}\mathcal{K}_N^{(\pm)}(\lambda)=:\mathcal{K}_N^{(\pm)}(\lambda_\pm)\in\Sigma_q$$

exists. Moreover,

$$\lim_{N\to\infty}\partial_q^{(\pm)}(\mathcal{K}_N^{(\pm)}(\lambda_\pm))=\partial_q^{(\pm)}(*),\tag{3.9±}$$
$$\lim_{N\to\infty}\Delta_1^{(\mp)}(\mathcal{K}_N^{(\pm)}(\lambda_\pm))=0.$$

We have $\mathcal{K}_N^{(\pm)}(\lambda_\pm)=\widetilde{\zeta}_N\mathcal{K}_N^{(\pm)}(\lambda_\pm)\widetilde{\zeta}_N$.

5. The operator $M_N(\lambda)$ in the limit $N \to \infty$ gives no contribution to the asymptotics.

PROPOSITION 3.7($\pm$). *We have*

$$M_N(\lambda) = \mathcal{M}_N^{(\pm)}(\lambda) + \widehat{M}_N^{(\pm)}(\lambda),$$

where $\operatorname{rank} \widehat{M}_N^{(\pm)}(\lambda) \le 2m_\pm$, *the limit*

$$\text{(u)-}\lim_{\lambda \to \lambda_\pm} \mathcal{M}_N^{(\pm)}(\lambda) =: \mathcal{M}_N^{(\pm)}(\lambda_\pm) \in \Sigma_q$$

exists, and

$$\lim_{N \to \infty} \Delta_q(\mathcal{M}_N^{(\pm)}(\lambda_\pm)) = 0. \tag{3.10$\pm$}$$

6. Now everything is prepared for applying the general method of Subsection 1. The operator $X(\lambda)$ is represented in the form (3.3$\pm$) with the operators

$$\Gamma_N^{(\pm)}(\lambda) := \mathcal{L}_N^{(\pm)}(\lambda) + \mathcal{K}_N^{(\pm)}(\lambda) + 2\mathrm{Re}\,\mathcal{M}_N^{(\pm)}(\lambda),$$

$$Y_N^{(\pm)}(\lambda) := \widehat{L}_N^{(\pm)}(\lambda) + \widehat{K}_N^{(\pm)}(\lambda) + 2\mathrm{Re}\,\widehat{M}_N^{(\pm)}(\lambda)$$

in the role of $\Gamma_\pm(\lambda)$, $Y_\pm(\lambda)$. Herewith, $\operatorname{rank} Y_N^{(\pm)}(\lambda) \le 6m_\pm$ and the limit

$$\text{(u)-}\lim_{\lambda \to \lambda_\pm} \Gamma_N^{(\pm)}(\lambda) =: \Gamma_N^{(\pm)}(\lambda_\pm) = \mathcal{L}_N^{(\pm)}(\lambda_\pm) + \mathcal{K}_N^{(\pm)}(\lambda_\pm) + 2\mathrm{Re}\,\mathcal{M}_N^{(\pm)}(\lambda_\pm) \tag{3.11$\pm$}$$

exists. We write relations (3.4$\pm$) for $\Gamma_N^{(\pm)}(\lambda_\pm)$:

$$\begin{aligned} \partial_q^{(+)}(\lambda_\pm; A, V) &= \partial_q^{(+)}(\Gamma_N^{(\pm)}(\lambda_\pm)), \quad \partial = \Delta, \delta, \\ \partial_q^{(-)}(\lambda_\pm; A, V) &= \partial_q^{(-)}(\Gamma_N^{(\pm)}(\lambda_\pm)), \quad \partial = \Delta, \delta. \end{aligned} \tag{3.12$\pm$}$$

Since the left-hand sides in (3.12$\pm$) are independent of N, so are the right-hand sides. We pass to the limit in (3.12$\pm$) as $N \to \infty$. Relations (3.5$\pm$), (3.9$\pm$), (3.10$\pm$), (3.11$\pm$) and (3.12$\pm$) with $q > 1$ imply that

$$\partial_q^{(\pm)}(\lambda_\pm; A, V) = \partial_q^{(\pm)}(*), \quad \partial = \Delta, \delta, \tag{3.13$\pm$}$$

$$\Delta_q^{(\mp)}(\lambda_\pm; A, V) = 0. \tag{3.14$\pm$}$$

If $q = 1$, the same relations yield the following equalities:

$$\partial_1^{(+)}(\lambda_+; A, V) = J(V, g) + \partial_1^{(+)}(*), \quad \partial = \Delta, \delta, \tag{3.15+}$$

$$\Delta_1^{(+)}(\lambda_-; A, V) = \delta_1^{(+)}(\lambda_-; A, V) = J(V, g), \tag{3.15$-$}$$

$$\Delta_1^{(-)}(\lambda_+; A, V) = 0, \tag{3.16+}$$

$$\partial_1^{(-)}(\lambda_-; A, V) = \partial_1^{(-)}(*), \quad \partial = \Delta, \delta. \tag{3.16$-$}$$

Now it is easy to complete the proof of Theorem 2.2, applying a similar regularization for the model operators and comparing the results with formulas (3.13$\pm$)–(3.16$\pm$).

7. Scheme of the proof of Theorem 2.4. Below $\partial = \Delta, \delta$. The additional Condition 2.3 on V allows us to calculate $\partial_q^{(\pm)}(*)$ (cf. (3.8±)). Under this condition, we have

$$(\mathcal{T}_{jN}^{(\pm)}(\lambda_\pm)\Psi_\pm)(\mathcal{T}_{kN}^{(\pm)}(\lambda_\pm)\Psi_\pm)^* \in \Sigma_q^0, \quad j \neq k.$$

Then $\partial_q^{(\pm)}(*)$ is equal to the sum of the corresponding quantities for the operators $(\mathcal{T}_{jN}^{(\pm)}(\lambda_\pm)\Psi_\pm)(\mathcal{T}_{jN}^{(\pm)}(\lambda_\pm)\Psi_\pm)^*$. In [BLSu, §3], it was shown that these functionals are related to the auxiliary problem on the semi-axis. They coincide with $\partial_q^{(\pm)}(V, \beta_j^{(\pm)}, \varphi_j^{(\pm)})$. Condition 2.3 allows us to eliminate (cf. [BLSu, §7]) the functions $\varphi_j^{(\pm)}$ from the asymptotic coefficients. We have $\partial_q^{(\pm)}(V, \beta_j^{(\pm)}, \varphi_j^{(\pm)}) = \partial_q^{(\pm)}(V, \beta_j^{(\pm)}, 1)$. This leads to the asymptotic formulas (2.21)–(2.24).

References

[B1] Birman M. Sh., *Discrete spectrum in the gaps of a continuous one for perturbations with large coupling constant*, Estimates and Asymptotics for Discrete Spectra of Integral and Differential Equations, Adv. Soviet Math. **7** (1991), Amer. Math. Soc., Providence, RI, 57-73.

[B2] Birman M.Sh., *Discrete spectrum of the periodic Schrödinger operator perturbed by a decaying potential*, Algebra i Analiz **8** (1996), no. 1, 3-20; English transl., St. Petersburg Math. J. **8** (1997), no. 1, 1–14.

[B3] Birman M.Sh., *The discrete spectrum in gaps of the perturbed Schrödinger operator. I. Regular perturbations*, Boundary value problems, Schrödinger operators, deformation quantization, Math. Top. **8** (1995), Akademie Verlag, Berlin, 334-352.

[B4] Birman M.Sh., *Discrete spectrum in the gaps of a perturbed periodic Schrödinger operator.* II. *Nonregular perturbations*, Algebra i Analiz **9** (1997), no. 6, 62-89; English transl., St. Petersburg Math. J. **9** (1998), no. 6, 1073–1095.

[BL] Birman M.Sh., Laptev A., *The negative discrete spectrum of a two-dimensional Schrödinger operator*, Comm. Pure Appl. Math. **49** (1996), 967-997.

[BLSu] Birman M.Sh., Laptev A., Suslina T. A., *Discrete spectrum of a two-dimensional periodic elliptic second order operator perturbed by a decaying potential.* I. *Semibounded gap*, Algebra i Analiz **12** (2000), no. 4, 36-78; English transl., St. Petersburg Math. J. **12** (2001), no. 4, 535–567.

[BS] Birman M.Sh., Solomyak M.Z., *Estimates for the number of negative eigenvalues of the Schrödinger operator and its generalizations*, Estimates and Asymptotics for Discrete Spectra of Integral and Differential Equations, Adv. Soviet Math. **7** (1991), Amer. Math. Soc., Providence, RI, 1-55.

[Iv] Ivrii V., *Accurate spectral asymptotics for periodic operators*, Journées "Equations aux derivees partielles" (1999), Saint-Jean-de-Monts, 1-11.

[S] Solomyak M. Z., *Piecewise-polynomial approximation of functions from* $H^l((0,1)^d)$, $2l = d$, *and applications to the spectral theory of the Schrödinger operator*, Israel J. Math. **86** (1994), 253–275.

St. Petersburg State University, Physics Department, Ul'yanovskaya 1, Petrodvorets, 198904, St. Petersburg, Russia

E-mail address: tanya@petrov.stoic.spb.su

Contemporary Mathematics
Volume **339**, 2003

On the Principal Eigenvalue of the Schrödinger Operator with a Scaled Random Potential

V.V.Yurinsky

ABSTRACT. This article considers asymptotic behavior of the principal eigenvalue $\widehat{\Lambda}_t$ of the operator $-\frac{1}{2}\Delta\phi + (\ln t)^{-2/d}\beta V\phi$ under zero Dirichlet condition on the boundary of the cube $(-t/2, t/2)^d$. Here Δ is the Laplacian, t the large parameter of the problem, $\beta > 0$ one more parameter, and $V = \sum_{X\in\mathbb{X}} W(x - X)$ is a non-negative random potential generated by a Poisson point process $\mathbb{X}$ with constant intensity μ.

It was established by F. Merkl and M. Wütrich that the normalized principal eigenvalue $(\ln t)^{2/d}\widehat{\Lambda}_t$ converges in probability to a non-random limit as $t \to \infty$. The present article provides a new characterization of the limit.

1. The main result

This article deals with the asymptotic behavior of the principal eigenvalue of a random Schrödinger operator $\widehat{\mathcal{A}}_t$ under the zero Dirichlet condition on the boundary of the large cube $\widehat{Q}_t = (-t/2, t/2)^d$. The operator is self-adjoint in $L^2(\widehat{Q}_t)$ and acts on $C_0^\infty(\widehat{Q}_t)$ as

$$(1.1)\qquad \widehat{\mathcal{A}}_t\phi = -\tfrac{1}{2}\Delta\phi + (\ln t)^{-2/d}\beta V\phi,\ V(x) = \textstyle\sum_{X\in\mathbb{X}} W(x - X) \geq 0.$$

Here Δ is the Laplacian, t the large parameter of the problem, $\beta > 0$ one more parameter, $(\ln t)^{-2/d}$ the small scaling factor, and V the non-negative random potential. By a known variational principle [**2**, §8.12], the principal egenvalue $\widehat{\Lambda}_t$ of operator (1.1) is

$$(1.2)\qquad \widehat{\Lambda}_t = \inf_{\phi\neq 0}\left\{|\phi|_2^{-2}\left(\tfrac{1}{2}|\nabla\phi|_2^2 + \beta(\ln t)^{-2/d}\left|V\phi^2\right|_1\right)\right\},$$

where the infimum is over all non-zero test functions $\phi \in H_0^1(\widehat{Q}_t)$, and $|\cdot|_p$ is the norm in $L^p(\widehat{Q}_t)$. In the setting of the present article, $\widehat{\Lambda}_t$ converges in probability to a non-random limit under appropriate normalization:

$$(1.3)\qquad \Lambda_t = (\ln t)^{2/d}\widehat{\Lambda}_t \xrightarrow{\mathbf{P}} \lambda(\beta, d, \mu) \text{ as } t \to \infty.$$

2000 *Mathematics Subject Classification.* Primary 60H25, 60K40, 60K35; Secondary 35P15, 82D30.

Key words and phrases. Poisson cloud, soft obstacles, principal eigenvalue, random potential, scaled random potential.

Investigation of asymptotic behavior of the principal eigenvalue for operators with random elements originated in physics of disordered media. The principal eigenvalue of a Schrödinger operator with non-negative random potential (or its counterpart for the Laplacian in a domain with random fine-grained boundary carrying zero Dirichlet condition) characterizes, e.g., the rate at which diffusing particles are absorbed by randomly positioned traps. There is a considerable body of work on its deterministic asymptotic behavior and related problems. Much of it is covered by the recent monograph of A.-S. Sznitman [**9**].

Below, the set $\mathbb{X}$ in (1.1) is a configuration of a Poisson point process [**4**] with constant intensity μ. Namely, the number of its points $N(A) = \text{card}\,(A \cap \mathbb{X})$ in a set $A \subset \mathbb{R}^d$ of Lebesgue measure $|A| < \infty$ assumes a given value k with probability

$$(1.4) \qquad \mathbf{P}\{N(A) = k\} = (\mu |A|)^k \exp\{-\mu |A|\} / k!, \; k = 0, 1, 2, ...,$$

and the random variables $N(A_j)$ are independent for disjoint sets A_j.

The function W in (1.1) is a bounded probability density with compact support:

$$(1.5) \qquad 0 \le W(x) \le W_\infty, \quad W(x) = 0 \text{ for } x \notin (-W_0/2, W_0/2)^d, \quad |W|_1 = 1.$$

Earlier results on asymptotic behavior of the principal eigenvalue referred to the Schrödinger operator without the small scaling factor $(\ln t)^{-2/d}$ of (1.1) in the random potential. It was established (see [**9**]) that the principal eigenvalue, normalized as in (1.3), converges in probability to a non-random limit under fairly general assumptions about the function W. The limit depends only on d and μ of (1.4).

The influence of the small scaling factor in the random potential was analyzed in [**6, 7**]. It was proved in [**7**] that the principal eigenvalue of (1.1) obeys (1.3); it converges to zero at a faster rate if the scaling factor of the potential has higher order than in (1.1). Article [**7**] provides a variational characterization of the limit in (1.3): for $\mu = 1$

$$\lambda(\beta, d, 1) = \inf\left\{\tfrac{1}{2}|\nabla\phi|_2^2 + \beta\kappa : \kappa \in (0,1),\ \phi \in \Psi,\ \Lambda_\phi^*(\kappa) \le d\right\},$$

where $\Lambda_\phi^*(\kappa) = \sup_{\sigma\in\mathbb{R}}\left\{\kappa\sigma + \int_{\mathbb{R}^d}\left(1 - e^{\sigma\phi^2(x)}\right)dx\right\}$ and

$$(1.6) \qquad \Psi = \left\{\phi \in H^1(\mathbb{R}^d) : |\phi|_2 = 1, \quad \text{supp}\,(\phi) \Subset \mathbb{R}^d\right\}.$$

Varying the additional parameter β in (1.1), it proved possible [**7**] to trace the passage of $\lambda(\beta, d, 1)$ from the universal limit of [**9**] to values that correspond to a constant average potential.

The main result of the present paper is Theorem 1.1 below, which provides a different variational characterization of $\lambda(\beta, d, \mu)$. For a number $D > 0$ and a function $\phi \in \Psi$ define

$$(1.7) \quad \mathcal{M}(\phi; D, \mu) = \sup_{h>0}\frac{M(h;\phi,\mu) - D}{h}, \quad M(h;\phi,\mu) = \mu\int_{\mathbb{R}^d}\left(1 - e^{-h\phi^2(x)}\right)dx.$$

THEOREM 1.1. *Under conditions* (1.4) *and* (1.5), *the random principal eigenvalue* (1.2) *of operator* (1.1) *satisfies* (1.3), *where*

$$(1.8) \qquad \lambda(\beta, d, \mu) = \inf_{\phi\in\Psi}\left(\tfrac{1}{2}|\nabla\phi\,|_2^2 + \beta\mathcal{M}(\phi; d, \mu)\right).$$

To illustrate their utility, formulae (1.7)-(1.8) are applied in §4.4 to derive the above-mentioned results of [**7**] on asymptotic behavior of $\lambda(\beta, d, \mu)$ as function of β.

The proof of Theorem 1.1 and the necessary preparations occupy Sections 2–4. The key to it is the observation of [**6, 7**] that the limit behavior of the normalized principal eigenvalue (1.3) is determined by those values of Rayleigh ratios in (1.2) which are attained with high probability in sufficiently large families of translates of an individual test function. Although based on the same idea, the distinction between typical and anomalous values of Rayleigh ratios is described in §2.2 using functional (1.7) rather than the characterization of [**7**]. The upper bound on Λ_t is then established in Lemma 2.2.

The derivation of the lower bound on Λ_t is completed in §4.3. The method used to reduce the problem to that of analyzing Rayleigh ratios for a finite family of test functions differs from that of [**7**] by the use of techniques of [**10, 11**] and a large deviation estimate for the volume covered by very high peaks of the random potential. The latter are exposed in Section 3.

2. Feasibility of individual Rayleigh ratios

For a given test function$\int V\phi^2$ is rarely very small. Using this circumstance, Lemma 2.2 provides the upper bound of Theorem 1.1 for the principal eigenvalue.

2.1. Notation and change of scale. For a vector $x \in \mathbb{R}^d$, $x^{(j)}$ is its j-th coordinate, $x \cdot y = \sum_j x^{(j)} y^{(j)}$ is the scalar product, $|x| = \sqrt{x \cdot x}$ is the Euclidean norm, and $|x|_* = \max_j |x^{(j)}|$. As usual, $\nabla\phi = (\partial\phi/\partial x^{(j)})$ is the gradient of a function $\phi : \mathbb{R}^d \to \mathbb{R}$. Notation used to refer to scaling and translation of sets in $\mathbb{R}^d$ is $\gamma a + \delta A = \{x : x = \gamma a + \delta y,\ y \in A\}$. The unit ball is $B = \{x : |x| < 1\}$, and $Q = (-1/2, 1/2]^d$ is the unit cube.

The Lebesgue measure of a set $A \subset \mathbb{R}^d$ is $|A|$. For a finite set J, the number of elements is card(J). Probability and expectation are denoted $\mathbf{P}$ and $\mathbf{E}$.

The spaces of summable functions over a set G are denoted $L^p(G)$, $1 \le p < \infty$, and the corresponding norms are $|\phi|_p = \left(\int_G |\phi(x)|^p \, dx\right)$. Whenever this is unlikely to cause misunderstanding, notation of integrals and related norms is abbreviated: $\int_G \psi$ may stand for $\int_G \psi(x)\, dx$, $\int \psi = \int_{\mathbb{R}^d} \psi$, etc.

For the Sobolev spaces, notation follows [**2**]. For an open set G, $C_0^\infty(G)$ is the space of infinitely differentiable functions with compact support in G, and $H_0^1(G)$ is the closure of $C_0^1(G)$ with respect to the norm $(|\nabla\phi|_2^2 + |\phi|_2^2)^{1/2}$. The Sobolev space of functions defined on the interior of $a + rQ$ is denoted $H_0^1(a + rQ)$.

It proves convenient to change from the spatial variable $x \in \widehat{Q}_t$ of (1.1) to

$$y = \tau x \in Q_t = (-\tau t/2, \tau t/2)^d, \quad \tau = (\ln t)^{-1/d}. \tag{2.1}$$

This change of scale transforms operator (1.1) to one acting as

$$\mathcal{A}_t\phi = -\tfrac{1}{2}\Delta\phi + \beta V_t \phi \tag{2.2}$$

on functions from $C_0^\infty(Q_t)$. The random potential is

$$V_t(y) = V\left(\tfrac{1}{\tau} y\right) = \textstyle\sum_{\xi \in \mathbb{X}_t} W\left(\tfrac{1}{\tau}(y - \xi)\right), \quad \mathbb{X}_t = \{\xi = \tau X : X \in \mathbb{X}\}, \tag{2.3}$$

where the set $\mathbb{X}$ is a configuration of the point process in (1.4), so the new Poisson point process $\mathbb{X}_t$ has intensity μ/τ^d.

The principal eigenvalue of operator (2.2) is the normalized eigenvalue in (1.3); the variational characterization of (1.2) changes to

$$\Lambda_t = \inf\left\{|\phi|_2^{-2}\left(\tfrac{1}{2}|\nabla\phi|_2^2 + \beta\textstyle\int V_t\phi^2\right) : \phi \in H_0^1(Q_t), |\phi|_2 > 0\right\}, \tag{2.4}$$

where the set of test functions could be restricted to $C_0^\infty(Q_t)$ (see [**2**, §7.5–§7.6] on smooth approximations of functions with weak derivatives).

2.2. The upper bound on Λ_t. The key to fine localization of Λ_t is the analysis of the probability that $\int \beta V_t\phi^2$ is anomalously small for a given test function. The necessary estimates are collected in Proposition 2.1 below. They follow from classical theorems on large deviations (see, e.g., [**3**] or [**8**]). The calculations are sketched in §2.3 for the reader's convenience. Throughout the paragraph, the function $M(h) = M(h;\phi,\mu)$ is defined as in (1.7) using a function $\phi \in H^1(\mathbb{R}^d)$ which is not identically zero and has compact support.

PROPOSITION 2.1. *Let $h > 0$ be a fixed number.*
(a) *If $x + M'(h;\phi,\mu) < 0$, then*

$$\underline{\lim}_{t\to\infty}(\ln t)^{-1}\ln \mathbf{P}\left\{-\textstyle\int V_t\phi^2 > x\right\} \ge -\left(M(h;\phi,\mu) - hM'(h;\phi,\mu)\right).$$

(b) *If $x + M'(h;\phi,\mu) \ge 0$, then*

$$\overline{\lim}_{t\to\infty}(\ln t)^{-1}\ln \mathbf{P}\left\{-\textstyle\int V_t\phi^2 > x\right\} \le -\left(M(h;\phi,\mu) - hM'(h;\phi,\mu)\right).$$

It is easy to verify that the function $M(h)$ is increasing and its derivative $M'(h)$ decreasing for $h > 0$. Moreover,

$$M(0;\phi,\mu) = 0,\ h^{-1}M(h;\phi,\mu) \le \mu\textstyle\int\phi^2,\ 0 < M(h;\phi,\mu) < \mu|\{\phi \ne 0\}|. \tag{2.5}$$

The asymptotic behavior of M and M' is described by the relations

$$\begin{aligned} &\lim_{h\searrow 0} M(h;\phi,\mu) = 0, && \lim_{h\searrow 0} M'(h;\phi,\mu) = \mu|\phi|_2^2,\\ &\lim_{h\to\infty} M(h;\phi,\mu) = \mu|\{\phi > 0\}|, && \lim_{h\to\infty} hM'(h;\phi,\mu) = 0. \end{aligned} \tag{2.6}$$

It is immediate from (2.6) that $\mathcal{M}(\phi; D,\mu)$ of (1.7) is non-negative and finite.

LEMMA 2.1. *For each test function $\mathcal{M}(D;\phi)$ is a non-increasing and continuous function of $D > 0$ and*

$$\begin{aligned} &\mathcal{M}(D;\phi) > 0 \quad \text{for} \quad D < \mu|\{\phi \ne 0\}|,\\ &\mathcal{M}(D;\phi) = 0 \quad \text{for} \quad D \ge \mu|\{\phi \ne 0\}|. \end{aligned} \tag{2.7}$$

Proof of Lemma 2.1. The derivative of $\frac{1}{h}(M(h) - D)$ equals $h^{-2}U(h)$, where $U(h) = hM'(h) - (M(h) - D)$. The function U has negative derivative $U'(h) = hM''(h) < 0$ and varies from $U(0) = D$ to $\lim_{h\to\infty} U(h) = D - \mu|\{\phi \ne 0\}|$ according to (2.6).

If $D < \mu|\{\phi \ne 0\}|$, then the upper bound in (1.7) is attained at a finite value of $h > 0$ for which $U(h) = 0$, that is,

$$M(h) - hM'(h) = D. \tag{2.8}$$

Equation (2.8) defines a strictly increasing function $h(D)$ for $D < \mu|\{\phi \ne 0\}|$; since $M''(h) < 0$, this is ensured by the implicit function theorem. Moreover, $h(D) \nearrow \infty$ as $D \nearrow \mu|\{\phi \ne 0\}|$, so $\lim_{D\nearrow\mu|\{\phi\ne 0\}|}\mathcal{M}(D;\phi) = 0$.

If $D \ge \mu|\{\phi \ne 0\}|$, then $U(h) > 0$ for all finite h, which shows that $\mathcal{M}(D;\phi) = \lim_{h\to\infty}\frac{1}{h}(M(h;\phi) - D) = 0$. Since M is increasing,

$$\forall h > 0\ hM'(h;\phi) > 0 > M(h;\phi) - D. \tag{2.9}$$

The above also shows that whatever the value of $D > 0$ and arbitrarily small numbers ε, δ, one can find a finite $h > 0$ such that

$$\begin{aligned} hM'(h) &\geq (M(h) - D), \quad hM'(h) \leq (M(h) - D) + \varepsilon, \\ M'(h) &\leq \tfrac{1}{h}(M(h) - D) + \varepsilon, \quad 0 \leq \mathcal{M}(D;\phi) - h^{-1}(M(h) - D) \leq \delta. \end{aligned} \tag{2.10}$$

Indeed, for $D \geq \mu |\{\phi \neq 0\}|$ it suffices to choose $h > 0$ sufficiently large; for $D < \mu |\{\phi \neq 0\}|$ the desired value of h solves (2.8). □

LEMMA 2.2. *Under the assumptions of Theorem* 1.1

$$\forall \varepsilon > 0 \quad \lim_{t\to\infty} \mathbf{P}\{\Lambda_t \leq \lambda(\beta, d, \mu) + \varepsilon\} = 1.$$

This lemma is a corollary of the next one, which is stated in terms of the quantity (see (1.6) and (1.7))

$$\lambda_\beta(D;\phi) = \tfrac{1}{2}|\nabla\phi|_2^2 + \beta\mathcal{M}(\phi; D, \mu), \quad D > 0, \ \phi \in \Psi. \tag{2.11}$$

LEMMA 2.3. *Consider translations $S_\xi\phi(x) = \phi(x - \xi)$ of a non-zero function ϕ by vectors $\xi \in \Xi_t$, choosing the set Ξ_t so that* $\lim_{t\to\infty}(\ln t)^{-1}\operatorname{card}(\Xi_t) = D > 0$.

(a) *If $0 < \lambda < \lambda_\beta(D;\phi)$, then*

$$\lim_{t\to\infty} \mathbf{P}\left\{\lambda \geq \min_{\xi\in\Xi_t} |S_\xi\phi|_2^{-2}\left(\tfrac{1}{2}|\nabla(S_\xi\phi)|_2^2 + \textstyle\int \beta V_t (S_\xi\phi)^2\right)\right\} = 0.$$

(b) *If the random variables $\int V_t (S_\xi\phi)^2$ are independent and $\lambda > \lambda_\beta(D;\phi)$, then*

$$\lim_{t\to\infty} \mathbf{P}\left\{\lambda \geq \min_{\xi\in\Xi_t} |S_\xi\phi|_2^{-2}\left(\tfrac{1}{2}|\nabla(S_\xi\phi)|_2^2 + \textstyle\int \beta V_t (S_\xi\phi)^2\right)\right\} = 1.$$

Proof of Lemma 2.3. Below, the test function is normalized so that $|\phi|_2 = 1$. By invariance of the properties of a Poisson point process under translations, it suffices to estimate the probability $\mathbf{P}\left\{\lambda \geq \frac{1}{2}|\nabla\phi|_2^2 + \int\beta V_t\phi^2\right\} = \mathbf{P}\left\{-\int V_t\phi^2 \geq x\right\}$ with $x = \beta^{-1}(\frac{1}{2}|\nabla\phi|_2^2 - \lambda)$. Notation is abbreviated to $\lambda_\beta(D) = \lambda_\beta(D;\phi)$ and $\mathcal{M}(D,\mu) = \mathcal{M}(\phi; D, \mu)$.

(a) Take a small number $\delta \in \left(0, \frac{1}{10}\right)$ for which $\lambda_\beta(D) - \lambda > 2\delta$ and select $\varepsilon \in \left(0, \frac{1}{10}\right)$ so small that $\mathcal{M}(D) < \mathcal{M}(D+\varepsilon) + \delta/\beta$. By (2.10), there exists $h > 0$ satisfying the inequalities $\frac{1}{h}(M(h;\phi) - (D+\varepsilon)) \geq \mathcal{M}(D+\varepsilon) - \delta/\beta$ and

$$hM'(h) \geq M(h) - (D+\varepsilon), \quad hM'(h) \leq (M(h) - (D+\varepsilon)) + \varepsilon\delta.$$

It follows from the above that $x + M'(h) > 0$, because $M'(h) \geq \frac{1}{h}(M(h) - (D+\varepsilon))$ and so $x + M'(h) \geq x + \mathcal{M}(D) - 2\delta/\beta = \beta^{-1}(\lambda_\beta(D) - \lambda - 2\delta)$.

Moreover, $M(h) - hM'(h) \geq M(h) - (M(h) - (D+\varepsilon)) - \varepsilon\delta = D + \varepsilon(1-\delta)$ for this value of h. Consequently, it follows from Proposition 2.1 that

$$\varlimsup_{t\to\infty} (\ln t)^{-1} \ln \mathbf{P}\left\{\lambda > \tfrac{1}{2}|\nabla\phi|_2^2 + \textstyle\int\beta V_t\phi^2\right\} \leq -(D + \varepsilon(1-\delta)).$$

Combined with the bound on $\operatorname{card}(\Xi_t)$, this inequality implies assertion (a).

(b) Consider first the case $D \leq \mu|\{\phi \neq 0\}|$. Take a small number $\delta < \lambda - \lambda_\beta(D)$ and choose $\varepsilon > 0$ so small that $\lambda_\beta(D-\varepsilon) \leq \lambda_\beta(D) + \delta$.

There exists $h = h(\varepsilon) > 0$ for which $M'(h) = \frac{1}{h}(M(h) - (D-\varepsilon)) = \mathcal{M}(D-\varepsilon)$, and consequently $x + M'(h;\phi) = \frac{1}{\beta}(\lambda_\beta(D-\varepsilon) - \lambda) < 0$.

Moreover, $M(h) - hM'(h) = M(h) - (M(h) - (D - \varepsilon)) = D - \varepsilon$. For this reason $q_t = \mathbf{P}\left\{\lambda \geq \frac{1}{2}|\nabla\phi|_2^2 + \int \beta V_t \phi^2\right\}$ satisfies the relation

$$\text{(2.12)} \qquad \underline{\lim}_{t\to\infty} (\ln t)^{-1} \ln (q_t \operatorname{card}(\Xi_t)) \geq D - (D - \varepsilon) = \varepsilon > 0,$$

so $\underline{\lim}_{t\to\infty} q_t \operatorname{card}(\Xi_t) = \infty$. This proves assertion (b) because the probability in question is $1 - (1 - q_t)^{\operatorname{card}(\Xi_t)} \geq 1 - \exp\{-q_t \operatorname{card}(\Xi_t)\}$.

If $D > \mu|\{\phi \neq 0\}|$, then $\mathcal{M}(D) = 0$. Choose $\varepsilon > 0$ so small that

$$D - \varepsilon > \mu|\{\phi \neq 0\}|, \quad \mathcal{M}(D - \varepsilon) = 0$$

and $\delta \in (0, \lambda - \lambda_\beta(D - \varepsilon))$. It follows from (2.10) that one can choose $h > 0$ so as to satisfy the inequalities $x + M'(h) < \frac{1}{\beta}(\lambda_\beta(D - \varepsilon; \phi) - \lambda + \delta) < 0$ and $M(h) - hM'(h) \leq M(h; \phi) - (M(h) - (D - \varepsilon)) = D - \varepsilon$. This proves (2.12) for $D > \mu|\{\phi \neq 0\}|$ and the lemma. $\square$

Proof of Lemma 2.2. For each $\varepsilon > 0$ one can choose a smooth function ϕ with compact support so that $|\phi|_2 = 1$ and

$$\lambda_\beta(d; \phi) = \tfrac{1}{2}|\nabla\phi|_2^2 + \beta\mathcal{M}(\phi; d, \mu) < \lambda(\beta, d, \mu) + \varepsilon.$$

Choose a set of translations S_ξ with $\operatorname{card}(\Xi_t) = (d + o(1)) \ln t$ for which all the translated functions $S_\xi\phi$ belong to $H_0^1(Q_t)$, and the random variables $\int V_t (S_\xi\phi)^2$ are independent. For this choice of ϕ, outside an event of infinitesimal probability

$$\lambda(\beta, d, \mu) + \varepsilon \geq \min_{\xi\in\Xi_t} |S_\xi\phi|_2^{-2} \left(\tfrac{1}{2}|\nabla(S_\xi\phi)|_2^2 + \int \beta V_t (S_\xi\phi)^2\right),$$

so $\Lambda_t \leq \lambda(\beta, d, \mu) + \varepsilon$. $\square$

2.3. Appendix: large deviations for potential term. By infinite divisibility, the distribution of $\mathbb{X}_t$ in (2.3) can for each n be represented as one of a sum of independent Poisson point processes of intensity $\widehat{\mu} = \frac{1}{n}\tau^{-d}\mu$. Consequently $-\int V_t\phi^2$ is distributed as the mean $\overline{\xi} = \frac{1}{n}(\xi_{1,n} + \ldots + \xi_{n,n})$ of i.i.d. random variables $\xi_{k,n} = -n\int V_t^{(k,n)}\phi^2$, each generated by a process of intensity $\widehat{\mu}$.

Below, this representation is used with $n = [\ln t]$. In terms of (1.7), the exponential moment of $\xi_{1,n}$ is given [**4**, Example 1.5] by the formula

$$\ln \mathbf{E} \exp\{h\xi_{1,n}\} = -M_t(h) \equiv -(\tau^d n)^{-1} M(\tau^d nh; \phi_\tau),$$

where $\phi_\tau^2(x) = \int \phi^2(x + \tau y) W(y)\, dy$, and it is easily seen that M_t, M_t' converge to M, M' as $t \to \infty$.

To derive the estimates of Proposition 2.1, one can use standard arguments based on the use of the Cramér transform (see, e.g., [**3**, Ch. 8] or [**8**, Ch. 8]). This procedure yields the representation

$$\mathbf{P}\left\{\overline{\xi} > x\right\} = \exp\{-n(M_t(h) - h\, M_t'(h))\} \mathbf{E} \exp\{-nh\overline{\xi_h^0}\}\mathbf{1}\{\overline{\xi_h^0} > y\},$$

where $y = x + M_t'(h)$ and $\overline{\xi_h^0}$ is the average of i.i.d. random variables with the distribution obtained from that of $\xi_{k,n}$ by Cramér's transform and centering. Proposition 2.1 follows from this formula using the central limit theorem for the average $\overline{\xi_h^0}$, and the evident estimate $\left(\exp\{-nh\overline{\xi_h^0}\} - 1\right)\operatorname{sign}(\overline{\xi_h^0}) < 0$.

3. Pits and peaks of potential

The random potential (2.3) is rarely very large or very small over large spaces within Q_t. The present section contains large deviation estimates for the pertinent probabilities. Lemma 3.5 yields a rough lower bound for Λ_t derived by techniques of [**10, 11**].

3.1. Large deviations for block concentrations. Below, *cells* are small cubes $C_z^0 = H_0(z+Q)$, $H_0 = \tau W_0$, $z \in \mathbb{Z}^d$, and *blocks* are cubes $C_z^1 = H_1(z+Q)$ of size $H_1 = TH_0$ (notation is that of (1.5) and (2.1)). The large number T is for the sake of commodity supposed to be natural, odd, and a multiple of three. Thus, the scales used in the construction are

$$\tau = (\ln t)^{-1/d}, \quad T = T(t) \in \mathbb{N}, \quad T = 1 \,(\mathrm{mod}\, 2), \; T = 0 \,(\mathrm{mod}\, 3). \tag{3.1}$$

Each block consists of T^d cells; notation $z_0 \prec z_1$ indicates that $C_{z_0}^0 \subset C_{z_1}^1$. The "core" of the block $\mathring{C}_z^1 = H_1 z + (H_1 - 2H_0) Q$ consists of $(T-2)^d$ cells. It follows from (1.5) that for $x \in \mathring{C}_z^1$ the value of $V_t(x)$ depends only on the configuration of $\mathbb{X}_t$ on C_z^1.

The following lemma adapts classical large deviation estimates [**3, 8**] to show that the average of V_t over a large block is rarely small.

LEMMA 3.1. *There exists* $\varepsilon_0 > 0$ *such that*

$$\mathbf{P}\left\{\overline{V_{t,z}} \equiv \left|C_z^1\right|^{-1} \int_{C_z^1} V_t(x)\,dx \le (1-\varepsilon_*)\,\mu\right\} \le \exp\left\{-\tfrac{1}{5}\varepsilon^2 \mu W_0^d T^d\right\}$$

for $\varepsilon \in (0, \varepsilon_0)$, $(1-2/T)^d > \frac{1}{2}$, *and* $1 - \varepsilon_* = (1-2/T)^d (1-\varepsilon)$.

Proof of Lemma 3.1. Set $N_z = \mathrm{card}(\mathring{C}_z^1 \cap \mathbb{X}_t)$. The distribution of N_z is Poissonian with mean $\Lambda = (T-2)^d \mu W_0^d$ (see (1.4)). The average admits the estimate $\overline{V_{t,z}} \ge (W_0 T)^{-d} N_z$ because $\int_{C_z^1} W\left(\frac{1}{\tau}(x-\xi)\right) dx = \tau^d$ if $\xi \in \mathring{C}_z^1$ by (1.5).

An exponential Chebyshev inequality shows that

$$\begin{aligned}\mathbf{P}\left\{\overline{V_{t,z}} \le (1-\varepsilon)\,\mu\right\} &\le \inf\nolimits_{h>0} \mathbf{E} \exp\left\{h\left((1-\varepsilon_*)\,\mu W_0^d T^d - N_z\right)\right\} \\ &= \exp\left\{-\mu W_0^d T^d (1-2/T)^d \sup\nolimits_{h>0}\left(1 - e^{-h} - h(1-\varepsilon)\right)\right\}.\end{aligned}$$

The inequality of the lemma follows for small ε since

$$\lim_{\varepsilon \to 0} \varepsilon^{-2} \sup_{h>0}\left\{1 - e^{-h} - (1-\varepsilon)\,h\right\} = \lim_{\varepsilon \to 0} \varepsilon^{-2}\left(\varepsilon + (1-\varepsilon)\ln(1-\varepsilon)\right) = \tfrac{1}{2}. \quad \square$$

Another essential feature of the random potential V_t is that high peaks contribute little to its averages over large blocks.

LEMMA 3.2. *For* $K \in \mathbb{N}$ *such that* $K \ge \ell_* e^{1+\ell_*}$ *with* $\ell_* = 2^d W_0^d \mu$,

$$\mathbf{P}\left\{\overline{(V_t - W_\infty K)_{+,z}} \ge 3^d W_\infty / \sqrt{\ln K}\right\} \le \exp\left\{-T^d\left(\sqrt{\ln K} - 1\right)\right\},$$

where $\overline{(V_t - W_\infty K)_{+,z}} = |C_z^1|^{-1} \int_{C_z^1} (V_t(x) - W_\infty K)_+ \, dx$ *and* $a_+ = \max\{a, 0\}$.

Proof of Lemma 3.2. It suffices to derive the estimate for $z = 0$. Because of (1.5), the potential is determined on C_z^0 by the configuration of $\mathbb{X}_t$ in the double-sized cell $C_z^+ = H_0 z + 2H_0 Q$, and for each cell

$$\forall x \in C_z^0 \quad V_t(x) \le W_\infty N_z^+(x), \; N_z^+ = \mathrm{card}\left(\mathbb{X}_t \cap C_z^+\right).$$

Consequently, N_z^+ should be large in order that a cell C_z^0 contain points with large values of V_t, and $|C_z^0|^{-1}\int_{C_z^0}(V_t(x)-W_\infty K)_+\,dx \le W_\infty (N_z^+ - K)_+$, while for the larger block C_0^1

$$\overline{(V_t - W_\infty K)_{+,0}} \le W_\infty T^{-d}\mathcal{N},\ \mathcal{N} = \textstyle\sum_{z\prec 0}\left(N_z^+ - K\right)_+. \tag{3.2}$$

The random variable N_z^+ has Poisson distribution with mean ℓ_*. Since $K!/m! \le 1/(m-K)!$ for $m \ge K$, the exponential moment of $(N_z^+ - K)_+$ admits the estimate

$$\begin{aligned}\mathbf{E}\exp\left\{h\left(N_z^+ - K\right)_+\right\} &\le \sum_{m=0}^{K-1}\frac{\ell_*^m e^{-\ell_*}}{m!} + \delta(K)\sum_{n=0}^{\infty}\frac{\left(e^h\ell_*\right)^n e^{-\ell_*}}{n!}\\ &\le \exp\left\{\delta(K)\exp\left\{\ell_*\left(e^h-1\right)\right\}\right\},\end{aligned} \tag{3.3}$$

where by the Stirling formula

$$\delta(K) \equiv \ell_*^K/K! \le (\ell_* e/K)^K. \tag{3.4}$$

If $C_z^0 \subset C_0^1$, then z belongs to one of the 3^d disjoint sets

$$J_f = \left\{z = f + 3z' : z' \in \mathbb{Z}^d,\ f + 3z' \prec 0\right\} \subset \mathbb{Z}^d,$$

that correspond to $f \in F = \mathbb{Z}^d \cap \{|f|_* \le 1\}$. Since T is supposed to be a multiple of 3, $\operatorname{card}(J_f) = \left(\frac{1}{3}T\right)^d$ for each f.

The random variables N_z^+ are independent for $z \in J_f$, so by (3.3) the exponential moment of $\widehat{N}_f = \sum_{z\in J_f} h\,(N_z^+ - K)_+$ admits the estimate

$$\mathbf{E}\exp\left\{h\widehat{N}_f\right\} \le \exp\left\{(T/3)^d\,\delta(K)\exp\left\{\ell_*\left(e^h-1\right)\right\}\right\}. \tag{3.5}$$

Since $\mathcal{N} = \sum_{|f|_*\le 1}\widehat{N}_f$, and all summands have the same distribution, the above formula combined with the Hölder inequality yields the estimate

$$\begin{aligned}\mathbf{E}\exp\{h\mathcal{N}\} &= \mathbf{E}\exp\left\{h\textstyle\sum_{f\in F}\widehat{N}_f\right\} \le \prod_{f\in F}\left(\mathbf{E}\exp\left\{3^d h\widehat{N}_f\right\}\right)^{1/3^d}\\ &\le \exp\left\{\left(\tfrac{1}{3}T\right)^d\delta(K)\exp\left\{\ell_*\left(e^{3^d h}-1\right)\right\}\right\}.\end{aligned} \tag{3.6}$$

It follows from (3.6) that for $\alpha > 0$ and $U = 3^d h > 0$

$$\begin{aligned}\mathbf{P}\left\{\overline{(V_t - W_\infty K)_{+,0}} \ge \alpha W_\infty\right\} &\le \mathbf{E}\exp\left\{3^{-d}U\left(\mathcal{N} - \alpha T^d\right)\right\}\\ &\le \exp\left\{T^d\left(-3^{-d}\alpha U + 3^{-d}\delta(K)\exp\left\{\ell_*\left(e^U-1\right)\right\}\right)\right\}.\end{aligned}$$

To make small the right hand side of this inequality, one should choose U large and use the fact that $\delta(K)$ is very small for large K. A possible choice is $\alpha = 3^d(\ln K)^{-1/2}$ and $U = \ln K$. If $K \ge \ell_* e^{1+\ell_*}$, then by (3.4)

$$\delta(K)\,3^{-d}\exp\left\{\ell_*\left(e^U - 1\right)\right\} \le \exp\left\{K\ell_* + K\ln(\ell_* e/K)\right\} \le 1.$$

This yields the inequality of the lemma. □

3.2. A rough lower bound on Λ_t. In notation of Lemma 3.1 and Lemma 3.2, integer points labeling blocks (3.1) with irregular behavior of the potential belong to the sets

$$\begin{aligned} \mathbb{L}_t^\varepsilon &= \left\{z \in \mathbb{Q}_t : \overline{V_{t,z}} \le (1-2\varepsilon)\,\mu\right\}, \\ \mathbb{M}_t^K &= \left\{z \in \mathbb{Q}_t : \overline{(V_t - W_\infty K)_{+,z}} \ge 3^d W_\infty / \sqrt{\ln K}\right\}, \end{aligned} \tag{3.7}$$

where $\mathbb{Q}_t = \left\{z \in \mathbb{Z}^d : C_z^1 \cap \hat{Q}_t \neq \emptyset\right\}$. One can choose the block size $T = T(t) \in \mathbb{N}$ so as to guarantee, with probability near to one, the absence of blocks with pathologically low average potential or too many high peaks for large t.

LEMMA 3.3. *Suppose that the function $T : \mathbb{R}_+ \to \mathbb{N}$ satisfies condition* (3.1) *and as $t \to \infty$*

$$H_1(t) = W_0 (\ln t)^{-1/d}\, T(t) \to H_*, \quad 0 < H_* < \infty. \tag{3.8}$$

Suppose moreover that the parameters $\varepsilon > 0$ and $K \in \mathbb{N}$ satisfy the conditions of Lemmas 3.1 *and* 3.2.

(a) *If $\frac{1}{6}\varepsilon^2 \mu H_*^d > d$, then $\lim_{t\to\infty} \mathbf{P}\left\{\mathbb{L}_t^\varepsilon \neq \emptyset\right\} = 0$.*

(b) *If $\left(\sqrt{\ln K} - 1\right) H_*^d / W_0^d > d$, then $\lim_{t\to\infty} \mathbf{P}\left\{\mathbb{M}_t^K \neq \emptyset\right\} = 0$.*

Proof of Lemma 3.3. For large t, the number of blocks C_z^1 of size H_1 that have common points with Q_t does not exceed $\exp\{(\ln t)\, d\,(1 + o(1))\}$. By the choice of T in the estimate of Lemma 3.1 ε_* converges to ε, so the probability that the number of a given block belongs to $\mathbb{L}_t^\varepsilon$ is majorized by $\exp\left\{-\frac{1}{6}\varepsilon^2 \mu H_*^d \ln t\,(1 + o(1))\right\}$. Hence the probability $\mathbf{P}\left\{\mathbb{L}_t^\varepsilon \neq \emptyset\right\}$ does not exceed $\exp\left\{-\left(\frac{1}{6}\varepsilon^2 \mu H_*^d - d\right)\ln t\right\}$. This proves assertion (a).

A similar argument based on Lemma 3.2 yields assertion (b). □

For a block whose number is neither in $\mathbb{L}_t^\varepsilon$ nor in $\mathbb{M}_t^K$, the random potential is large on an appreciable part of its volume.

LEMMA 3.4. *If $\overline{V_{t,z}} > (1-2\varepsilon)\,\mu$ and $\overline{(V_t - W_\infty K)_{+,z}} < 3^d W_\infty / \sqrt{\ln K}$, then*

$$\left|\left\{V_t > \tfrac{1}{2}(1-2\varepsilon)\,\mu\right\}\right| \ge A\left|C_z^1\right|,$$

where $A(\varepsilon, K) = \left(W_\infty K - \frac{1}{2}(1-2\varepsilon)\,\mu\right)^{-1}\left(\frac{1}{2}(1-2\varepsilon)\,\mu - 3^d W_\infty (\ln K)^{-1/2}\right) > 0$ for each $\varepsilon \in \left(0, \frac{1}{2}\right)$ if $K > K_0(\varepsilon, \mu, W_\infty)$.

Proof of Lemma 3.4. Set $\xi = \left|C_z^1\right|^{-1}\left|\left\{V_t > \frac{1}{2}(1-2\varepsilon)\,\mu\right\}\right|$. The assumptions of the lemma permit one to conclude that

$$(1-2\varepsilon)\,\mu < \left|C_z^1\right|^{-1} \int_{C_z^1} V_t \le \tfrac{1}{2}(1-2\varepsilon)\,\mu\,(1-\xi) + W_\infty K \xi + 3^d W_\infty / \sqrt{\ln K}.$$

This yields the inequality of the lemma. □

LEMMA 3.5. *Suppose that the block size satisfies* (3.8) *and the parameters $\varepsilon > 0$ and $K \in \mathbb{N}$ are such that $A(\varepsilon, K) > 0$ in Lemma* 3.4.

If $z \notin \mathbb{L}_t^\varepsilon \cup \mathbb{M}_t^K$ and t is large enough, then for each function $\phi \in H_0^1(Q_t)$

$$\int_{C_z^1} \phi^2 \le \widehat{H}^2 \int_{C_z^1} \left(\tfrac{1}{2}\left|\nabla\phi\right|^2 + \beta V_t \phi^2\right), \tag{3.9}$$

where for some constants $c = c(d) > 0$ and $C = C(\varepsilon, K, \mu) > 0$

$$\widehat{H}^2 = cA^{-1} \max\left\{(1-2\varepsilon)^{-1}(\beta\mu)^{-1}, H_*^2\right\} \le C \max\left\{\beta^{-1}, H_*^2\right\}. \tag{3.10}$$

If t is large enough and $\mathbb{L}_t^\varepsilon = \mathbb{M}_t^K = \emptyset$, then for each function $\phi \in H_0^1(Q_t)$

$$|\phi|_2^2 \leq \widehat{H}^2 \left(\tfrac{1}{2} |\nabla \phi|_2^2 + \textstyle\int_{Q_t} \beta V_t \phi^2 \right). \tag{3.11}$$

Inequality (3.11) is a rough lower bound for the normalized principal eigenvalue Λ_t of (2.4) with correct dependence on β (cf. Corollary 4.2). As mentioned after Lemma 3.3, the probability that both $\mathbb{L}_t^\varepsilon$ and $\mathbb{M}_t^K$ are empty converges to one if the limit block size H_* of (3.8) is large enough. Hence, whatever the choice of ε and K, the bound of (3.11) is valid with probability convergent to one for $H_* \geq h(\varepsilon, K)$. The proof of Lemma 3.5 uses a version of the Poincaré–Friedrichs inequality (see [**10**] or [**11**]).

PROPOSITION 3.1. *Consider a convex set Q^+ of diameter ρ and its subsets Q^0, Q^1 of positive Lebesgue measure. If ϕ is a smooth function defined on Q^+, and $|Q^1| / |Q^0| \leq \alpha^d$, then*

$$\textstyle\int_{Q^1} \phi^2(x)\, dx \leq 2\alpha^d \int_{Q^0} \phi^2(x)\, dx + \frac{2}{d-1} \left((1+\alpha)^d - 1 - \alpha^d \right) \rho^2 \int_{Q^+} |\nabla \phi(x)|^2 \, dx.$$

Proof of Lemma 3.5. Consider an arbitrary block C_z^1, $z \in \mathbb{Q}_t$. By the assumption, conditions of Lemma 3.4 are satisfied, so for

$$Q^0 = C_z^1 \cap \left\{ V_t \geq \tfrac{1}{2}(1-\varepsilon)\mu \right\}, \quad Q^1 = C_z^1 \backslash Q^0$$

the ratio of volumes admits an estimate that depends on A alone: $|Q^1| / |Q^0| \leq \alpha^d \stackrel{def}{=} (1-A)/A$. By the choice of parameters,

$$\textstyle\int_{Q^0} \phi^2 \leq \left(\frac{1}{2}\beta(1-\varepsilon)\mu \right)^{-1} \int_{Q^0} \beta V_t \phi^2,$$

and (3.9) follows from Proposition 3.1.

If inequality (3.9) applies to each block intersecting the interior of Q_t, then summation over all such blocks yields (3.11). □

4. Lower bound on principal eigenvalue

The main aim of this section is to derive the lower bound of Theorem 1.1. In the calculations to follow, the sizes of blocks (3.1) satisfy (3.8). When using the estimates of §3.2, it is supposed that sets (3.7) satisfy the condition

$$\mathbb{L}_t^{\varepsilon_0} = \mathbb{M}_t^K = \mathbb{M}_t^{K_0} = \emptyset, \tag{4.1}$$

At the final stage of proof of Theorem 1.1 in §4.3, it will be shown that $T(t)$ and the fixed parameters $\varepsilon_0 > 0$, $K_0 \in \mathbb{N}$, and $K \in \mathbb{N}$ can be chosen so as to fulfill the condition

$$\lim_{t\to\infty} H_1 = H_*, \quad \lim_{t\to\infty} \mathbf{P} \left\{ \mathbb{L}_t^{\varepsilon_0} = \mathbb{M}_t^K = \mathbb{M}_t^{K_0} = \emptyset \right\} = 1, \tag{4.2}$$

and make non-trivial the estimate of (3.10).

No attempt is made at keeping track of numerical values of constants. Only their dependence on K, L, and other variable parameters introduced below is specified.

4.1. Localization. Below the parameters τ and T are as in (2.1), (3.1), and (3.8). Choose a large natural number L and put

$$Q_z^{(L)} = LH_1 \left(z + 2Q\right), \quad Q_{z-}^{(L)} = LH_1 \left(z + \tfrac{19}{10}Q\right). \tag{4.3}$$

The principal eigenvalue of the scaled operator (2.2) restricted to $H_0^1(Q_{z-}^{(L)})$ is

$$\lambda_z^{(L)} = \inf \left\{ |\phi|_{L^2}^{-2} \left(\tfrac{1}{2} |\nabla\phi|_2^2 + \beta |V_t\phi^2|_1\right) : \phi \in H_0^1(Q_{z-}^{(L)}) \right\}. \tag{4.4}$$

LEMMA 4.1. *If condition* (4.2) *is satisfied, then with* $\widehat{H}$ *of* (3.10)

$$\Lambda_t \geq \left(1 + \hat{c}L^{-2}(\widehat{H}/H_1)^2\right)^{-1} \min_{z\in\mathbb{J}_t} \lambda_z^{(L)}, \quad \mathbb{J}_t = \{z \in \mathbb{Z}^d : Q_z^{(L)} \cap Q_t \neq \emptyset\}.$$

Proof of Lemma 4.1. Following [**7**], choose a smooth cutoff function $\zeta(x) \in [0,1]$ such that $\zeta(x) = 1$ for $|x|_* \leq 1$ and $\zeta(x) = 0$ for $|x|_* > \frac{19}{10}$, while in all the space $\sum_{z\in\mathbb{Z}^d} \zeta^2(x-z) = 1$ and $|\nabla\zeta(x)| \leq c$. Define $\zeta_z(x) = \zeta\left((LH_1)^{-1}x - z\right)$. Evidently, $\psi_z(x) = \psi(x)\zeta_z(x) \in H_0^1(Q_{z-}^{(L)})$ for each function $\psi \in H_0^1(Q_t)$, and

$$\textstyle\sum_{z\in\mathbb{Z}^d} \psi_z^2(x) = \psi^2(x), \quad \sum_{z\in\mathbb{Z}^d} |\nabla\psi_z(x)|^2 = |\nabla\psi(x)|^2 + \psi^2 \sum_{z\in\mathbb{J}_t} |\nabla\zeta_z|^2,$$

because $\sum_{z\in\mathbb{J}_t} \nabla\zeta_z\psi \cdot \zeta_z\nabla\psi = \nabla\left(\frac{1}{2}\sum_{z\in\mathbb{J}_t}\zeta_z^2\right) \cdot \nabla\left(\frac{1}{2}\psi^2\right) = 0$. Consequently

$$\textstyle\sum_{z\in\mathbb{J}_t} |\nabla\psi_z|_2^2 = |\nabla\psi|_2^2 + \int_{Q_t} \psi^2 \sum_{z\in\mathbb{J}_t} |\nabla\zeta_z|^2 \leq |\nabla\psi|_2^2 + c(LH_1)^{-2}|\psi|_2^2. \tag{4.5}$$

For each z, one can use the estimate

$$\lambda_z^{(L)} |\psi_z|_2^2 \leq \tfrac{1}{2}|\nabla\psi_z|_2^2 + \textstyle\int_{Q_z^{(L)}} \beta V_t \psi_z^2,$$

Estimate (4.5), (4.4) and (3.11) show that for each $\psi \in H_0^1(Q_t)$

$$\begin{aligned}\left(\min_{z\in\mathbb{J}_t}\lambda_z^{(L)}\right)|\psi|_2^2 &= \left(\min_{z\in\mathbb{J}_t}\lambda_z^{(L)}\right)\sum_{z\in\mathbb{J}_t}|\psi_z|_2^2 \leq \sum_{z\in\mathbb{J}_t}\left(\tfrac{1}{2}|\nabla\psi_z|_2^2 + \textstyle\int_{Q_z^{(L)}}\beta V_t\psi_z^2\right)\\ &\leq \tfrac{1}{2}|\nabla\psi|_2^2 + \left|\beta V_t\psi^2\right|_1 + c\left(LH_1\right)^{-2}|\psi|_2^2\\ &\leq \left(1 + c\widehat{H}^2\left(LH_1\right)^{-2}\right)\left(\tfrac{1}{2}|\nabla\psi|_2^2 + |\beta V_t\psi^2|_1\right). \quad \square\end{aligned}$$

4.2. Reduction to individual Rayleigh ratios. Using (4.2) and the upper bound of Lemma 2.2, the class of test functions in (4.4) can be restricted to

$$\Phi_z^{(L)} = \left\{\phi \in H_0^1\left(Q_{z-}^{(L)}\right) : |\nabla\phi|_2^2 \leq \lambda_+, |\phi|_2^2 = 1\right\}, \tag{4.6}$$

where $\lambda_+ = 3\lambda(\beta,d,\mu)$ and $\lambda(\beta,d,\mu)$ is defined in Theorem 1.1. Indeed, the exclusion of test functions with large gradients can only influence the value of $\lambda_z^{(L)}$ when it is greater than Λ_t. If the inequality in (4.6) is violated, then $\frac{1}{2}|\nabla\phi|_2^2 + \beta|V_t\phi^2|_1 > \frac{3}{2}\lambda(\beta,d,\mu)|\phi|_2^2$ while $\mathbf{P}\left\{\Lambda_t > \frac{3}{2}\lambda(\beta,d,\mu)\right\} \to 0$ by Lemma 2.2. Hence with probability convergent to one

$$\forall z \in \mathbb{J}_t \quad \lambda_z^{(L)} = \inf\left\{\tfrac{1}{2}|\nabla\phi|_2^2 + \beta|V_t\phi^2|_1 : \phi \in \Phi_z^{(L)}\right\}. \tag{4.7}$$

Note that the function family (4.6) is obtained from $\Phi_0^{(L)}$ by the translation

$$\phi(x) \mapsto S_z\phi(x) = \phi(x - 2LH_1z). \tag{4.8}$$

All the random variables $\lambda_z^{(L)}$ have the same distribution.

LEMMA 4.2. *If condition* (4.2) *is satisfied, then there exist positive constants* $c_i > 0$ *such that whenever positive numbers* δ, B *satisfy the inequality* $1/\delta + B \geq c_0$ *one can select a finite set* $\widehat{G}_0 = \widehat{G}_0(\delta, K, B) \subset H_0^1(Q_0^{(L)})$ *so that its number of elements admits the estimate* $\operatorname{card}(\widehat{G}_0) \leq N(\delta, K)$ *and*

$$\min_z \lambda_z^{(L)} \geq (1 - c_1\varepsilon_1) \min_{z \in \mathbb{J}_t} \min_{\widehat{G}_z} |\psi|_2^{-2} \left(\tfrac{1}{2} |\nabla \psi|_2^2 + \beta |V_t \psi^2|_1\right) - c_2\varepsilon_2,$$

where the sets $\widehat{G}_z$ *are obtained from* $\widehat{G}_0$ *by translations* (4.8), $\varepsilon_1 = \delta + B^{-2/(d-1)}$, *and* $\varepsilon_2 = \delta + B^2 L^d H_*^d / \sqrt{\ln K}$.

Proof of Lemma 4.2. Take a test function $\phi \in \Phi_0^{(L)}$ and for $\delta \in (0,1)$ consider its convolution $\phi_\delta(x) = \phi * n_\delta(x) = \int \phi(x + \delta y)\, n(y)\, dy$ with a C^∞-smooth kernel $n(x) \geq 0$ such that $\int n(x)\, dx = 1$ and $n(x) = 0$ for $|x| \geq 1$. It follows from (4.6) that ϕ_δ vanishes outside $Q_0^{(L)}$. Well-known estimates for convolutions show that

$$(4.9) \qquad |\phi - \phi_\delta|_2 \leq c\sqrt{\lambda_+}\delta\, |\phi|_2, \quad |\phi_\delta|_2 \leq |\phi|_2, \quad |\nabla \phi_\delta|_2 \leq |\nabla \phi|_2.$$

Moreover, the functions ϕ_δ and all their derivatives stay uniformly bounded as long as δ remains bounded from zero. Consequently, the class $\Phi_0^{(L,\delta)} = \{\phi_\delta : \phi \in \Phi_0^{(L)}\}$ is compact in $C_0^m(Q_0^{(L)})$ for each m and $\delta \in \left(0, \tfrac{1}{2}\right)$. Combined with the invariance of L^p norms with respect to translations, this fact makes it possible to estimate $\lambda_z^{(L)}$ from below using a finite family of test functions — at the price of truncating the potential.

Set $V_t^{[K]}(x) = \min\{V_t(x), W_\infty K\}$. Using the fact that (4.8) preserves L^2 norms, it is easy to see that for each fixed pair of $\delta > 0$ and $K \in \mathbb{N}$ there exists a finite set $G(\delta, K) \subset \Phi_0^{(L)}$ such that for each $z \in \mathbb{J}_t$

$$(4.10) \qquad \lambda_z^{(L)} \geq \min_{\phi \in G_z} \left\{\tfrac{1}{2} |\nabla \phi_\delta|_2^2 + \beta |V_t^{[K]} \phi_\delta^2|\right\} - c_1\delta, \quad c_1 = 2c^* \beta W_\infty \sqrt{\lambda_+},$$

where $G_z = \{S_z \phi_\delta : \phi \in G(\delta, K)\}$. Moreover, by (4.9) there exists a constant c_2 such that $1 - c_2\delta \leq |\phi_\delta|_2^2 \leq 1$ for each z and $\phi \in G_z$.

Truncating functions ϕ_δ proves practical on the return from $V_t^{[K]}$ of (4.10) to the original potential. The truncated function $\phi_{\delta,B}(x) = \min\{|\phi_\delta(x)|, B\} \operatorname{sign}(\phi_\delta(x))$ does not differ much from the original one ϕ_δ for large B. Indeed, it follows from (4.6), (4.9), and the classical multiplicative inequality for embedding of H^1 in L^q (see [**5**, Th.II.2.1]) that for $q = 2d/(d-1)$ with a constant only dependent on d

$$(4.11) \qquad |\phi_\delta|_{L^q\left(Q_z^{(L)}\right)} \leq c\lambda_+^{1/4}, \quad \int_{\{|\phi_\delta| > B\}} |\phi_\delta|^2 \leq c\lambda_+^{q/4} B^{-(q-2)},$$

$$(4.12) \qquad 0 \leq |\phi_\delta|_2^2 - |\phi_{\delta,B}|_2^2 \leq B^{-(q-2)} \hat{c} \lambda_+^{q/4}, \quad |\nabla \phi_{\delta,B}|_2 \leq |\nabla \phi_\delta|_2.$$

Obviously, $\int \beta V_t \phi_{\delta,B}^2 - \int \beta V_t^{[K]} \phi_{\delta,B}^2 \leq B^2 \int_{Q_z^{(L)}} \beta (V_t - W_\infty K)_+$. Lemma 3.4 permits one to conclude that $\int \beta V_t \phi_{\delta,B}^2 - \int \beta V_t^{[K]} \phi_{\delta,B}^2 \leq c\varepsilon_2$. Combined with (4.10), this last inequality yields the estimate for $\lambda_z^{(L)}$ and that of the lemma because $|\phi_{\delta,B}|_2^2 \geq 1 - c_3(\delta + B^{-2/(d-1)} \hat{c} \lambda_+^{d/(2d-2)})$ by (4.12). The test functions $\phi_{\delta,B}$ used to estimate $\lambda_z^{(L)}$ belong to $\widehat{G}_z = \{S_z \phi_{\delta,B} : \phi \in G(\delta, K)\}$, which proves the lemma.
$\square$

4.3. Proof of Theorem 1.1: the final lower bound on Λ_t. Fix some numbers $\varepsilon > 0$ and $K_0 \in \mathbb{N}$ such that $A(\varepsilon, K_0) > 0$ in the estimate of Lemma 3.4. Given K_0 and ε, one can chose the block size $T(t)$ so as to satisfy (3.8) with the limit block size H_* so large that condition (4.2) holds true provided that the inequality $K \geq K_0$ is satisfied. For the configurations of the potential obeying condition (4.1), the rough lower bound of (3.11) for Λ_t is a finite constant that depends on ε and K_0, and the calculations to follow are justified for $K \geq K_0$.

Given an arbitrary small fixed $\delta > 0$ and sufficiently large L, B, and K, the estimates of Lemmas 4.1 and 4.2 yield, for large t and potential configurations satisfying condition (4.1), the inequality

$$\Lambda_t \geq (1 + c\varkappa)^{-1} \min_{z \in \mathbb{J}_t} \min_{\psi \in \hat{G}_z} |\psi|_2^{-2} (\tfrac{1}{2} |\nabla \psi|_2^2 + \beta |V_t \psi^2|_1) - c\varkappa, \tag{4.13}$$

where $\varkappa = L^{-2} + \delta + B^{-2/(d-1)} + B^2 L^d / \sqrt{\ln K}$, and the choice of the parameters B, K, L is still free.

Choose B and L as functions of the natural parameter K in such a way that $B = L \sim (\ln K)^\gamma$ for a small exponent $\gamma \in \left(0, \frac{1}{2d+4}\right)$. Given $\delta > 0$, the parameter $K = K(\delta)$ can be chosen so large that $\kappa \leq \delta$ in (4.13) for the above choice of B and L.

Since K is chosen as a function of δ, the cardinality of $\widehat{G}_0$ is, for a given δ, a fixed finite number. In Lemma 4.2, card $(\mathbb{J}_t)$ is also determined by δ and satisfies the relation $\lim_{t \to \infty} (\ln t)^{-1} \ln \operatorname{card}(\mathbb{J}_t) = d$. Lemma 2.3 and (4.13) show that $\lim_{t \to \infty} \mathbf{P}\left\{\Lambda_t > (1 + c\delta)^{-1} \lambda(\beta, d, \mu) - c\delta\right\} = 1$ in notation of Theorem 1.1. The choice of $\delta > 0$ being arbitrary, this proves the lower bound of the theorem. □

4.4. Asymptotics of principal eigenvalue for large and small β. The formulae discussed below were derived in [**7**]. This subsection is intended to show that representation (1.7)-(1.8) proves efficient in the study of asymptotic behavior of $\lambda(\beta, d, \mu)$ as a function of β.

Case of large β. Denote by $\mathfrak{L}$ the principal eigenvalue of the Laplace operator in the solid sphere of unit volume under zero Dirichlet condition on the boundary. It was found by A.-S. Sznitman (see [**9**]) that for the Laplacian with unscaled potential the limit of the normalized principal value is $(d/\mu)^{-2/d} \mathfrak{L}$.

COROLLARY 4.1. [**7**, Theorem 0.2] *For large β, the limit value of the normalized principal eigenvalue* (2.4) *approaches the corresponding limit for the unscaled case:*

$$\lim_{\beta \to \infty} \lambda(\beta, d, \mu) = (d/\mu)^{-2/d} \mathfrak{L}. \tag{4.14}$$

Proof of Corollary 4.1 By Lemma 2.1 below, $\mathcal{M}(d; \phi, \mu) = 0$ if $\mu |\{\phi \neq 0\}| \leq d$. To get an upper bound, one can use the Faber–Krahn inequality (see [**1**], §VI.2):

$$\lambda(\beta, d, \mu) \leq \inf \left\{ \tfrac{1}{2} |\nabla \phi|_2^2 / |\phi|_2^2, \ \mu |\{\phi \neq 0\}| \leq d \right\} = (d/\mu)^{-2/d} \mathfrak{L}. \tag{4.15}$$

The derivation of the lower bound makes use of sorting test functions from Ψ of (1.6) in several disjoint classes. For a large value of β the sets $\Psi_i^c = \Psi \backslash \Psi_i$ are the complements of

$$\Psi_1 = \{\phi \in \Psi : |\{(\ln \beta) \phi^2 > 1\}| \leq d/\mu + (\ln \beta)^{-1}\},$$

$$\Psi_2 = \{\phi \in \Psi : |\phi_\beta|_2^2 \leq (\ln \beta)^{-1/2}\}, \ \phi_\beta(x) = \operatorname{sign}(\phi(x)) \min\{|\phi(x)|, (\ln \beta)^{-1/2}\}.$$

For $\phi \in \Psi_1 \cap \Psi_2$, one can disregard $\beta\mathcal{M}(\phi)$ and use the evident estimate $\phi^2 \geq \phi_\beta^2 + (\phi - \phi_\beta)^2$, estimates of §II.3 in [**5**] for truncated functions (see also remark on local Sobolev spaces in §7.5 of [**2**]), and the Faber-Krahn inequality to conclude that

$$\begin{aligned}(4.16)\qquad \tfrac{1}{2}|\nabla\phi|_2^2 &\geq \tfrac{1}{2}|\nabla\phi - \nabla\phi_\beta|_2^2 \geq (d/\mu + 1/\ln\beta)^{-2/d}\,\mathfrak{L}\,|\phi - \phi_\beta|_2^2 \\ &\geq (d/\mu + 1/\ln\beta)^{-2/d}\,\mathfrak{L}(1 - 2/\sqrt{\ln\beta}).\end{aligned}$$

It remains to show that for test functions outside $\Psi_1 \cap \Psi_2$ the value of $\mathcal{M}(\phi)$ is very large in comparison with $1/\beta$, so (4.16) is a lower bound for $\lambda(\beta, d, \mu)$.

Assume that $\phi \in \Psi_1^c$. For large β and $h = (\ln\beta)^2$, the estimate $1 - e^{-h\phi^2} \geq 1 - 1/\beta$ holds on the set where $(\ln\beta)\phi^2(x) \geq 1$. This yields the inequality

$$(4.17)\qquad \mathcal{M}(\phi) \geq (\ln\beta)^{-2}\mu\left((1 - 1/\beta)|\{(\ln\beta)\phi^2 > 1\}| - d/\mu\right) \geq c\mu(\ln\beta)^{-3}.$$

For $\phi \in \Psi_1 \cap \Psi_2^c$, take $h = \ln\beta$. The inequality $h\phi_\beta^2(x) \leq 1$ holds for all β and x, so $1 - \exp\{-h\phi^2(x)\} \geq 1 - \exp\left\{-h\phi_\beta^2(x)\right\} \geq Ch\phi_\beta^2(x)$, and consequently $\mu\int\left(1 - e^{-h\phi^2}\right) \geq C\mu h\,|\phi_\beta|_2^2$. Hence $\mathcal{M}$ is also large in this case:

$$(4.18)\qquad \mathcal{M}(\phi) \geq C\mu\,|\phi_\beta|_2^2 - d/h \geq C\mu(\ln\beta)^{-1/2} - (\ln\beta)^{-1}d.$$

Estimates (4.15), (4.16), (4.17), and (4.18) prove (4.14). □

Case of small β. It follows from (1.7) and the inequality $1 - e^{-x} \leq x$ for $x \geq 0$ that $\mathcal{M} \leq \mu|\phi|_2^2$. Scaling $\phi \in \Psi$ to $(\beta\mu)^{-d/2}\hat{\phi}(y) = \phi(y/\sqrt{\beta\mu})$, one derives from the above the inequality

$$\lambda(\beta, d, \mu) \leq \inf_{\phi\in\Psi}\left\{\tfrac{1}{2}|\nabla\phi|_2^2 + \beta\mu\right\} = \beta\mu \inf_{\phi\in\Psi}\left\{\tfrac{1}{2}|\nabla\phi|_2^2 + 1\right\} = \beta\mu.$$

It is immediate from this rough estimate that the pertinent test functions in (1.8) are ones satisfying the condition

$$(4.19)\qquad |\phi|_2^2 = 1, \quad |\nabla\phi|_2^2 \leq c\beta\mu.$$

Finer analysis of asymptotic behavior of $\lambda(\beta, d, \mu)$ for small β is based on a more exact approximation of $\exp\{-h\phi^2\}$ in the integral defining $\mathcal{M}(\phi)$ by an initial segment of its Taylor expansion.

COROLLARY 4.2. [**7**, Theorems 0.3–0.4] *The quantity* $\beta^{-1}\lambda(\beta, d, \mu)$ *converges to a value that corresponds to constant potential equal to* $\mathbf{E}V_t$ *as* $\beta \searrow 0$.

(a) *For* $d = 2, 3$ *there exists a positive constant* $\beta_0(d, \mu)$ *such that if* $0 < \beta \leq \beta_0(d, \mu)$, *then* $|\lambda(\beta, d, \mu) - \beta\mu| \leq c(d)\left(\beta\sqrt{\mu}\right)^{4/(4-d)}$.

(b) *If* $d \geq 4$, *then there exists a positive constant* $\beta_0(d, \mu)$ *such that*

$$\lambda(\beta, d, \mu) = \beta\mu \quad \text{for} \quad 0 < \beta \leq \beta_0(d, \mu).$$

Proof of Corollary 4.2. (a) The test functions below have unit L^2 norm, so the multiplicative inequalities in §II.2 of [**5**] provide the estimates

$$(4.20)\qquad |\phi|_4^4 \leq C\,|\nabla\phi|_2^d,\ |\phi|_6^6 \leq C\,|\nabla\phi|_2^{2d},\ d = 2, 3.$$

A lower bound for $\mathcal{M}(\phi)$ follows from the inequality $1 - e^{-x} \geq x - \frac{1}{2}x^2$:

$$\tfrac{1}{h}(M(h) - d) \geq -d/h + \mu\int\left(\phi^2 - \tfrac{1}{2}h\phi^4\right) \geq \mu\left(1 - \tfrac{1}{2}Ch\,|\nabla\phi|_2^d\right) - d/h.$$

The maximum of the right-hand side is attained at $h = \left(\frac{1}{2}C\,|\nabla\phi|_2^d\,\mu/d\right)^{-1/2}$. It is positive for small β and test functions satisfying (4.19), which yields

$$\lambda(\beta,d,\mu)=\inf\left\{\tfrac12|\nabla\phi|_2^2+\beta\mathcal{M}(\phi)\right\}\geq\inf\left\{\tfrac12|\nabla\phi|_2^2+\beta\left(\mu-\sqrt{2C\mu/d}\,|\nabla\phi|_2^{d/2}\right)\right\}$$
$$\geq\inf_{\varepsilon>0}\left\{\tfrac12\varepsilon^2+\beta\mu-\beta\sqrt{2C\mu/d}\varepsilon^{d/2}\right\}=\beta\mu-c'\left(\beta\sqrt{\mu}\right)^{4/(4-d)}.$$

One can derive an upper bound on $\mathcal{M}(\phi)$ by a similar reasoning. Indeed, retaining one more term in the Taylor expansion, one concludes that

$$\tfrac1h(M(h)-d)\leq\mu\textstyle\int\left(\phi^2-\frac12h\phi^4+\frac16h^2\phi^6\right)-d/h\leq Q(h)+\frac16h^2\mu|\phi|_6^6,$$
$$Q(h)=\mu\left(1-\tfrac12h\,|\phi|_4^4-\tfrac1h(d/\mu)\right).$$

The maximum of Q is attained with $h=h_0=\sqrt{2d/\mu}\,|\phi|_4^{-2}$. The choice of h_0 guarantees that $(M(h)-d)/h\leq\mu Q(h_0)+\frac16h_0^2\,|\phi|_6^6$ for $h\leq h_0$, and $M(h_0)/h_0\leq\mu Q(h_0)+\frac16h_0^2\,|\phi|_6^6+d/h_0$. Since $M(h)/h$ is a decreasing function, this implies

$$\forall h>0\ \ \tfrac1h(M(h)-d)\leq\mu Q(h_0)+\tfrac16h_0^2\,|\phi|_6^6+d/h_0\leq1-\tfrac12h_0\,|\phi|_4^4+\tfrac16h_0^2\,|\phi|_6^6$$

(Monotonicity of $M(h)/h$ is verified by derivation using the inequality $e^{-x}(e^x-1-x)>0$.) Note that $Q(h_0)=\mu\left(1-\sqrt{2d/\mu}\,|\phi|_4^2\right)$.

It follows from the above and (4.20) that

$$\sup_{h>0}\tfrac1h(M(h)-d)\leq\mu\left(1-\sqrt{\tfrac{d}{2\mu}}\,|\phi|_4^2+\tfrac{d}{3\mu}\,|\phi|_6^6\,|\phi|_4^{-4}\right).$$

To get an upper bound on $\lambda(\beta,d,\mu)$, consider the special test function $\phi_\varepsilon(x)=\varepsilon^{d/2}\psi\left(\varepsilon^2|x|^2\right)$, where the smooth function $\psi>0$ with compact support satisfies condition $|\psi|_2=1$. It is easily seen that there exist constants $C_i>0$ such that

$$|\nabla\phi_\varepsilon|_2^2=C_1\varepsilon^2,\quad|\phi_\varepsilon|_4^4=C_2\varepsilon^d,\quad|\phi_\varepsilon|_6^6=C_3\varepsilon^{2d}.$$

Consequently, the above estimate for $\mathcal{M}(\phi)$ yields the inequality

$$\lambda(\beta,d,\mu)=\inf_{\phi\in\Psi}\left\{\tfrac12|\nabla\phi|_2^2+\beta\mathcal{M}(\phi)\right\}\leq\inf_{\varepsilon>0}\left\{\tfrac12|\nabla\phi_\varepsilon|_2^2+\beta\mathcal{M}(\phi_\varepsilon)\right\}\leq L(\varepsilon),$$

where $L(\varepsilon)=\hat C_1\varepsilon^2+\beta\mu-\hat C_2\beta\sqrt{\mu}\varepsilon^{d/2}+\hat C_3\beta\varepsilon^d$, the constants $\hat C_i$ being determined by the choice of ψ. The function $L(\varepsilon)$ is strictly decreasing for small $\varepsilon>0$ — its derivative is negative because the two positive terms in the expression

$$L'(\varepsilon)=2\hat C_1\varepsilon-\tfrac d2\beta\sqrt{\mu}\hat C_2\varepsilon^{d/2-1}+d\beta\mu\hat C_3\varepsilon^{d-1}<0$$

are infinitesimals of higher order that the negative one. Choosing ε of order $\left(\beta\sqrt{\mu}\right)^{1/(2-d/2)}$, one gets the upper estimate which yields that of the corollary: $\lambda(\beta,d,\mu)\leq\mu\beta-c_1\left(\beta\sqrt{\mu}\right)^{4/(4-d)}+c_2\beta^{(4+d)/(4-d)}\mu^{d/(4-d)}$.

(b) For $d\geq4$, the multiplicative inequalities in §II.2 of [**5**] provide the estimate

$$|\phi|_{2d/(d-2)}\leq C\,|\nabla\phi|_2\,,\ d\geq4. \tag{4.21}$$

One can get a lower bound for $\mathcal{M}(\phi)$ using the inequality $1-e^{-x}\geq x-\frac{d-2}{d}\,|x|^{d/(d-2)}$ and (4.21): since $\frac1hM(h)\geq\mu\int\left(\phi^2-\frac{d-2}{d}h^{2/(d-2)}\,|\phi|^{2d/(d-2)}\right)$,

$$\mathcal{M}(\phi)\geq\mu\max_{h>0}\left(1-C_1h^{2/(d-2)}\,|\nabla\phi|_2^{2d/(d-2)}-\tfrac1h(d/\mu)\right).$$

The maximum is no less than the value of the right-hand side at $h_0 = |\nabla\phi|_2^{-2}$, so for small $|\nabla\phi|_2$ there is a non-trivial bound $\beta\mathcal{M}(\phi) \geq \beta\mu - \beta(c_1\mu + d)|\nabla\phi|_2^2$.

Consequently, $\lambda(\beta, d, \mu) \geq \inf\left\{\frac{1}{2}|\nabla\phi|_2^2 + \beta\mu - \beta(c_1\mu + d)|\nabla\phi|_2^2\right\}$. For small enough β, the right-hand side of the latter inequality is a growing function of $|\nabla\phi|_2^2$. Since one can choose a test function with $|\phi|_2^2 = 1$ and an arbitrarily small $|\nabla\phi|_2^2$, the above inequality implies that $\lambda(\beta, d, \mu) = \beta\mu$ for $\beta \leq \beta_0(d, \mu)$. □

References

[1] Chavel I. *Eigenvalues in Riemannian Geometry.* Academic Press, Orlando (Florida), 1984.

[2] Gilbarg D., Trudinger N. S. *Differential Equations of Second Order*, 2nd ed. Springer-Verlag, Berlin, 1983.

[3] Ibragimov I.A., Linnik Yu.V. *Independent and Stationary Sequences of Random Variables.* Wolters-Noordhoff, Groningen, 1971.

[4] Karr A. F. *Point Processes and Their Statistical Inference.* Marcel Dekker, New York – Basel, 1986.

[5] Ladyzhenskaya, O. A., Uraltseva, N. N. *Linear and Quasilinear Elliptic Equations, Mathematics in Science and Engineering*, vol. 46. Academic Press, New York, 1968 (in Russian: 2nd ed., Nauka, Moscow, 1973).

[6] Merkl F., Wütrich M.V. Phase transition of the principal Dirichlet eigenvalue in a scaled Poissonian potential. *Probab. Theory Relat. Fields*, 2001, v.119, 475-507.

[7] Merkl F., Wütrich M. Infinite volume asymptotics of the ground state energy in a scaled Poissonian potential. *Ann. Inst. H.Poincaré. Probability and Statistics*, 2002, v.38, No 3, 2002.

[8] Petrov V.V. *Sums of Independent Random Variables.* Springer-Verlag, Berlin New York, 1975.

[9] Sznitman A.-S. *Brownian Motion, Obstacles and Random Media.* Springer-Verlag, New York, 1998.

[10] Beliaev A.Yu., Yurinsky V.V. Bottom of Laplacian's spectrum in Poisson cloud, *Siberian Adv. Math.*, 1995, v.5, No 4, 113–150.

[11] Yurinsky V.V. Spectrum bottom and largest vacuity. *Probab. Theory Relat. Fields*, 1999, v.114, 151-175.

Departamento de Matemática, Universidade da Beira Interior, Rua Marquês d'Ávila e Bolama, 6201-001 Covilhã, Portugal

Current address: Departamento de Matemática, Universidade da Beira Interior, Rua Marquês d'Ávila e Bolama, 6201-001 Covilhã, Portugal

E-mail address: yurinsky@ubi.pt

Titles in This Series

339 **Peter Kuchment, Editor,** Waves in periodic and random media, 2003

338 **Pascal Auscher, Thierry Coulhon, and Alexander Grigor'yan, Editors,** Heat kernels and analysis on manifolds, graphs, and metric spaces, 2003

337 **Krishan L. Duggal and Ramesh Sharma, Editors,** Recent advances in Riemannian and Lorentzian geometries, 2003

336 **José González-Barrios, Jorge A. León, and Ana Meda, Editors,** Stochastic models, 2003

335 **Geoffrey L. Price, B. Mitchell Baker, Palle E.T. Jorgensen, and Paul S. Muhly, Editors,** Advances in quantum dynamics, 2003

334 **Ron Goldman and Rimvydas Krasauskas, Editors,** Topics in algebraic geometry and geometric modeling, 2003

333 **Giovanni Alessandrini and Gunther Uhlmann, Editors,** Inverse problems: Theory and applications, 2003

332 **John Bland, Kang-Tae Kim, and Steven G. Krantz, Editors,** Explorations in complex and Riemannian geometry, 2003

331 **Luchezar L. Avramov, Marc Chardin, Marcel Morales, and Claudia Polini, Editors,** Commutative algebra: Interactions with algebraic geometry, 2003

330 **S. Y. Cheng, C.-W. Shu, and T. Tang, Editors,** Recent advances in scientific computing and partial differential equations, 2003

329 **Zhangxin Chen, Roland Glowinski, and Kaitai Li, Editors,** Current trends in scientific computing, 2003

328 **Krzysztof Jarosz, Editor,** Function spaces, 2003

327 **Yulia Karpeshina, Günter Stolz, Rudi Weikard, and Yanni Zeng, Editors,** Advances in differential equations and mathematical physics, 2003

326 **Kenneth D. T-R McLaughlin and Xin Zhou, Editors,** Recent developments in integrable systems and Riemann-Hilbert problems, 2003

325 **Seok-Jin Kang and Kyu-Hwan Lee, Editors,** Combinatorial and geometric representation theory, 2003

324 **Caroline Grant Melles, Jean-Paul Brasselet, Gary Kennedy, Kristin Lauter, and Lee McEwan, Editors,** Topics in algebraic and noncommutative geometry, 2003

323 **Vadim Olshevsky, Editor,** Fast algorithms for structured matrices: theory and applications, 2003

322 **S. Dale Cutkosky, Dan Edidin, Zhenbo Qin, and Qi Zhang, Editors,** Vector bundles and representation theory, 2003

321 **Anna Kamińska, Editor,** Trends in Banach spaces and operator theory, 2003

320 **William Beckner, Alexander Nagel, Andreas Seeger, and Hart F. Smith, Editors,** Harmonic analysis at Mount Holyoke, 2003

319 **W. H. Schikhof, C. Perez-Garcia, and A. Escassut, Editors,** Ultrametric functional analysis, 2003

318 **David E. Radford, Fernando J. O. Souza, and David N. Yetter, Editors,** Diagrammatic morphisms and applications, 2003

317 **Hui-Hsiung Kuo and Ambar N. Sengupta, Editors,** Finite and infinite dimensional analysis in honor of Leonard Gross, 2003

316 **O. Cornea, G. Lupton, J. Oprea, and D. Tanré, Editors,** Lusternik-Schnirelmann category and related topics, 2002

315 **Theodore Voronov, Editor,** Quantization, Poisson brackets and beyond, 2002

314 **A. J. Berrick, Man Chun Leung, and Xingwang Xu, Editors,** Topology and Geometry: Commemorating SISTAG, 2002

313 **M. Zuhair Nashed and Otmar Scherzer, Editors,** Inverse problems, image analysis, and medical imaging, 2002

For a complete list of titles in this series, visit the AMS Bookstore at **www.ams.org/bookstore/**.